Molecular Basis
of Morphogenesis

The Fifty-First Annual Symposium of
the Society for Developmental Biology
University of Washington, June 25–27, 1992

Executive Committee
1991–1992
Merton Bernfield, Harvard Medical School, President
Allan C. Spradling, Carnegie Institution of Washington, Past President
David R. McKay, Duke University, President-Elect
Cynthia Kenyon, University of California, San Francisco, Secretary
Susan Strome, University of Indiana, Treasurer
Peter J. Bryant, University of California, Irvine, Editor-in-Chief
JoAnn Render, University of Illinois, Urbana, Member-at-Large
Rachel Fink, Mt. Holyoke College, Massachusetts, Member-at-Large
Charles D. Little, University of Virginia Medical School, Member-at-Large
Gerald P. Schatten, University of Wisconsin, Member-at-Large
Samuel Ward, University of Arizona, Member-at-Large
Alexandra L. Joyner, Samuel Lunefeld Research Institute, Member-at Large
Stephen D. Hauschka, University of Washington, Member-at-Large
Holly Schauer, Business Manager

1992–1993

David R. McClay, Duke University, President
Merton Bernfield, Harvard Medical School, Past President
Charles Kimmel, University of Oregon, President-Elect
Cynthia Kenyon, University of California, San Francisco, Secretary
Susan Strome, University of Indiana, Treasurer
Peter J. Bryant, University of California, Irvine, Editor-in-Chief
Celeste A. Berg, University of Washington, Member-at-Large
Rachel Fink, Mt. Holyoke College, Massachusetts, Member-at-Large
Charles D. Little, University of Virginia Medical School, Member-at-Large
Gerald P. Schatten, University of Wisconsin, Member-at-Large
Samuel Ward, University of Arizona, Member-at-Large
Victor D. Vacquier, University of California, San Diego, Member-at-Large
Alexandra L. Joyner, Samuel Lunefeld Research Institute, Member-at-Large
Stephen D. Hauschka, University of Washington, Member-at-Large
Ellen J. Henderson, Georgetown University, Member-at-Large
Holly Schauer, Business Manager

Society for Developmental Biology. Symposium (51st: 1992: University of Washington)

Molecular Basis of Morphogenesis

Editor
Merton Bernfield
**Joint Program in Neonatology
Harvard Medical School
Boston, Massachusetts**

QH491
M635
1992

WILEY-LISS
A JOHN WILEY & SONS, INC., PUBLICATION
New York • Chichester • Brisbane • Toronto • Singapore

Address all Inquiries to the Publisher
Wiley-Liss, Inc., 605 Third Avenue, New York, NY 10158-0012

Copyright © 1993 Wiley-Liss, Inc.

Printed in the United States of America.

Under the conditions stated below the owner of copyright for this book hereby grants permission to users to make photocopy reproductions of any part or all of its contents for personal or internal or organizational use, or for personal or internal use of specific clients. This consent is given on the condition that the copier pay the stated per-copy fee through the Copyright Clearance Center, Incorporated, 27 Congress Street, Salem, MA 01970, as listed in the most current issue of "Permissions to Photocopy" (Publisher's Fee List, distributed by CCC, Inc.), for copying beyond that permitted by sections 107 or 108 of the US Copyright Law. This consent does not extend to other kinds of copying, such as copying for general distribution, for advertising or promotional purposes, for creating new collective works, or for resale.

Library of Congress Cataloging-in-Publication Data

Molecular basis of morphogenesis / editor, Merton Bernfield.
    p. cm.
  "The Fifty-first Annual Symposium of the Society for Developmental Biology, University of Washington, June 25–27, 1992"—P.
  Includes bibliographical references and index.
  ISBN 0-471-30515-4
  1. Morphogenesis—Molecular aspects—Congresses. I. Bernfield, Merton. II. Society for Developmental Biology. Symposium (51st : 1992 : University of Washington)
QH491.M635 1993
574.3'32—dc20   93-2500
                                                               CIP

The text of this book is printed on acid-free paper.

Contents

Contributors vii

Preface
Merton Bernfield xi

Young Investigators Awards, 1992 xv

Abstract of the First Place Young Investigator Award xvi

I. GAMETES

Chapter 1. Role in Gametogenesis of *c-kit* Encoded at the *W* Locus of Mice
Rosemary F. Bachvarova, Katia Manova, and Peter Besmer 1

Chapter 2. Regulation of Expression of *mZP3*, the Sperm Receptor Gene, During Mouse Development
Ross A. Kinloch, Sergio A. Lira, Steven Mortillo, Michael Schickler, Richard J. Roller, and Paul M. Wassarman 19

II. EARLY ORGANOGENESIS

Chapter 3. Building a Frog: The Role of *Xenopus Wnts* in Patterning the Embryonic Mesoderm
Jan L. Christian and Randall T. Moon 35

Chapter 4. Components of Intercellular Adhesive Junctions and Their Roles in Morphogenesis
Pamela Cowin and Anthony M.C. Brown 49

Chapter 5. Eyes, Ears, and Homeobox Genes in Zebrafish Embryos
Monte Westerfield, Marie-Andrée Akimenko, Marc Ekker, and Andreas Püschel 69

Chapter 6. Cardiac Morphogenesis in the Zebrafish, Patterning the Heart Tube Along the Anteroposterior Axis
Didier Y.R. Stainier and Mark C. Fishman 79

Chapter 7. Genetic Hierarchy Controlling Flower Development
Detlef Weigel and Elliot M. Meyerowitz 93

III. CENTRAL NERVOUS SYSTEM

Chapter 8. The Role of Induction in Cell Choice and Cell Cycle in the Developing *Drosophila* Retina
Ross L. Cagan, Barbara J. Thomas, and S. Lawrence Zipursky 109

vi   Contents

Chapter 9.   Neurogenesis, Determination, and Migration During Cerebral
              Cortical Development
              Susan K. McConnell, Christine E. Kaznowski, Nancy A. O'Rourke,
              Michael E. Dailey, and Jennifer S. Roberts 135

IV. SOMITES

Chapter 10.   Myogenic Factor Gene Expression in Mouse Somites and
              Limb Buds
              Gary E. Lyons and Margaret E. Buckingham 155

Chapter 11.   Myogenic Lineages Within the Developing Somite
              Charles P. Ordahl 165

Chapter 12.   Positional Specification During Muscle Development
              Uta Grieshammer, David Sassoon, and Nadia Rosenthal 177

V. SKELETON

Chapter 13.   From Cartilage to Bone—The Role of Collagenous Proteins
              Phyllis Lu Valle, Olena Jacenko, Masahiro Iwamoto,
              Maurizio Pacifici, and Bjorn R. Olsen 189

Chapter 14.   Pattern Formation and Limb Morphogenesis
              Lewis Wolpert and Cheryll Tickle 207

Chapter 15.   The Bone Morphogenetic Proteins in Cartilage and Bone
              Development
              John M. Wozney, Joanna Capparella, and Vicki Rosen 221

VI. EPITHELIA

Chapter 16.   Homeobox Genes and Epithelial Patterning in *Hydra*
              M.A. Shenk and R.E. Steele 231

Chapter 17.   Human Epidermal Keratinocytes in Culture: Role of Integrins in
              Regulating Adhesion and Terminal Differentiation
              Fiona M. Watt 241

Chapter 18.   Differentiation of the Endodermal Epithelium During Gastrulation
              in the Sea Urchin Embryo
              Gary M. Wessel 255

Index 267

Contributors

Marie-Andrée Akimenko, Institute of Neuroscience, University of Oregon, Eugene, OR 97403 [69]

Rosemary F. Bachvarova, Department of Cell Biology, Cornell University Medical College, New York, NY 10021 [1]

Merton Bernfield, Joint Program in Neonatology, Harvard Medical School, Boston, MA 02115 [xi]

Peter Besmer, Molecular Biology Program, Sloan Kettering Institute, New York, NY 10021 [1]

Anthony M.C. Brown, Department of Cell Biology and Anatomy, Cornell University Medical College, New York, NY 10021 [49]

Margaret E. Buckingham, Department of Molecular Biology, Pasteur Institute, 75724 Paris, France [155]

Ross L. Cagan, Howard Hughes Medical Institute and Department of Biological Chemistry, University of California, Los Angeles, CA 90024-1662 [109]

Joanna Capparella, Genetics Institute, Inc., Cambridge, MA 02140 [221]

Jan L. Christian, Department of Pharmacology, University of Washington School of Medicine, Seattle, WA 98195 [35]

Pamela Cowin, Departments of Cell Biology and Dermatology, New York University Medical Center, New York, NY 10016 [49]

Michael E. Dailey, Department of Biological Sciences, Stanford University, Stanford, CA 94305 [135]

Marc Ekker, Institute of Neuroscience, University of Oregon, Eugene, OR 97403 [69]

Mark C. Fishman, Cardiovascular Research Center, Massachusetts General Hospital, Charlestown, MA 02129 [79]

Uta Grieshammer, Department of Biochemistry, Boston University School of Medicine, Boston, MA 02118 [177]

Masahiro Iwamoto, Department of Anatomy and Histology, University of Pennsylvania, School of Dental Medicine, Philadelphia, PA 19104 [189]

Olena Jacenko, Department of Anatomy and Cellular Biology, Harvard Medical School, Boston, MA 02115 [189]

The numbers in brackets are the opening page numbers of the contributors' articles.

Contributors

Christine E. Kaznowski, Department of Biological Sciences, Stanford University, Stanford, CA 94305 **[135]**

Ross A. Kinloch, Department of Cell and Developmental Biology, Roche Institute of Molecular Biology, Roche Research Center, Nutley, NJ 07110 **[19]**

Sergio A. Lira, Department of Cell and Developmental Biology, Roche Institute of Molecular Biology, Roche Research Center, Nutley, NJ 07110 **[19]**

Phyllis Lu Valle, Department of Anatomy and Cellular Biology, Harvard Medical School, Boston, MA 02115 **[189]**

Gary E. Lyons, Department of Anatomy, University of Wisconsin Medical School, Madison, WI 53706 **[155]**

Katia Manova, Department of Cell Biology, Cornell University Medical College, New York, NY 10021 **[1]**

Susan K. McConnell, Department of Biological Sciences, Stanford University, Stanford, CA 94305 **[135]**

Elliot M. Meyerowitz, Division of Biology 156-29, California Institute of Technology, Pasadena, CA 91125 **[93]**

Randall T. Moon, Department of Pharmacology, University of Washington School of Medicine, Seattle, WA 98195 **[35]**

Steven Mortillo, Department of Cell and Developmental Biology, Roche Institute of Molecular Biology, Roche Research Center, Nutley, NJ 07110 **[19]**

Bjorn R. Olsen, Department of Anatomy and Cellular Biology, Harvard Medical School, Boston, MA 02115 **[189]**

Charles P. Ordahl, Department of Anatomy, Cardiovascular Research Institute, University of California, San Francisco, CA 94143-0452 **[165]**

Nancy A. O'Rourke, Department of Biological Sciences, Stanford University, Stanford, CA 94305 **[135]**

Maurizio Pacifici, Department of Anatomy and Histology, University of Pennsylvania, School of Dental Medicine, Philadelphia, PA 19104 **[189]**

Andreas Püschel, Institute of Neuroscience, University of Oregon, Eugene, OR 97403 **[69]**

Jennifer S. Roberts, Department of Biological Sciences, Stanford University, Stanford, CA 94305 **[135]**

Richard J. Roller, Department of Cell and Developmental Biology, Roche Institute of Molecular Biology, Roche Research Center, Nutley, NJ 07110 **[19]**

Vicki Rosen, Genetics Institute, Inc., Cambridge, MA 02140 **[221]**

Nadia Rosenthal, Department of Biochemistry, Boston University, School of Medicine, Boston, MA 02118 **[177]**

David Sassoon, Department of Biochemistry, Boston University School of Medicine, Boston, MA 02118 **[177]**

Michael Schickler, Department of Cell and Developmental Biology, Roche Institute of Molecular Biology, Roche Research Center, Nutley, NJ 07110 **[19]**

M.A. Shenk, Department of Biological Chemistry and the Developmental Biology Center, University of California, Irvine, CA 92717-1700 **[231]**

Didier Y.R. Stainier, Cardiovascular Research Center, Massachusetts General Hospital, Charlestown, MA 02129 **[79]**

R.E. Steele, Department of Biological Chemistry and the Developmental Biology Center, University of California, Irvine, CA 92717-1700 **[231]**

Barbara J. Thomas, Howard Hughes Medical Institute and Department of Biological Chemistry, University of California, Los Angeles, CA 90024-1662 **[109]**

Cheryll Tickle, Department of Anatomy and Developmental Biology, University College and Middlesex School of Medicine, London W1P 6DB, UK **[207]**

Paul M. Wassarman, Department of Cell and Developmental Biology, Roche Institute of Molecular Biology, Roche Research Center, Nutley, NJ 07110 **[19]**

Fiona M. Watt, Keratinocyte Laboratory, Imperial Cancer Research Fund, London WC2A 3PX, UK **[241]**

Detlef Weigel, Division of Biology 156-29, California Institute of Technology, Pasadena, CA 91125 **[93]**

Gary M. Wessel, Division of Biology and Medicine, Brown University, Providence, RI 02912 **[255]**

Monte Westerfield, Institute of Neuroscience, University of Oregon, Eugene, OR 97403 **[69]**

Lewis Wolpert, Department of Anatomy and Developmental Biology, University College and Middlesex School of Medicine, London W1P 6DB, UK **[207]**

John M. Wozney, Genetics Institute, Inc., Cambridge, MA 02140 **[221]**

S. Lawrence Zipursky, Howard Hughes Medical Institute and Department of Biological Chemistry, University of California, Los Angeles, CA 90024-1662 **[109]**

Preface

The 51st Symposium of the Society for Developmental Biology, "Molecular Basis of Morphogenesis," was held June 24–27, 1992 at the University of Washington in Seattle. The symposium explored how various cells and tissues come together to generate the specific form and shape of organs. Each of the nine sessions focused on a distinct organ system and each session was chaired by a well-known investigator of that system. The speakers examined how molecular processes generate specific tissue and organ forms in order to highlight possible common (and unique) mechanisms. Several organisms were represented, each exploiting a developmental paradigm relevant to a specific organ system. Emphasis was placed on molecules that influence cell behavior, including extracellular matrix components, signaling peptides, degradative enzymes and their inhibitors, cell adhesion molecules and cell junction-associated molecules, and various intracellular organelles, including the actin-containing cytoskeleton, intermediate filaments, and microtubules.

Inquiry on how tissues and organs acquire their shape was a major focus of the earliest practitioners in our discipline, as witnessed by the controversy between proponents of pre-formation and adherents to the countervailing view. With advances in technology, in this case microscopy, it became clear that organization of the body form involved a sequential and complex set of structural changes. These observations led embryologists to devise sophisticated hypotheses to explain the generation of organ form, including the mathematically reasonable but biologically contrived approaches of D'Arcy Thompson. The problems underlying the stepwise construction of an organism were defined by Wilson in his classic text, *The Cell in Development and Inheritance*. Genetic influences on organ formation were recognized, especially from the work of Morgan on *Drosophila* in the teens and twenties, but detailed mechanisms awaited advances in the cell biology of embryos ushered in during the thirties, especially with the work of Paul Weiss. Our understanding of morphogenetic interactions came from a variety of organisms, including Wilson's turn-of-the-century work on the dissociation and reorganization of sponge cells, Spemann's transplantation of portions of amphibian embryos, and Grobstein's use of membrane filters to separate interacting mouse embryonic tissues, each of these milestones occurring at about 25-year intervals. With the discovery in the forties that environmental factors could be human teratogens, embryonic morphogenesis became increasingly relevant to clinical problems.

No previous technological advance, however, has been as dramatic in providing mechanistic understanding as the application of molecular biology and the use of genetic approaches to identify key genes and their functions in specific morphogenetic processes. Molecular approaches to morphogenesis have provided the best evidence for the unity of mechanisms in developing systems. As emphasized in this symposium, regardless of whether plants, invertebrates, or vertebrates are being studied, similar processes control the organization of cells into discrete forms. Organisms use both genetic and cellular hierarchies to establish these forms. Certain genes, when activated, coordinate the sequential activation of other genes, ultimately resulting in the gene products that dictate cellular behavior. At the same time, cells become stably committed to distinct lineages with specific developmental fates, yet can stop their level of commitment at many stages along the path to their ultimate fate.

Although several classes of molecules dictate the selective organization of cells into distinct tissues, the large functional repertoire of molecules within each class has only recently become apparent. One lesson of this symposium is that the expression and interactions of these molecules are regulated exquisitely. Another lesson is that the physiological implications of findings obtained by study of molecules and isolated cells must be confirmed at the organ level and, indeed, in the whole organism. While we have some insights into the molecules involved in controlling morphogenetic processes, we have less information on how these molecular events are integrated. This coordination must have its own regulation with a complex set of controls.

The molecular basis of morphogenesis has become increasingly relevant to clinical problems. Birth defects are now the prevalent cause of infant mortality in this country, accounting for about 20% of infant deaths and for many handicapped children. Indeed, the frequency of birth defects has been unchanged for decades while the mortality and disability rates from other causes have been reduced. The vast bulk of these problems, possibly greater than 90%, are due to non-environmental causes. Thus, the most reasonable way to prevent these defects is to decipher the mechanisms of morphogenesis, identify the molecules involved and their genes, and use this information to prevent the defects.

The chapters in this volume attest to the increased understanding of recent years. Exciting progress was palpable at this meeting and I hope that these chapters convey some of the enthusiasm. This symposium was the final meeting of the type, begun in 1939 by the symposia of the Growth Society, in which a single topic was explored. Subsequent annual meetings of our Society will be more eclectic in the topics covered, reflecting the extensive and successful incursions of our discipline into nearly every aspect of modern biology. Some of this eclecticism was reflected in the two mini-symposia that were held prior to this year's symposium: "Inheritance

Patterns During Fertilization," arranged by Gerald Schotten (University of Wisconsin, Madison), focused on non-traditional mechanisms of inheritance in a wide variety of organisms, and "Molecular Embryology and the Study of Lung Development," arranged by David Warburton (University of Southern California), Mary Ann Berberich (National Heart, Lung, and Blood Institute), and myself, focused on research tools of developmental biology that can be used to study the lung (summarized in the *American Journal of Respiratory Cell and Molecular Biology* 9:5–9, 1993). In addition, the meeting provided an opportunity for hands-on participation in a plant development workshop organized and directed by Judith E. Heady (University of Michigan, Dearborn).

The meeting was organized through the University of Washington's Office of Convention Services, and the special assistance of Celeste Berg, Cynthia Finch, Stephen Hauschka, and Merrill Hille helped to ensure that the symposium and minisymposia were successes. The meeting was funded, in part, by a grant from the National Science Foundation to assist with the travel expenses of the speakers. On behalf of the Society members and symposium participants, I thank Dr. Judith Plesset, Program Administrator for Developmental Biology, for this support.

Merton Bernfield

Young Investigators Awards, 1992

First Place

Mary Constance Lane
Department of Molecular and Cell Biology
University of California
Berkeley, California

**Regulated Secretion of Extracellular Matrix Drives Invagination
of the Vegetal Plate During Sea Urchin Gastrulation**

Second Place

Marnie E. Halpern
Institute of Neuroscience
University of Oregon
Eugene, Oregon

**The Notochord Induces Somite Patterning
in the Zebrafish Embryo**

Abstract of the First Place Young Investigator Award

Regulated Secretion of Extracellular Matrix Drives Invagination of the Vegetal Plate During Sea Urchin Gastrulation

Mary Constance Lane, F. Wilt, and R. Keller
*Department of Molecular and Cell Biology,
University of California, Berkeley, California*

Invagination of the vegetal plate in the gastrulating sea urchin is dependent on the hyaline layer (HL), a complex, stratified extracellular matrix (ECM) that surrounds the embryo. Invagination is inhibited by sulfate deprivation and exogenous xyloside, treatments that perturb proteoglycan synthesis in the embryo. Chondroitin sulfate proteoglycan has been localized to the HL of the vegetal plate at gastrulation using a monoclonal antibody (CS-56) to native chondroitin sulfate. Exposure to the calcium ionophore A23187 just before gastrulation (23 hr) stimulates precocious invagination in embryos, while exposure of mid-mesenchyme blastulae (21 hr) stimulates precocious exocytosis and results in a visible thickening of the matrix on the apical side of the vegetal plate epithelium. This apical matrix contains agarose, which provides a mechanical resistance to the swelling, secreted matrix, causes an inpocketing of the plate that appears similar to primary invagination in normal embryos. Neither normal nor precocious invagination is inhibited by treatment with cytochalasin D, but both are disrupted in embryos pretreated with monensin, which blocks secretion. We propose that the selective synthesis and directed secretion of matrix molecules by certain cells within an epithelium may provide the mechanical force to produce epithelial folding. The effects of proteoglycan secretion and deposition within a complex ECM during epithelial folding will be compared with a biophysical model in which thermal expansion in bimetallic strips produces a curved structure analogous to an invaginating epithelium.

1. Role in Gametogenesis of c-*kit* Encoded at the *W* Locus of Mice

Rosemary F. Bachvarova, Katia Manova, and Peter Besmer

Department of Cell Biology, Cornell University Medical College, New York, New York 10021 (R.B., K.M.) Molecular Biology Program, Sloan Kettering Institute, New York, New York 10021 (P.B.)

INTRODUCTION

The mechanisms by which mutations at the *white spotting locus (W)* of mice exert their effects on several unrelated cell types has long been of interest to classical geneticists and developmental biologists. Three major targets of *W* mutations have been defined: (1) hematopoietic cells, particularly erythroid progenitors and mast cells, (2) melanocytes and their precursors, and (3) germ cells (Russell, 1979; Silvers, 1979; Besmer, 1991). Homozygotes for the original *W* allele die of anemia, usually in the fetal-neonatal period (de Aberle, 1927), are black-eyed whites, and have few or no germ cells in the gonads. This mutation is effectively a null mutation (Nocka et al., 1990a). A mutation at another locus called *steel (Sl)* (see Silvers, 1979) gives rise to a very similar phenotype. A variety of studies have indicated that for hematopoietic cells and melanocytes, *W* mutations are cell autonomous: that is, the gene functions within the target cell, while *Sl* mutations affect the environment of target cells. Analysis of chimeric embryos has shown that germ cells carrying severe *Sl* genotypes can nevertheless form functional gametes (Nakayama et al., 1988; Kuroda et al., 1988), a result which suggests that for germ cells also, *Sl* acts through the environment.

The three major affected cell types share the property of migration within the embryo. Melanocytes migrate from the neural crest to their final destination in the skin, and hematopoietic progenitor cells travel from yolk sac to fetal liver (Toles et al., 1989) and other sites of hematopoiesis. The germ cells are found first in the allantois and later in the hindgut, from which they migrate through the dorsal mesentery to the gonadal ridge. The three cell types also proliferate rapidly early in development. These properties are shared to varying degrees with other cell types, so that there is no obvious feature that explains why these three cell types are affected in common.

A major step foward occurred when it was discovered that the *W* locus encodes c-*kit*, a protooncogene corresponding to the oncogene of a feline sarcoma virus (Chabot et al., 1988; Geissler et al., 1988; Nocka et al., 1989). Its sequence showed that c-*kit* encodes a tyrosine kinase receptor in the PDGF (platelet-derived growth factor) and CSF-1 (colony stimulating factor-1) receptor family (Besmer et al., 1986; Qui et al., 1988). The gene is located adjacent to the α subunit of the PDGF receptor on chromosome 5 of the mouse (Chabot et al., 1988; Stephenson et al., 1991). It seemed likely that the *Sl* locus would encode its ligand, and indeed the factor activating processes mediated by c-*kit* was isolated and shown to map to the *Sl* locus (Copeland et al., 1990; Nocka et al., 1990c; Zsebo et al., 1990). The kit-ligand is a membrane-spanning protein synthesized from two alternatively spliced forms of the message. One of these encodes a protein that is more resistant to proteolytic cleavage and, therefore, is predominantly membrane-associated (Flanagan et al., 1991; Huang et al., 1992). Activation of the c-*kit* receptor promotes survival and proliferation of mast cells (Fujita et al., 1988; Nocka et al., 1990b), and may have a similar action on other cells expressing c-*kit*. Additional activities may include stimulation of motility (Blume-Jensen et al., 1991) and of differentiated functions. Signal transduction by the activated receptor is apparently mediated by phosphorylation of downstream targets (e.g., Rottapel et al., 1991).

A large number of alleles are available at both the *W* and *Sl* loci, and their phenotypes vary widely. Heterozygotes for null mutations such as *Sl*/+ and *W*/+ display a white belly spot, slight anemia, and are fertile. Some mutations have dominant negative effects, that is, heterozygotes are more severely affected than *W*/+. Most severe is the W^{42} allele for which heterozygotes are white and anemic, and show reduced fertility (Geissler et al., 1981). The W^{42} mutation is a missense mutation in the kinase domain that eliminates c-*kit* function (Tan et al., 1990). The dominant negative effect is presumably due to the fact that dimerization of the receptor is necessary for normal function, and heterodimers consisting of mutant and normal c-*kit* receptors are nonfunctional.

ROLE OF c-*kit* IN THE DEVELOPMENT OF PRIMORDIAL GERM CELLS

The function of c-*kit* and its ligand in germ cells occurs at two major phases of development: during proliferation and migration of primordial germ cells in the 8- to 13-day mouse embryo, and during postnatal gametogenesis. The classic studies of Bennett (1956) and Mintz and Russell (1957) on *Sl* and *W* mutant mice, respectively, followed primordial germ cells from their time of appearance in the 7½-day (E7½) embryo, to their arrival in the gonads. Germ cells can be located by their expression of alkaline phosphatase, which provides a convenient histochemical marker. Germ cells are first identified in the allantois soon after they are generated from the epiblast, and move through the primitive streak into the extraembryonic mesoderm (Ginsburg et al., 1990). Mutations at *W* and *Sl* apparently do not affect the number of germ cells generated by segregation from somatic cells. Thus, c-*kit* is not involved in determination of primordial germ cells, but in some later process. In normal embryos, the number of germ cells increases from about 40 at E7½ to about 25,000 at E13½ (Tam and Snow, 1981). In severely affected mutant mice, the number of germ cells does not increase, but they persist and may reach the gonad (Coulombre and Russell, 1954; McCoshen and McCallion, 1975). Consistent with this phenotype, c-*kit* is expressed in primordial germ cells from 7½ days; expression continues throughout the period of proliferation to E13½, and ceases at E13½ to E14½, as seen by in situ hybridization (Fig. 1) (Manova and Bachvarova, 1991) and by immunocytochemistry (Fig. 2). Expression of c-*kit* during the germ cell cycle is summarized in Figure 3.

Interestingly, the first expression of c-*kit* in embryo-derived cells is in parietal endoderm, starting at 5½ days (see Fig. 4), and precedes expression in germ cells. c-*kit* is also expressed in uterine tissues during and after implantation (Orr-Urtreger et al., 1990; Horie et al., 1992). Both of these cases are examples of the widespread expression of c-*kit* in tissues other than those known to be affected by *W* and *Sl* mutations.

Throughout the period of germ cell proliferation, kit-ligand is produced in cells around the primordial germ cells, starting in the allantois and endoderm, and continuing in hindgut mesoderm, dorsal mesentery, and gonadal ridges (see Figs. 5 and 6D) (Matsui et al., 1990; Keshet et al., 1991, Motro et al., 1991). Expression of kit-ligand continues in the fetal gonad (Manova and Bacharova unpublished data; Motro et al., 1991), while c-*kit* expression ceases in male germ cells as they enter a quiescent period or in female germ cells as they enter meiotic prophase. Thus, the pattern of expression is consistent with a role of c-*kit* in promoting germ cell proliferation. Complicating the picture is the fact that c-*kit* is also expressed in tissues that the germ cells migrate through, that is, gut mesoderm and dorsal mesentery (Orr-Urtreger et al., 1990; Manova and Bachvarova, 1991; Keshet et al., 1991), tissues that also express the ligand. This expression of c-*kit* in the environment of target cells probably reduces the amount of available ligand, and a balance of c-*kit* and ligand expression may regulate germ cell proliferation. A 50% reduction of ligand in *Sl*/+ embryos appears to reduce the number of primordial germ cells (Bennett, 1956), and the larger reduction that probably occurs in Sl^{pan}/Sl^{pan} embryos results in a larger decrease in the number of germ cells (see below). The small size of $W^{42}/+$ gonads (Geissler et al., 1981) suggests that primordial germ cell number is also reduced by decreased levels of functional c-*kit*.

In vitro studies have shown that kit-ligand is a survival factor for primordial germ cells (Godin et al., 1991; Dolci et al., 1991) and also a proliferation factor acting in synergy with other factors (Matsui et al., 1991). While primordial germ cell motility is stimulated by the ligand, it does not act as a

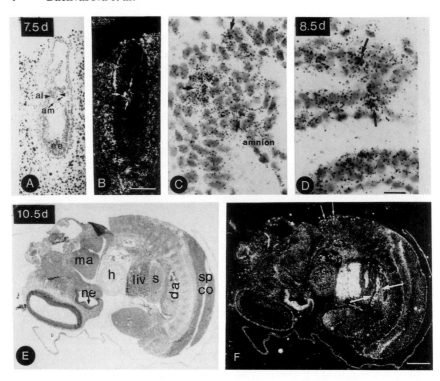

Fig. 1. Expression of c-*kit* RNA in embryonic germ cells. **A,B:** Bright-field and corresponding dark-field images of an E7½ embryo. Germ cells (arrows in B) in the allantois (al) are labeled; amnion (am). Scale bar = 200 μm. Labeling outside the embryo is an artifact. **C:** High-magnification view of the labeled germ cells (arrows) in (B). **D:** Labeled germ cells (arrows) in the hindgut of an E8½ embryo. The gut endoderm and mesenchyme are also labeled. Scale bar = 20 μm for (C) and (D). **E,F:** Labeling of germ cells in the gonadal ridge (large arrows) below the dorsal aorta (da), and of melanoblasts (small arrows) at E10½. The liver (liv) and nasal epithelium (ne) are also highly labeled. Abbreviations: ma, mandible; h, heart; s, stomach. (Panels A–D reprinted from Manova and Bacharova, 1991, with permission of the publisher.)

chemotactic factor (Dolci et al., 1991). Ligand emanating from the gonad in vivo probably does not attract germ cells, since some germ cells arrive in the gonad even in *W/W* embryos (Coulombre and Russell, 1954).

The role of c-*kit* in melanoblasts may be similar to its role in primordial germ cells. Melanocyte precursors first appear at about E9½ as cells derived from the neural crest and are sparsely distributed along a dorsal-lateral pathway over the somites, while most cells derived from the neural crest move ventrally within or medial to the somites (Serbedzija et al., 19901). From day 9½ to day 12½, the melanoblasts spread through the dermis, covering the whole body, and then enter the epidermis (Rawles, 1947; Mayer, 1973). Expression of c-*kit* in this lineage is first detected in cells that are on the dorsal-

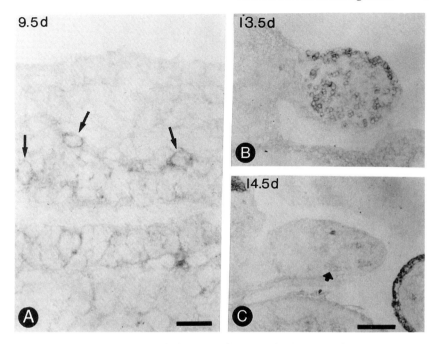

Fig. 2. Expression of c-*kit* protein in embryonic germ cells. Frozen sections incubated with anti-c-*kit* antiserum (A) or monoclonal antibody, ACK2 (B,C), and stained with an indirect horseradish peroxidase system. **A:** Three stained germ cells in the hindgut of a E9½ embryo (arrows). Scale bar = 20 μm. **B:** Stained germ cells in the ovary at E13½. **C:** A few lightly stained germ cells (arrow) remain in the E14½ ovary (arrow). Scale bar = 100 μm.

lateral path above the somites (Fig. 6C,E; Fig. 1F) (Manova and Bachvarova, 1991). The first source of kit-ligand along the migratory path is the dorsal somite, followed by the dermatome that develops from the somite (Fig. 6A,G) (Matsui et al., 1990). The kit-ligand is likely to be involved in survival and/or proliferation of melanoblasts migrating over the somite and through the dermis. Whether c-*kit* might also play a function in determination of melanoblasts is difficult to evaluate; upregulation of c-*kit* may reflect one step in determination, and stimulation by ligand could provide a second determining signal only to cells on the dorsal-lateral path.

ROLE OF c-*kit* IN THE POSTNATAL OVARY

A variety of *W* and *Sl* genotypes have mild effects on target cells, and thus it is possible to study the function of the c-*kit* receptor-ligand system at developmental stages subsequent to the first point at which it is required. Also, in some cases differential effects can be seen on the three major targets, which suggests that different aspects of the system are important in different cell types.

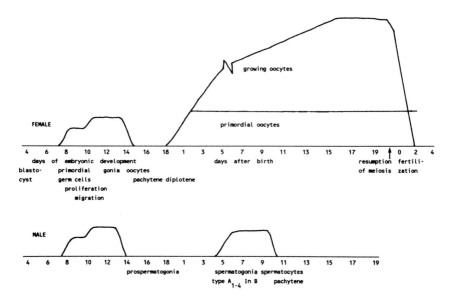

Fig. 3. Expression of c-*kit* during the germ cell cycle in females and males. This diagrammatic representation is not quantitative. Expression is shown for primordial oocytes and for the first wave of growing oocytes appearing at birth and for the first wave of developing male germ cells.

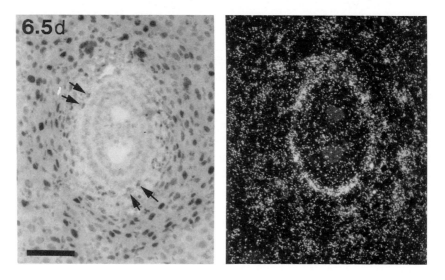

Fig. 4. Expression of c-*kit* RNA in parietal endoderm in a cross section of an E6½ embryo. Arrows in left panel indicate some of the labeled cells shown in the dark-field image (right panel). Scale bar = 100 μm.

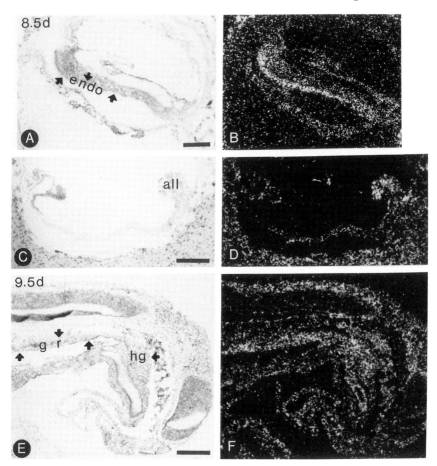

Fig. 5. Expression of kit-ligand RNA at E8½ and E9½. **A,B:** Labeled endoderm (endo) at E8½. **C,D:** Labeled allantois at E8½. **E,F:** Labeled gonadal ridge (gr) bordering on the mesonephric region; and hindgut (hg) mesenchyme at E9½. The meninges around the spinal cord are also labeled. Scale bars = 100 μm.

For germ cells, certain mutant alleles of W and Sl affect postnatal gonads and fertility differentially in males and females (see Geissler et al., 1981; Beechey and Searle, 1983; Copeland et al., 1990). If one sex is fertile, clearly germ cells reached the gonad in the embryo, while if the other sex is infertile, then the receptor-ligand system must function at some point after sexual differentiation.

In order to determine at what stage and in what cell types c-*kit* might be functioning in the ovary, in situ hybridization and immunohistochemical studies have been carried out. Kit is highly expressed on the surface of all oocytes of the postnatal ovary (Fig. 7) (Manova et al., 1990; Horie et al., 1991). In

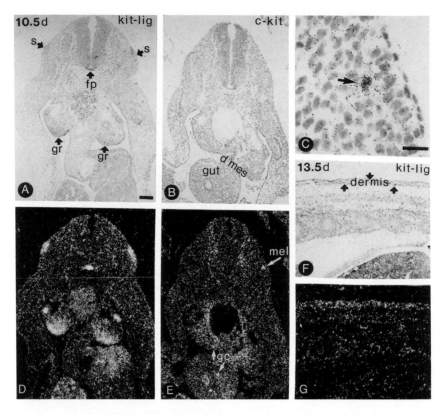

Fig. 6. Expression of kit-ligand and c-*kit* RNA at E10½ and E13½. **A,D:** Expression of kit-ligand in somites (s), floorplate (fp), and gonadal ridges (gr) at E10½. A composite from two adjacent sections is presented. Scale bar = 100 μm for panels A, B, D, E, F, and G. **B,E:** Expression of c-*kit* in germ cells (gc) and in the gut and dorsal mesentery that they migrate through. One labeled melanocyte precursor (mel) is seen dorsolateral to the somite. **C:** The melanocyte precursor seen in (E), shown at high magnification. Scale bar = 20 μm. **F,G:** Kit-ligand RNA in the dermis at E13½.

mice, oocytes progress through meiotic prophase in fetal life, enter the diplotene stage near the time of birth, and then become surrounded by follicle cells to form primordial follicles (see Bachvarova, 1985). Growth of oocytes and follicles commences almost immediately in the central region of perinatal ovaries. This wave of growing oocytes reaches half-size in one- to two-layered cuboidal follicles at about 8 days of age, and reaches full-size in in follicles with three or more layers at about 16 days of age. Growing oocytes continue to be generated from primordial oocytes throughout the fertile life of the animal. The onset of c-*kit* expression in oocytes can be traced back to the diplotene stage in late fetal life, before primordial follicles form. It continues in

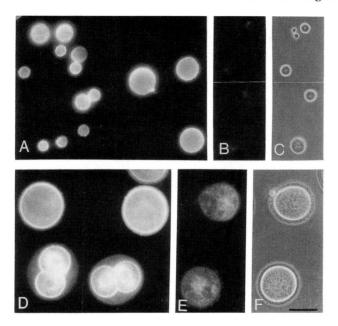

Fig. 7. Expression of c-*kit* protein on the surface of oocytes from postnatal ovary, and on 2-cell embryos. Unfixed oocytes and embryos were incubated with anti-c-*kit* antiserum which was detected with fluorescein-labeled secondary antibody. **A:** Primordial and growing oocytes. **B:** Same as (A), incubated with preimmune serum. **C:** Phase contrast image of (B). **D:** Full-grown oocytes and 2-cell embryos incubated with immune serum absorbed with *W/W* mast cells specifically lacking c-*kit*. **E:** Full grown oocytes incubated with immune serum absorbed with +/+ mast cells expressing c-*kit*. **F:** Corresponding phase micrograph. Scale bar = 50 μm for all panels.

primordial, growing, and full-grown oocytes, and through meiotic maturation and into early embryonic development (see Fig. 3).

The kit-ligand is expressed in a pattern consistent with a role in oocyte growth (see Fig. 8). It is expressed in the gonad during fetal development, and by birth is present in high amounts in cords and clusters of somatic cells in the central region of the ovary (Manova et al., 1993). Such cells contribute to follicles that form in this region and start to grow without delay. In one- and two-layered cuboidal follicles, expression is low and variable, and most of the protein detected in 5- to 8-day-old mice is found within the oocytes, presumably taken up by receptor-mediated endocytosis. Expression is clearly evident in follicle cells of three-layered follicles during late oocyte growth. In follicles with full-grown oocytes, expression is restricted to the outer layer, and ligand within the oocyte is reduced or absent. Thus, the ligand is apparently available from nearby follicle cells throughout the period of oocyte growth, but much less is available both before and after oocyte growth. It is likely that a

kit-ligand in ovary

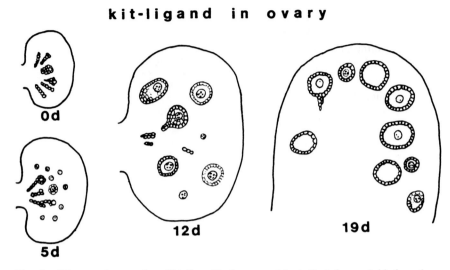

Fig. 8. Diagram of expression of kit-ligand in the ovary at the indicated ages. Aside from the outline of the ovaries and of oocytes, the drawing indicates kit-ligand expression as defined with antibodies direct against the soluble protein and against a peptide located in the cytoplasmic domain (Manova et al., 1993).

soluble factor diffusing from two to three layers of follicle cells mediates more of the response than a membrane-bound factor, since follicle cells contact growing oocytes only at the ends of processes projecting through the zona pellucida.

In functional studies, the effect of added ligand on the growth of oocytes cultured over a 6-day period in the presence of follicle cells has been monitored. In the range of 5 to 100 ng/ml, the ligand was able to stimulate growth significantly, and at 50 ng/ml it more than doubled the rate of increase in diameter which was 0.5 μm per day in controls. Moreover, addition of c-*kit* blocking antibody (ACK2 obtained from S.-I. Nishikawa) to cultures of dispersed newborn ovaries significantly inhibited oocyte growth (Packer AI, Hsu YC, Besmer P, Bachvarova RF, manuscript submitted).

The Sl^t and Sl^{pan} mutations are of particular interest here, because female homozygotes are sterile, while males are fertile. In Sl/Sl^t females, oocyte and follicle development is arrested at the one-layered cuboidal stage (Kuroda et al., 1988). For Sl^{pan}, the mutation does not affect the amino acid sequence of the ligand, but in homozygotes reduces the level of its expression in several tissues to about 20% of normal (Huang et al., 1993). The Sl^{pan}/Sl^{pan} ovary has a greatly reduced number of germ cells relative to normal, indicating an effect during primordial germ cell development. In the postnatal ovary, follicle development proceeds only to the one-layered cuboidal stages, as in Sl/Sl^t females. These results suggest that kit-ligand is required at a level greater than 20% of normal at this stage. The fact that the follicle cells, in turn, do not develop suggests that the normal growing oocyte emits factors that maintain follicle cell proliferation (see Vanderhyden et al., 1992).

For comparison, in $W^{42}/+$ mice, c-*kit* function is reduced by a dominant negative effect. Nevertheless, oocyte and follicle development proceeds normally in postnatal ovaries (Manova K, Bachvarova RF, unpublished data). Also, follicle and oocyte development can be quite normal in W/W^v and W^v/W^v mice (Coulombre and Russell, 1954), having limited c-*kit* function (Nocka et al., 1990a). Thus, it appears that during oocyte growth the receptor is present in greater excess than the ligand.

In conclusion, kit-ligand appears to be a growth factor for oocytes. Its action may be in many ways analogous to that of growth factors such as EGF (epidermal growth factor) and PDGF, which promote growth and proliferation of various somatic cells by activation of their tyrosine kinase receptors. The primordial oocyte in the diplotene stage of prophase of meiosis is arrested in the cell cycle, and stimulation may be required to trigger reentry into a growing state. Kit-ligand maintains growth and may be involved in an initial triggering event as well. Further movement in the cell cycle is represented by acquisition of competence to resume meiosis and to enter M phase; these are acquired near the end of oocyte growth, and not expressed in vivo until the surge of luteinizing hormone (Schultz, 1986; Motlik and Kubelka, 1990). There is no evidence for involvement of c-*kit* in the late phase of oocyte development.

ROLE OF c-*kit* IN POSTNATAL TESTIS

Prospermatogonia resume proliferation in early postnatal testes, and type A spermatogonia are produced by 6 days of age. Intermediate and type B spermatogonia and early meiotic spermatocytes appear by 10 days, and spermatids by 20 days (Sutcliffe and Burgoyne, 1989). Expression of c-*kit* reappears in the most advanced spermatogonia between 4 and 6 days of age, and remains in differentiating type A, intermediate and type B spermatogonia, and at a lower level in early meiotic spermatocytes (Fig. 9A–E); this pattern is maintained in the adult (Manova et al., 1990; Yoshinaga et al., 1991). By in situ hybridization, kit-ligand is expressed in Sertoli cells during the fetal period and in juvenile mice (Fig. 9F,G) (Manova et al., 1993). Immunohistochemistry demonstrates that the protein is present at all ages and is distributed throughout the tubule surrounding most of the germ cells (e.g., Fig. 10). The protein appears to be concentrated basally during the portion of the cycle of the seminiferous epithelium when type A_{2-4} spermatogonia are proliferating in the basal layer. Analysis of the forms of kit-ligand RNA present suggests that more of the membrane-bound protein is present in testes than in ovary.

Two lines of evidence demonstrate that c-*kit* is required for proliferation of type A spermatogonia. Heterozygous $Sl/+$ mice show reduced efficiency of production of type B from type A spermatogonia (Nishimune et al., 1980). In

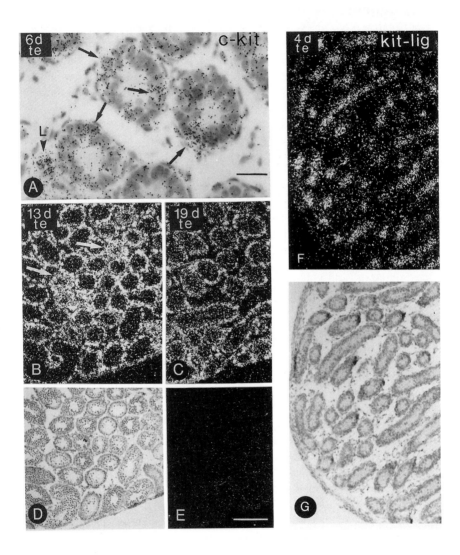

Fig. 9. Expression of c-*kit* RNA and kit-ligand RNA in testes; c-*kit* probe (A–D). **A:** Day 6 testis with labeled spermatogonia (arrows). **B,D:** Day 13 testis with labeled spermatogonia or early spermatocytes (arrows) in all tubules. **C:** Day 19 testis with labeled cells located basally at the outside of the tubule. **E:** Control day 13 testis hybridized to the sense strand probe. **F,G:** kit-ligand probe showing labeled Sertoli cells in day 4 testis.

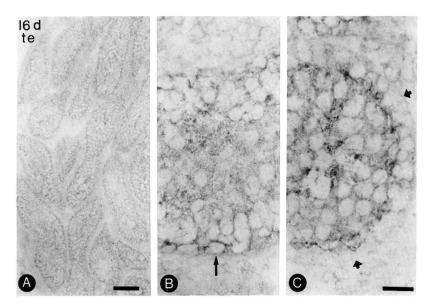

Fig. 10. Expression of kit-ligand in day 16 testis. **A:** Frozen section of testis incubated with antiserum directed against the soluble form of the ligand. Reaction was detected with an indirect horseradish peroxidase system. Scale bar = 100 μm. **B:** Tubule with staining in the basal layer and staining in a putative type A spermatogonium (arrow). Presumably the staining represents uptake of ligand. **C:** Tubule at about stage VII with little staining in the basal layer of the tubule. Arrows indicate the outer limit of the tubule. Scale bar = 20 μm for (B) and (C).

Sl/Sl^d mice, which synthesize reduced levels and only the soluble form of the ligand (Brannan et al., 1991), only a few undifferentiated type A spermatogonia are found, and these do not progress further (Nakayama et al., 1988). Most dramatic, injection of a monoclonal c-*kit* blocking antibody into male mice arrested development of, or eliminated, differentiating type A spermatogonia (Yoshinaga et al., 1991). Injection of the antibody from day 0 to day 5 of age had no effect on ongoing mitosis, injection from day 6 to day 14 blocked spermatogonial development, while injection from day 8 allowed some gonia to develop and produce spermatocytes and spermatids. Thus, the most advanced spermatogonia move through the phase when c-*kit* is required on days 6 and 7. These and other results showed that c-*kit* is necessary for proliferation of most, or all, of the four generations of type A spermatogonia, but not for the transition of type B spermatogonia to spermatocytes.

The ligand appears to be present much more widely within the tubule than necessary to serve this function and could be released either apically into the lumen of the tubule or basally to targets outside the tubule, to mediate other functions. The membrane-bound form of the ligand may be directed specifically toward differentiating type A spermatogonia at the appropriate stage of the cycle, or the onset of c-*kit* expression in spermatogonia may control the onset of function in this system. The function of c-*kit* is presumably to provide growth signals to promote the gonial cell cycle. It is interesting to note that the cell cycles of differentiating spermatogonia are limited to a given number, and other factors presumably mediate the variable proliferation of undifferentiated stem cell spermatogonia.

CONCLUSION

c-*kit* is involved in different aspects of cell cycle progression at several stages of germ cell development. In oocytes, it functions during oocyte growth, a substage of diplotene stage of meiotic prophase. Functional and cytological evidence suggests it functions throughout growth. In spermatogonia, it has been clearly shown that c-*kit* is required during approximately the first four cell cycles of differentiating spermatogonia, while c-*kit* continues to be expressed at both the RNA and protein levels through the last two cycles and into preleptotene spermatocytes. c-*kit* may also be required for a defined period during primordial germ cell development. In fact, primordial germ cells from E11½ embryos respond to the kit-ligand in vitro, but those from E12½ do not (Matsui et al., 1991), although they still express c-*kit*.

Another important function of the c-*kit* receptor may be to promote cell survival. Interaction with growth factors may be required quite generally to prevent the programmed cell death that would otherwise occur (Raff, 1992). The same growth factor may stimulate proliferation at one concentration and promote cell survival at a lower concentration. While not required for survival of primoridal germ cells in vivo, its demonstrated action in vitro (see above) suggests that the kit-ligand may be one of the factors contributing to their survival. Spontaneous cell death is a very dramatic feature of germs cells during fetal life, particularly during the meiotic prophase of female germ cells (e.g., Bakken and McClanahan, 1978). The apparent absence of functional c-*kit* receptor during this period could be related to the low survival rate. Similarly, primordial oocytes undergo atresia at a higher rate than growing oocytes (Faddy et al., 1983), and in this case the atresia is correlated with a lower availability of kit-ligand.

With regard to the role of c-*kit* in the cell cycle, we propose that the three cell types most severely affected by *W* mutations may have in common a proliferation phase that is closely linked to a differentiation program. For germ cells, the concept of differentiation has to be modified, since these are potentially totipotent. In most animals, the growing oocyte is a critical stage of "differentiation" in that information is stored to program early spatial development of the embryo. In mammals, this function is minimized, but a variety of steps must occur in preparation for meiotic maturation, fertilization, and early cleavage. Proliferating primordial germ cells proceed through several developmental changes, including a period of migration (e.g., Clark and Eddy, 1975), and perhaps a programmed number of cell cycles limited to the approximately eight that normally occur. As indicated above, spermatogonial proliferations are limited, the gonia go through a morphological transition, and perhaps a series of interactions with surrounding cells. It will be of great interest to understand how cell cycle and differentiation signals are linked at the molecular level.

REFERENCES

Bachvarova R (1985): Gene expression during oogenesis and oocyte development in the mammal. *In* Browder LW (ed): "Developmental Biology. A Comprehensive Synthesis," vol. 1. New York: Plenum, pp 453–524.

Bakken AH, McClanahan M (1978): Patterns of RNA synthesis in early meiotic prophase oocytes from fetal mouse ovaries. Chromosoma 67:21–30.

Beechey CV, Searle A (1983): Contrasted, a steel allele in the mouse with intermediate effects. Genet Res 42:183–191.

Bennett D (1956): Developmental analysis of a mutation with pleiotropic effects in the mouse. J Morphol 98:199–233.

Besmer P (1991): The *kit*-ligand encoded at the murine *steel* locus: A pleiotropic growth and differentiation factor. Curr Opin Cell Biol 3:939–946.

Besmer P, Murphy PC, George PC, Qiu F, Bergold PJ, Lederman L, Snyder HW, Brodeur D, Zuckerman EE, Hardy WD (1986): A new acute transforming feline retrovirus and relation of its oncogene v-*kit* with the protein kinase gene family. Nature 320:415–421.

Blume-Jensen P, Claesson-Welsh L, Siegbahn A, Zsebo KM, Westermark B, Heldin C-H (1991): Activation of the human c-*kit* product by ligand-induced dimerization mediates circular actin reorganization and chemotaxis. EMBO J 10:4121–4128.

Brannan CI, Lyman SD, Williams DE, Eisenman J, Anderson DM, Cosman D, Bedell MA, Jenkins NA, Copeland NG (1991): Steel-Dickie mutation encodes a c-kit ligand lacking transmembrane and cytoplasmic domains. Proc Nat Acad Sci USA 88:4671–4674.

Chabot B, Stephenson DA, Chapman VM, Besmer P, Bernstein A (1988): The protooncogene c-*kit* encoding a transmembrane tyrosine kinase receptor maps to the mouse *W* locus. Nature 335:88–89.

Clark JM, Eddy EM (1975): Fine structural observations on the origins and associations of primordial germ cells of the mouse. Dev Biol 47:136–155.

Copeland NG, Gilbert DJ, Cho BC, Donovan PJ, Jenkins NA, Cosman D, Anderson D, Lyman SD, Williams DE (1990): Mast cell growth factor maps near the steel locus on mouse chromosome 10 and is deleted in a number of steel alleles. Cell 63:175–183.

Coulombre JL, Russell ES (1954): Analysis of the pleiotropism at the W-locus in the mouse. The effects of W and W^v substitution upon postnatal development of germ cells. J Exp Zool 126:277–295.

de Aberle SB (1927): A study of the hereditary anemia of mice. Am J Anat 40:219–247.

Dolci S, Williams DE, Ernst MK, Resnick JL, Brannan CI, Lock LF, Lyman SD, Boswell HS, Donovan PJ (1991): Requirement for mast cell growth factor for primordial germ cell survival in culture. Nature 352:809–811.

Faddy MJ, Gosden RG, Edwards RG (1983): Ovarian follicle dynamics in mice: A comparative study of three inbred strains and an F_1 hybrid. J Endocrinol 96:23–34.

Flanagan JG, Chan DC, Leder P (1991): Transmembrane form of the *kit* ligand growth factor is determined by alternative splicing and is missing in the Sl^d mutant. Cell 64:1025–1035.

Fujita J, Nakayama H, Onoue H, Kanakura Y, Nakano T, Asai H, Takeda, S-I, Honjo T, Kitamura Y (1988): Fibroblast-dependent growth of mast cells in vitro: Duplication of mast cell depletion in mutant mice of W/W^v phenotype. J Cell Physiol 134:78–84.

Geissler EN, McFarland EC, Russell ES (1981): Analysis of pleiotropism at the dominant white-spotting *(W)* locus of the house mouse: A description of ten new *W* alleles. Genetics 97:337–361.

Geissler EN, Ryan MA, Housman DE (1988): The dominant-white spotting (W) locus of the mouse encodes the c-kit proto-oncogene. Cell 55:185–192.

Ginsburg M, Snow MHL, McLaren A (1990). Primordial germ cells in the mouse embryo during gastrulation. Development 110:521–528.

Godin I, Deed R, Cooke J, Zsebo K, Dexter M, Wylie CC (1991): Effects of the *steel* gene product on mouse primordial germ cells in culture. Nature 352:807–809.

Horie K, Takakura K, Taii S, Narimoto K, Noda Y, Nishikawa S, Nakayama H, Fujita J, Mori T (1991): The expression of c-kit protein during oogenesis and early embryonic development. Biol Reprod 45:547–552.

Horie K, Fujita J, Takakura K, Kanazaki H, Kaneko Y, Iwai M, Nakayama H, Mori T (1992): Expression of c-kit protein during placental development. Biol Reprod 47:614–620.

Huang EJ, Nocka KH, Buck J, Besmer P (1992): Differential expression and processing of two cell associated forms of the kit-ligand: KL-1 and KL-2. Mol Biol Cell 3:349–362.

Huang EJ, Manova K, Packer AI, Sanchez S, Bachvarova RF, Besmer P (1993): The murine steel panda mutation affects kit-ligand expression and growth of early ovarian follicles. Dev Biol 157:100–109.

Keshet E, Lyman SD, Williams DE, Anderson DM, Jenkins NA, Copeland NG, Parada LF (1991): Embryonic RNA expression patterns of the *c-kit* receptor and its cognate ligand suggest multiple functional role in mouse development. EMBO J 10:2425–2435.

Kuroda H, Terada N, Nakayama H, Matsumoto K, Kitamura Y (1988): Infertility due to growth arrest of ovarian follicles in Sl/Sl^t mice. Dev Biol 126:71–79.

Manova K, Bachvarova R (1991): Expression of c-*kit* encoded at the *W* locus of mice in developing embryonic germ cells and presumptive melanoblasts. Dev Biol 146:312–324.

Manova K, Nocka K, Besmer P, Bachvarova RF (1990): Gonadal expression of c-*kit* encoded at the *W* locus of mice. Development 110:1057–1069.

Manova K, Huang EJ, Angeles M, De Leon V, Sanchez S, Pronovost SM, Besmer P, Bachvarova RF (1993): The expression pattern of the c-*kit* ligand in gonads of mice supports a role for the c-*kit* receptor in oocyte growth and in proliferation of spermatogonia. Dev Biol 157:85–99.

Matsui Y, Zsebo KM, Hogan BLM (1990): Embryonic expression of SCF (stem cell factor), a haematopoietic growth factor encoded by the *Sl* locus and the ligand for c-kit. Nature 347:667–669.

Matsui Y, Toksoz D, Nishikawa S, Nishikawa S-I, Zsebo K, Hogan BLM (1991): Effect of *steel* factor and leukaemia inhibitory factor on murine primordial germ cells in culture. Nature 353:750–752.

Mayer TC (1973): The migratory pathway of neural crest cells into the skin of mouse embryos. Dev Biol 34:39–46.

McCoshen JA, McCallion DJ (1975): A study of the primordial germ cells during their migratory phase in steel mutant mice. Experientia 31:589–590.

Mintz B, Russell ES (1957): Gene-induced embryological modifications of primordial germ cells in the mouse. J Exp Zool 134:207–237.

Motlik J, Kubelka M (1990): Cell-cycle aspects of growth and maturation of mammalian oocytes. Mol Reprod Dev 27:366–375.

Motro B, Van der Kooy D, Rossant J, Reith A, Bernstein A (1991): Contiguous patterns of c-*kit* and *steel* expression: Analysis of mutations at the *W* and *Sl* loci. Development 113: 1207–1221.
Nakayama H, Kuroda H, Onoue H, Fujita J, Nishimune Y, Matsumoto K, Nagano T, Suzuki F, Kitamura Y (1988): Studies of $Sl/Sl^d \leftrightarrow +/+$ mouse aggregation chmieras, II. Effect of the steel locus on spermatogenesis. Development 102:117–126.
Nishimune Y, Haneji T, Kitamura Y (1980): The effects of Steel mutation on testicular germ cell differentiation. J Cell Physiol 105:137–141.
Nocka K, Majumder S, Chabot B, Ray P, Cervone M, Bernstein A, Besmer P (1989) Expression of c-*kit* gene products in known cellular targets of *W* mutations in normal and *W* mutant mice—evidence for an impaired c-*kit* in mutant mice. Genes Dev 3:816–826.
Nocka K, Tan JC, Chiu E, Chu TY, Ray P, Traktman P, Besmer P (1990a): Molecular bases of dominant negative and loss of function mutations at the murine c-*kit*/white spotting locus: W^{37}, W^v, W^{41}, W. EMBO J 9:1805–1813.
Nocka K, Buck J, Levi E, Besmer P (1990b). Candidate ligand for the c-*kit* transmembrane receptor: KL, a fibroblast-derived growth factor stimulates mast cells and erythroid progenitors. EMBO J 9:3287–3294.
Nocka K, Huang E, Beier DR, Chu T-Y, Buck J, Lahm H-W, Wellner D, Leder P, Besmer P (1990c): The hematopoietic growth factor KL is encoded by the *Sl* locus and is the ligand of the c-*kit* receptor, the gene product of the *W* locus. Cell 63:225–233.
Orr-Urtreger A, Avivi A, Zimmer Y, Givol D, Yarden Y, Lonai P (1990): Developmental expression of c-*kit,* a proto-oncogene encoded by the *W* locus. Development 109:911–923.
Qiu F, Ray P, Brown K, Barker PE, Jhanwar S, Ruddle FH, Besmer P (1988): Primary structure of c-*kit:* Relationship with the CSF-1/PDGF receptor kinase family—Oncogenic activation of v-*kit* involves deletion of extracellular domain and C terminus. EMBO J 7:1003–1011.
Raff MC (1992): Social controls on cell survival and cell death. Nature 356:397–400.
Rawles ME (1947): Origin of pigment cells from the neural crest in the mouse embryo. Physiol Zool 20:248–265.
Rottapel R, Reedijk M, Williams DE, Lyman SD, Anderson DM, Pawson T, Bernstein A (1991): The *Steel/W* transduction pathway: Kit autophosphorylation and its association with a unique subset of cytoplasmic signaling proteins is induced by the Steel factor. Mol Cell Biol 11:3043–3051.
Russell ES (1979): Hereditary anemias of the mouse. Adv Genet 20:357–459.
Schultz R (1986): Molecular aspects of mammalian oocyte growth and maturation. In Rossant J, Pedersen RA (eds): "Experimental Approaches to Mammalian Embryonic Development." New York: Cambridge University Press, pp 195–237.
Serbedzija GN, Fraser SE, Bronner-Fraser M (1990): Pathways of trunk neural crest cell migration in the mouse embryo as revealed by vital dye labeling. Development 108:605–612.
Silvers WK (1979). "The Coat Colors of Mice." New York: Springer-Verlag.
Stephenson DA, Mercola M, Anderson E, Wang C, Stiles CD, Bowen-Pope DF, Chapman VM (1991): Platelet-derived growth factor receptor α-subunit gene *(Pdgfra)* is deleted in the mouse Path *(Ph)* mutation. Proc Nat Acad Sci USA 88:6–10.
Sutcliffe MJ, Burgoyne PS (1989): Analysis of the testes of H-Y negative XOSxr^b mice suggests that the spermatogenesis gene *(Spy)* acts during the differentiation of the A spermatogonia. Development 107:373–380.
Tam PL, Snow MHL (1981): Proliferation and migration of primordial germ cells during compensatory growth in mouse embryos. J Embryol Exp Morph 64:133–147.
Tatn JC, Nocka K, Chiu E, Chu TY, Ray P, Traktman P, Besmer P (1990): The dominant W^{42} spotting phenotype results from a missense mutation in the c-*kit* receptor kinase. Science 247:209–212.

Toles JF, Chui DH, Belbeck LW, Starr E, Barker JE (1989): Hemopoietic stem cells in murine embryonic yolk sac and peripheral blood. Proc Natl Acad Sci USA 86:7456–7459.

Vanderhyden BC, Telfer EE, Eppig JJ (1992): Mouse oocytes promote proliferation of granulosa cells from preantral and antral follicles in vitro. Biol Reprod 46:1196–1204.

Yoshinaga K, Nishikawa S, Ogawa M, Hayashi SI, Kunisada T, Fujimoto T, Nishikawa S-I (1991): Role of c-*kit* in mouse spermatogenesis: Identification of spermatogonia as a specific site of c-*kit* expression and function. Development 113:689–699.

Zsebo KM, Williams DA, Geissler EN, Broudy VC, Martin FH, Atkins HL, Hsu R-Y, Birkett NC, Okino KH, Murdock DC, Jacobsen FW, Langley KE, Smith KA, Takeishi T, Cattanach BM, Galli SJ, Suggs SV (1990): Stem cell factor is encoded at the *Sl* locus of the mouse and is the ligand for the c-*kit* tyrosine kinase receptor. Cell 63:213–224.

Molecular Basis of Morphogenesis, pages 19–33
© 1993 Wiley-Liss, Inc.

2. Regulation of Expression of *mZP3*, the Sperm Receptor Gene, During Mouse Development

Ross A. Kinloch, Sergio A. Lira, Steven Mortillo, Michael Schickler, Richard J. Roller, and Paul M. Wassarman

Department of Cell and Developmental Biology, Roche Institute of Molecular Biology, Roche Research Center, Nutley, New Jersey 07110

INTRODUCTION

The mammalian egg extracellular coat, or zona pellucida (ZP) (Dietl, 1989), plays several important roles during development. Probably best known is the role it plays during fertilization (Gwatkin, 1977; Wassarman 1987a,b, 1990; Yanagimachi, 1988). Free-swimming sperm bind in a relatively species-specific manner to sperm receptors that are located in the ovulated egg ZP (Fig. 1). This leads ultimately to fusion of a sperm and egg (fertilization) and the onset of development. Shortly after gamete fusion, the ZP is modified by cortical granule enzymes in such a way that it prevents fertilization of the egg by more than one sperm (polyspermy). In this manner, the ZP regulates the process of fertilization in mammals.

The mouse sperm receptor, a ZP glycoprotein called mZP3 (Bleil and Wassarman, 1980a), was identified more than ten years ago (Bleil and Wassarman, 1980b) and in succeeding years has been characterized extensively (Wassarman, 1988a, 1990, 1993). There are more than 10^9 mZP3 molecules per ZP and they are located periodically (~ 150 Å repeat) along the interconnected filaments that constitute the ZP (Greve and Wassarman, 1985; Wassarman and Mortillo, 1991). The glycoprotein mZP3 ($M_r \sim 83,000$ kd) consists of a polypeptide ($M_r \sim 44,000$ kd), three or four complex-type N-linked asparagine-oligosaccharides, and an undetermined number of O-linked serine- or threonine-oligosaccharides (Salzmann et al., 1983; Wassarman, 1988a). The latter includes a class of oligosaccharides ($M_r \sim 3,900$ kd) that are recognized by free-swimming sperm when they bind to mZP3 (Florman and Wassarman, 1985; Bleil and Wassarman, 1988; Wassarman, 1989, 1991, 1992).

More than 10 years ago, the growing oocyte was identified as the site of mZP3 synthesis and secretion in mice (Bleil and Wassarman, 1980c; Salzmann et al., 1983; Roller and Wassarman, 1983; Wassarman et al., 1985; Wassarman, 1988a). Since then, molecular probes directed against the mZP3 gene (*mZP3*)

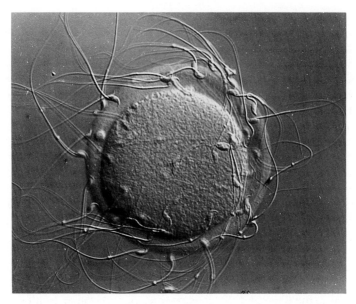

Fig. 1. Light micrograph of mouse sperm bound to the ZP of an unfertilized mouse egg in vitro. The micrograph was taken using Nomarski differential interference contrast (DIC) microscopy.

have become available (Kinloch et al., 1988, 1990; Kinloch and Wassarman, 1989a, 1992). These probes have permitted an analysis of *mZP3* expression during mouse development, as well as of *cis*-acting and *trans*-acting elements that regulate *mZP3* expression. The latter analyses have been facilitated by the use of transgenic mice (Lira et al., 1990, 1993; Schickler et al., 1992; Kinloch et al., 1992). Here, we summarize results of such studies that have been carried out in our laboratory.

EXPERIMENTAL PROCEDURES

For details of experimental procedures used in our experiments, see Kinloch et al. (1988, 1990, 1991, 1992), Kinloch and Wassarman (1989b), Roller et al. (1989), Lira et al. (1990, 1993), and Schickler et al. (1992).

OOGENESIS IN MICE AND PRODUCTION OF THE ZONA PELLUCIDA

Oogenesis is initiated during fetal development of female mice, with production of primordial germ cells (PGCs; ~7- to ~9-day fetus), conversion of PGCs into proliferating oogonia (~13-day fetus), and conversion of oogonia into oocytes (~12- to ~14-day fetus) that have progressed to the diplotene (dictyate) stage of meiosis by the time of birth (~21 days following fertilization) (Jones, 1978; Wassarman, 1988b). Thus, at birth, oocytes are arrested at the dictyate stage of meiosis, where they remain until stimulated to resume meiotic progression at the time of ovulation. This pool of small, nongrowing oocytes (~8,000 per ovary; ~15 μm in diameter), each contained within a cellular follicle consisting of a few epithelial-like cells, is the sole source of unfertilized eggs in sexually mature mice. Nongrowing oocytes do not have a ZP.

In ovaries of sexually mature mice, there is a continuous pool of growing oocytes. During a 2- to 3-week period, while arrested in the dictyate stage of meiosis, oocytes grow to a diameter of about 85 μm and acquire a ZP. The ZP increases in thickness (to ~7 μm) as oocytes increase in diameter. The period of oocyte growth is characterized by extensive transcription and translation, modification of existing organelles (e.g., Golgi apparatus and mitochondria), and production of new organelles (e.g., cortical granules) (Wassarman, 1983, 1988b). The major portion of follicular development takes place when oocyte growth has ceased, with production of a Graafian follicle (~600 μm in diameter) that consists of more than 50,000 cells and is characterized by a large cavity, or antrum. Fully grown oocytes in Graafian follicles resume meiosis and complete the first reductive division (meiotic maturation), with separation of homologous chromosomes and emission of the first polar body, just prior to ovulation. Meiotic maturation of oocytes occurs either in response to a surge in luteinizing hormone (LH) in vivo or by release of oocytes from follicles into a suitable culture medium in vitro. Oocytes that undergo meiotic maturation become unfertilized eggs, arrested at metaphase II, and only complete meiosis, with separation of chromatids and emission of a second polar body, upon fertilization or artificial activation of the eggs.

In summary, a morphologically definable ZP first appears during the oocyte-growth phase of oogenesis and its production is completed by the time oocytes are fully grown (~2–3 weeks). The ZP remains around cleavage-stage mouse embryos until about 4.5 days of development (expanded blastocyst stage), at which time embryos ''hatch'' from their ZP and implant in the uterus. Thus, the fully formed ZP is a transient participant in mouse development, existing for only a week or so before it is discarded. However, even in this relatively short period the ZP carries out several vital functions during oogenesis, fertilization, and embryogenesis.

SYNTHESIS AND SECRETION OF mZP3 DURING MOUSE DEVELOPMENT

The glycoprotein mZP3 is synthesized and secreted by growing mouse oocytes not, as thought for many years, by the follicle cells that surround the growing oocytes. Analysis of oocytes completely denuded of follicle cells, intact follicles, and dispersed follicle cells cultured in the presence of

[^{35}S]-methionine, under a variety of conditions, revealed that only oocytes synthesized and secreted ZP glycoproteins (Bleil and Wassarman, 1980c; Greve et al., 1982; Salzmann et al., 1983; Roller and Wassarman, 1983; Wassarman et al., 1985). Antisera directed against individual ZP glycoproteins detected mature glycoproteins and their precursors in oocytes, but not in follicle cells. Furthermore, electron micrographs of immunogold-labeled follicles in ovarian sections revealed the presence of nascent ZP3 in oocyte Golgi and secretory vesicles, but not in follicle cells (Wassarman, 1990; Wassarman and Mortillo, 1991) (Fig. 2). Since results of similar experiments revealed that fertilized eggs and cleavage-stage embryos did not synthesize mZP3 (or the two other ZP glycoproteins, mZP1 and mZP2; Wassarman, 1988a), these data suggested that expression of the mZP3 gene (*mZP3*) during mouse development might be restricted to oocytes (i.e., sex-specific).

CHARACTERISTICS OF THE mZP3 GENE (*mZP3*) AND POLYPEPTIDE

mZP3 is a single-copy gene that includes an 8.5-kb transcription unit, of which 1,271 nt are a coding sequence that is distributed in eight exons (Kinloch et al., 1988, 1990; Kinloch and Wassarman, 1989a,b) (Fig. 3). The *mZP3* transcription unit is not significantly similar to any other entries in DNA sequence data bases. *mZP3* has unusually short (Kozak, 1984) 5'- and 3'-noncoding regions, 29 nt and 16 nt, respectively. In fact, the termination codon (UAA) forms part of the poly(A)-addition signal (AATAAA) at the 3'-end of mZP3 messenger RNA. There is a potential TATA box located 29 nt upstream of the transcription start-site. The initiation codon (ATG) is surrounded by 10 nt which match those of a consensus sequence reported to be involved in initiation of translation in vertebrates (Kozak, 1987). There is a sequence found 26 nt downstream of the poly(A)-addition site that may be involved in cleavage and processing of nuclear transcripts (McLauchalan et al., 1985; Birnsteil et al., 1985). Consensus splice-site donor and acceptor sequences (Mount, 1982) are present at all 14 exon-intron junctions.

Nascent mZP3 polypeptide (424 amino acids) has a 22-amino-acid putative signal sequence (von Heijne, 1986) at its amino terminus that is not present on the mature polypeptide (Fig. 4). The mature mZP3 polypeptide primary structure includes 402 amino acids (M_r ~44,000 kd) and is not significantly similar to any other entries in either the GenBank or Protein Identification Resources data base. The mature polypepide has a nearly neutral isoelectric point (~6.7) and migrates as a discrete spot on high-resolution, two-dimensional gels. Such behavior is consistent with other evidence indicating that the low isoelectric point (~4.2–5.2) and substantial microheterogeneity of mature mZP3 observed on gels is due to its oligosaccharides, not to its polypeptide. The polypeptide is unusually rich in serine plus threonine (71 residues; 17%) and proline (27 resi-

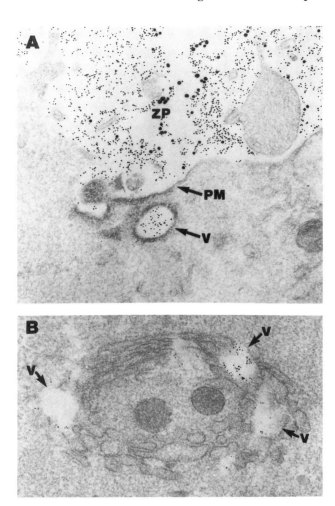

Fig. 2. Transmission electron micrograph of immunogold-labeled mouse oocytes in ovarian sections. **A:** Surface of a growing oocyte with the zona pellucida (ZP) labeled with immunogold, plasma membrane (PM), and a secretory vesicle (V) containing nascent ZP glycoproteins labeled with immunogold. **B:** Shown is the Golgi apparatus of a growing oocyte with vesicles (V) containing nascent ZP glycoproteins labeled with immunogold.

dues; 7%); possesses six potential N-linked glycosylation sites (i.e., consensus sequence–Asn–Xaa–Ser/Thr–; Kornfeld and Kornfeld, 1985); contains very little α-helix; and, overall, it is not particularly hydrophobic or hydrophilic. Interestingly, all 13 cysteine residues present in mZP3 are conserved in both hamster ZP3 (hZP3) and human ZP3 (huZP3) (Wassarman, 1993), and this, perhaps, suggests an important structural role for intermolecular disulfides.

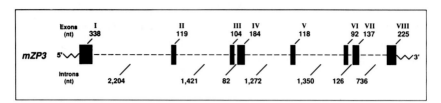

Fig. 3. Diagrammatic representation of the *mZP3* genomic locus. Shown are the positions and sizes (number of nt) of the 8 exons (black boxes) and 7 introns (dashed lines) that, together with the 5' and 3'-noncoding regions, constitute the *mZP3* transcription unit. For details see Kinloch et al., 1988 and Kinloch and Wassarman, 1989a,b.

PATTERN OF *mZP3* EXPRESSION DURING MOUSE DEVELOPMENT

Molecular probes specifically targeted for mZP3 messenger RNA have been employed (e.g., Northern gels, RNase protection, and in situ hybridization) to determine the pattern of *mZP3* expression during mouse development (Roller

```
MASSYFLFLCLLLCGGPELCNSQTLWLLPG    30
GTPTPVGSSSPVKVECLEAELVVTVSRDLF    60
GTGKLVQPGDLTLGSEGCQPRVSVDTDVVR    90
FNAQLHECSSRVQMTKDALVYSTFLLHDPR   120
PVSGLSILRTNRVEVPIECRYPRQGNVSSH   150
PIQPTWVPFRATVSSEEKLAFSLRLMEENW   180
NTEKSAPTFHLGEVAHLQAEVQTGSHLPLQ   210
LFVDHCVATPSPLPDPNSSPYHFIVDFHGC   240
LVDGLSESFSAFQVPRPRPETLQFTVDVFH   270
FANSSRNTLYITCHLKVAPANQIPDKLNKA   300
CSFNKTSQSWLPVEGDADICDCCSHGNCSN   330
SSSSQFQIHGPRQWSKLVSRNRRHVTDEAD   360
VTVGPLIFLGKANDQTVEGWTASAQTSVAL   390
GLGLATVAFLTLAAIVLAVTRKCHSSSYLV   420
SLPQ   424
```

Fig. 4. Primary structure of the mZP3 polypeptide using the single-letter amino acid code. The predicted potential signal sequence is italicized (residues 1–22), and locations of 13 cysteine residues (C) and potential N-linked glycosylation sites (N) are underlined. For details see Kinloch et al., 1988 and Kinloch and Wassarman, 1989a.

et al., 1989; Kinloch and Wassarman, 1989a; Kinloch et al., 1990). All experimental approaches revealed that no tissue other than ovary expressed *mZP3*, and within the ovary, mZP3 messenger RNA was detected only in growing oocytes (Fig. 5). Quantitative RNase protection assays provided a measure of the steady-state levels of mZP3 messenger RNA in oocytes, eggs, and cleavage stage embryos (Fig. 6). During 2 to 3 weeks, as oocytes increase from about 15 μm to about 85 μm in diameter, the amount of mZP3 messenger RNA increases from undetectable levels in nongrowing oocytes (<1,000 copies/oocyte) to about 250,000 copies per fully grown oocyte (~0.3% of total poly(A)+-RNA and ~1.8% of polysomal poly(A)+-RNA; Roller et al., 1989) (Fig. 6). During the relatively short period of ovulation (~10–12 hours), when fully grown oocytes become unfertilized eggs, the steady-state level of mZP3 messenger RNA falls to about 5,000 copies per egg (i.e., ~2% of mZP3 messenger RNA remains). mZP3 messenger RNA is undetectable in fertilized eggs, cleavage stage embryos, and adult tissues.

In summary, results of experiments that examined *mZP3* expression during mouse development are entirely consistent with results of experiments that examined mZP3 synthesis and secretion. *mZP3* is expressed exclusively by growing oocytes, for about 2 to 3 weeks, at a time in mouse development when the ZP is being laid down around growing oocytes. *mZP3* is not expressed by unfertilized and fertilized eggs, nor by preimplantation or postimplantation embryos. Ovary is the only tissue in adult animals that has detectable levels of mZP3 messenger RNA. Therefore, *mZP3* expression is, indeed, oocyte-specific and sex-specific.

CIS-ACTING ELEMENTS THAT REGULATE *mZP3* EXPRESSION DURING MOUSE DEVELOPMENT

Transgenic mice have been used to identify *cis*-acting DNA elements that are involved in regulating *mZP3* expression during mouse development. Several lines of transgenic mice were produced that harbored various DNA constructs as transgenes (Lira et al., 1990, 1993; Schickler et al., 1992; Kinloch et al., 1992; S. Lira and P. Wassarman, unpublished results). In most instances, the constructs consisted of various amounts of *mZP3* 5'-flanking sequence fused to the coding region of the firefly luciferase gene (*ZP3/LUC*), with the latter serving as a "reporter gene" (Fig. 7). Tissues excised from progeny of transgenic lines were assayed for luciferase activity and results were compared with assays using nontransgenic littermates.

Mice harboring *ZP3/LUC,* containing either 6,500 (Lira et al., 1990) or 470 nt (Schickler et al., 1992) of *mZP3* 5'-flanking sequence, expressed the reporter gene exclusively in ovaries (Table I), with about 10^5 to 10^6 light units (LU) of luciferase activity present per ovary [however, one of eight expressing transgenic lines did exhibit a low level of luciferase activity (1.8×10^3 LU)

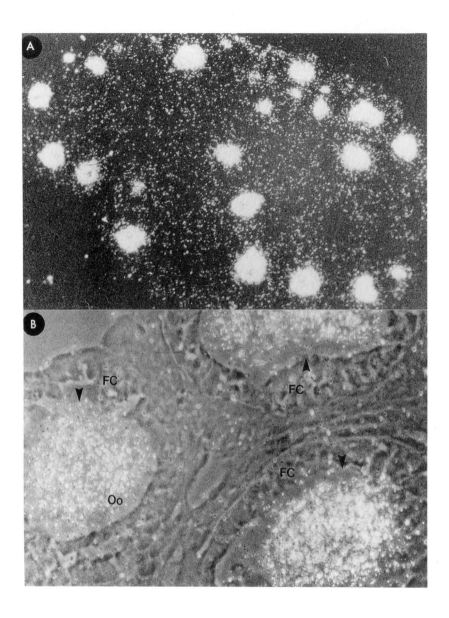

Fig. 5. In situ hybridization assay for mZP3 messenger RNA in mouse ovaries. Light micrographs of the same ovarian section hybridized with a radiolabeled mZP3 probe and subjected to autoradiography. **A:** Dark-field, low magnification. **B:** Bright-field, high magnification. Only the oocytes, not the surrounding follicle cels, display a concentration of silver grains above background. FC, follicle cells; Oo, oocyte. For details see Roller et al., 1989.

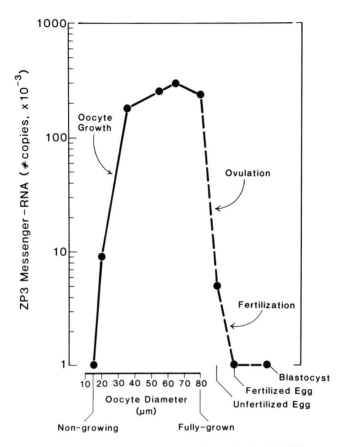

Fig. 6. Diagrammatic representation of the steady-state levels of mZP3 messenger RNA in mouse oocytes, eggs, and preimplantation embryos, as determined by quantitative RNase protection assays. The number of copies of mZP3 messenger RNA ($\times 10^3$) is plotted as a function of the stage of mouse development. For details see Roller et al., 1989.

in forebrain]. Within ovaries, luciferase messenger RNA and luciferase activity were restricted to growing oocytes. This was concluded on the basis of results from in situ hybridization analyses carried out with ovarian sections, and of luciferase activity measurements carried out with individual isolated oocytes and individual isolated follicles from which oocytes had been removed. In these experiments, 470 nt was found to be virtually as effective as 6,500 nt of *mZP3* 5'-flanking sequence with respect to both targeting expression to the ovary and the extent of expression of *ZP3/LUC* within the ovaries of transgenic mice (Table I). Results of recent experiments suggest that as little as 153 nt of *mZP3* 5'-flanking sequence will target expression of firefly luciferase to oocytes of transgenic mice, although levels of expression are re-

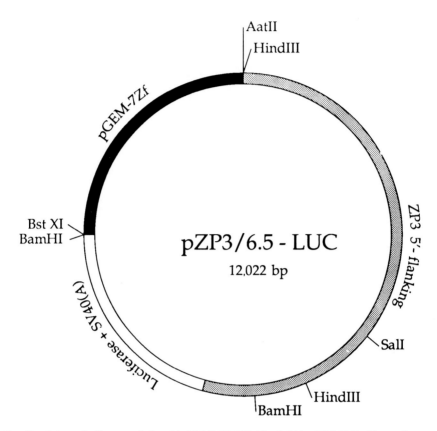

Fig. 7. Schematic diagram of plasmid pZP3/6.5-LUC. The 6.5-kb *mZP3* 5'-flanking region is indicated by the stippled region; the firefly luciferase sequences, simian virus 40 (SV40) polyadenylation signal, and SV40 splicing site are indicated by the open region; the pGEM-72f vector backbone is indicated by the solid region. Restriction enzymes and restriction sits used in constructing the plasmid and in isolation of the transgene *ZP3/LUC* are indicated. For details see Lira et al., 1990, 1993, and Schickler et al., 1992.

duced as compared to transgenes containing larger segments of *mZP3* 5'-flanking sequence (Lira et al., 1993).

These findings suggest that DNA element(s) located relatively close to the *mZP3* transcription start-site are responsible, at least in part, for oocyte-specific (sex-specific) expression of the gene. In this context, it should be noted that *mZP3* 5'-flanking sequence (3.3 kb) has been used to target expression of a foreign (hamster) sperm receptor (hZP3) in transgenic mice (Kinloch et al., 1992). As a result, oocytes of these transgenic mice synthesized and secreted hZP3, in addition to mZP3, and assembled a mosaic ZP.

TABLE I.
Luciferase Activities in Transgenic Mouse Tissues

Tissue	Mean Luciferase Activity, LU (range; number of lines)		
	6.5-kb ZP3/LUC[a]	0.47-kb ZP3/LUC[b]	0.15-kb ZP3/LUC[c]
Adrenal	56 (53–61; 3)	176 (160–200; 3)	191 (184–199; 3)
Brain	92 (51–169; 4)	—	—
Cerebellum	—	227 (167–313; 4)	242 (185–335; 3)
Epididymis	66 (40–103; 4)	—	—
Forebrain	—	600 (172–1,831; 4)[d]	2.4×10^3 (0.2–6×10^3; 3)
Gut	53 (51–54; 4)	186 (180–194; 4)	180 (174–188; 3)
Heart	56 (47–75; 4)	185 (161–233; 4)	193 (187–199; 3)
Kidney	51 (48–52; 4)	181 (171–201; 4)	189 (172–198; 3)
Liver	65 (48–85; 4)	185 (176–190; 4)	183 (179–186; 3)
Lung	54 (47–59; 4)	196 (166–278; 4)	181 (178–184; 3)
Ovary	3.98×10^5 (2.15–5.20×10^5; 4)	5.5×10^5 (0.99–11×10^5; 4)	1.6×10^4 (0.9–36×10^3; 3)
Pituitary	—	175 (164–194; 3)	184 (175–194; 3)
Spleen	50 (46–58; 4)	165 (150–173; 4)	178 (173–184; 3)
Testis	89 (61–141; 4)	197 (190–204; 2)	—
Thymus	54 (—; 1)	179 (172–191; 4)	193 (186–201; 3)
Background[e]	110	180	180

[a]Transgene containing 6,500 nt of *mZP3* 5'-flanking sequence.
[b]Transgene containing 470 nt of *mZP3* 5'-flanking sequence.
[c]Transgene containing 153 nt of *mZP3* 5'-flanking sequence.
[d]This relatively high value is due only to one line (206p), which had an unusually high level of luciferase activity in forebrain (1,831 LU).
[e]Homogenization buffer alone (background) subtracted from assays.
LU, light units.

TRANS-ACTING ELEMENTS THAT REGULATE mZP3 EXPRESSION DURING MOUSE DEVELOPMENT

DNase I footprinting and gel mobility-shift assays have been carried out to identify *trans*-acting factors involved in regulating *mZP3* expression during mouse development. Results of such experiments suggest the presence of an ovary-specific protein that binds to a small region (nt -99 to nt -86) within the 470-nt fragment of the *mZP3* promoter, with 5'-G(G/A)T(G/A)A-3' representing the minimal sequence required for binding (Schickler et al., 1992). This sequence, together with an additional base (T) at the 5' end, constitutes the consensus sequence WGATAR (W = T or A; R + G or A) that is the binding motif for the GATA multigene family of DNA-binding proteins (Yamamoto et al., 1990; Orkin, 1991). Southwestern gels revealed the presence of an oocyte-specific protein (OSP-1, M_r ~60,000 kd) that bound to the minimal sequence (Schickler et al., 1992). Changes in the levels of OSP-1 during oogenesis and early cleavage are consistent with the pattern of *mZP3* expression during these developmental stages in mice. Collectively, these results suggest that OSP-1 may be a mammalian oocyte-specific transcription factor involved in regulating oocyte-specific *mZP3* expression.

Like other members of the GATA family of proteins, OSP-1 probably directs expression of *mZP3* during oocyte growth by interacting with other DNA-binding proteins. In this context, the *mZP3* promoter contains other protein-binding domains. One domain (CACCC; Orkin, 1991) is only 7 nt downstream (nt -81 to nt -85) from the GATA sequence (Schickler et al., 1992; Wassarman, 1993), and it has been suggested that the protein that binds to this sequence may interact with GATA-binding protein (Orkin, 1991). Another *mZP3* sequence, CACGTG, located 181 nt to the 5' side of the transcription start-site, is identical to the consensus sequence required for binding of a number of proteins that belong to the helix-loop-helix family of DNA-binding proteins (Beckmann et al., 1990; Gregor et al., 1990; Blackwood and Eisenman, 1991; Cole, 1991). Thus, OSP-1 may interact with ubiquitous and/or cell-specific proteins to activate *mZP3* expression during oocyte growth.

CONCLUSIONS

The glycoprotein mZP3 from the ZP functions as sperm receptor during species-specific binding of free-swimming mouse sperm to ovulated eggs. It has a molecular weight of approximately 83,000 kd and consists of a polypeptide (M_r ~44,000 kd; 402 amino acids), three or four complex-type N-linked oligosaccharides, and an undetermined number of O-linked oligosaccharides. Certain of the latter are responsible for mZP3 sperm receptor activity. mZP3 is synthesized and secreted by oocytes during their growth phase (2–3 weeks), when the ZP is laid down. mZP3 is encoded by a single-copy gene (*mZP3*).

Expression of *mZP3* is stringently regulated during mouse development. The gene is silent throughout preimplantation and postimplantation fetal development. The gene is silent in all adult tissues except ovary (i.e., sex-specific). Within the ovary, *mZP3* is expressed only in growing oocytes during a 2- to 3-week period when the ZP is laid down. The gene is turned on when nongrowing oocytes begin to grow, and turned off when fully grown oocytes undergo meiotic maturation (ovulation). mZP3 messenger RNA accumulates to unusually high steady-state levels during oocyte growth ($\sim 250-300 \times 10^3$ copies/oocyte) and is rapidly degraded ($\sim 98\%$ during $\sim 10-12$ hours) during ovulation. Expression of *mZP3* is regulated by *cis*-acting DNA elements, certain of which are located close to the gene's transcription start-site (within 153 nt), and by *trans*-acting factors, one of which is an oocyte-specific protein (OSP-1; M_r 60,000 kd) that is a member of the GATA class of DNA-binding proteins. OSP-1 appears when oocytes begin to grow, increase in abundance during oocyte growth, and disappears when fully grown oocytes undergo meiotic maturation (ovulation).

The ZP is a highly specialized organelle only found on mammalian eggs and preimplantation embryos. To a certain extent, it combines in one extracellular coat the functions associated with the vitelline envelope, jelly coat, and fertilization envelope of many nonmammalian eggs (e.g., echinoderm eggs). Apparently, the three glycoproteins that comprise the mouse ZP are synthesized and secreted concomitantly during oogenesis. It is likely that common *cis*-acting and *trans*-acting elements are involved in regulating coordinate expression of the genes that encode these three glycoproteins.

ACKNOWLEDGMENTS

We thank all members of our laboratory for advice and assistance throughout the course of this research, members of Mel DePamphilis's laboratory for assistance with firefly luciferase constructs and assays, Sharon Herrmann and other members of Laboratory Animal Resources for assistance with transgenic animals, Barbara Rinde for providing synthetic oligonucleotides, and Alice O'Connor for assistance in preparation of manuscripts.

REFERENCES

Beckmann H, Su L-K, Kadesch T (1990): TFE3: A helix-loop-helix protein that activates transcription through the immunoglobulin enhancer μE3 motif. Genes Dev 4:167–179.

Birnsteil ML, Busslinger M, Strub K (1985): Transcription termination and 3' processing: The end is in site. Cell 41:349–359.

Blackwood EM, Eisenman RN (1991): Max: A helix-loop-helix zipper protein that forms sequence-specific DNA-binding complex with Myc. Science 251:1211–1217.

Bleil JD, Wassarman PM (1980a): Structure and function of the zona pellucida: Identification and characterization of the proteins of the mouse oocyte's zona pellucida. Dev Biol 76:185–203.

Bleil JD, Wassarman PM (1980b): Mammalian sperm-egg interaction: Identification of a glycoprotein in mouse egg zone pellucidae possessing receptor activity for sperm. Cell 20:873–882.

Bleil JD, Wassarman PM (1980c): Synthesis of zona pellucida proteins by denuded and follicle-enclosed mouse oocytes during culture in vitro. Proc Natl Acad Sci USA 77:1029–1033.

Bleil JD, Wassarman PM (1988): Galactose at the nonreducing terminus of O-linked oligosaccharides of mouse egg zona pellucida glycoprotein ZP3 is essential for the glycoprotein's sperm receptor activity. Proc Natl Acad Sci USA 85:6778–6782.

Cole MD (1991): Myc meets its Max. Cell 65:715–716.

Dietl J (ed) (1989): "The Mammalian Egg Coat: Structure and Function." Berlin: Springer-Verlag.

Florman HM, Wassarman PM (1985): O-Linked oligosaccharides of mouse egg ZP3 account for its sperm receptor activity. Cell 41:313–324.

Gregor PD, Sawadogo M, Roeder RG (1990): The adenovirus major late transcription factor USF is a member of the helix-loop-helix group of regulatory proteins and binds to DNA as a dimer. Genes Dev 4:1730–1740.

Greve JM, Salzmann GS, Roller RJ, Wassarman PM (1992): Biosynthesis of the major zona pellucida glycoprotein secreted by oocytes during mammalian oogenesis. Cell 31:749–759.

Greve JM, Wassarman PM (1985): Mouse egg extracellular coat is a matrix of interconnected filaments possessing a structural repeat. J Mol Biol 181:253–264.

Gwatkin RBL (1977): "Fertilization Mechanisms in Man and Mammals." New York: Plenum Press.

Jones R (ed) (1978): "The Vertebrate Ovary." New York: Plenum Press.

Kinloch RA, Wassarman PM (1989a): Profile of a mammalian sperm receptor gene. New Biol 1:232–238.

Kinloch RA, Wassarman PM (1989b): Nucleotide sequence of the gene encoding zona pellucida glycoprotein ZP3—the mouse sperm receptor. Nucleic Acids Res 17:2861–2863.

Kinloch RA, Wassarman PM (1992): Specific gene expression during oogenesis in mice. In Gwatkin RBL (ed): "Genes in Mammalian Reproduction." New York: Wiley-Liss, pp 27–43.

Kinloch RA, Roller RJ, Fimiani CM, Wassarman DA, Wassarman PM (1988): Primary structure of the mouse sperm receptor's polypeptide chain determined by genomic cloning. Proc Natl Acad Sci USA 85:6409–6413.

Kinloch RA, Roller RJ, Wassarman PM (1990): Organization and expression of the mouse sperm receptor gene. In Davidson EH, Ruderman JV, Posakony JW (eds): "Developmental Biology," vol 125 (UCLA Symposia, new series). New York: Wiley-Liss, pp 9–20.

Kinloch RA, Mortillo S, Stewart CL, Wassarman PM (1991): Embryonal carcinoma cells transfected with ZP3 genes differentially glycosylate similar polypeptides and secrete active mouse sperm receptor. J Cell Biol 115:655–664.

Kinloch RA, Mortillo S, Wassarman PM (1992): Transgenic mouse eggs with functional hamster sperm receptors in their zona pellucida. Development 115:937–946.

Kornfeld R, Kornfeld S (1985): Assembly of asparagine-linked oligosaccharides. Annu Rev Biochem 54:631–664.

Kozak M (1984): Complication and analysis of sequences upstream from the translational start site in eukaryotic mRNAs. Nucleic Acids Res 12:857–872.

Kozak M (1987): An analysis of 5'-noncoding sequences from 699 vertebrate messenger RNAs. Nucleic Acids Res 15:8125–8131.

Lira SA, Kinloch RA, Mortillo S, Wassarman PM (1990): An upstream region of the mouse ZP3 gene directs expression of firefly luciferase specifically to growing oocytes in transgenic mice. Proc Natl Acad Sci USA 87:7215–7219.

Lira SA, Schickler M, Wassarman PM (1993): Cis-acting DNA elements involved in oocyte-specific expression of mouse sperm receptor gene mZP3 are located close to the gene's transcription start-site. Mol Reprod Dev (in press).

McLauchalan J, Gaffney D, Whitten GL, Clements JB (1985): The consensus sequence

YGTGTTYY located downstream from the AATAAA signal is required for efficient formation of mRNA 3′ terminal. Nucleic Acids Res 13:1347–1368.

Mount SM (1982): A catalogue of splice junction sequences. Nucleic Acids Res 10:459–472.

Orkin SH (1991): Globin gene regulation and switching: Circa 1990. Cell 63:665–672.

Roller RJ, Wassarman PM (1983): Role of asparagine-linked oligosaccharides in secretion of glycoproteins of the mouse egg's extracullar coat. J Biol Chem 258:13243–13249.

Roller RJ, Kinloch RA, Hiroaka BY, Li SSL, Wassarman PM (1989): Gene expression during mammalian oogenesis and early embryogenesis: Quantification of three messenger-RNAs abundant in fully-grown mouse oocytes. Development 106:251–261.

Salzmann GS, Greve JM, Roller RJ, Wassarman PM (1983): Biosynthesis of the sperm receptor during oogenesis in the mouse. EMBO J 2:1451–1456.

Schickler M, Lira S, Kinloch R, Wassarman P (1992): A mouse oocyte-specific protein that binds to a region of mZP3 promoter responsible for oocyte-specific mZP3 gene expression. Mol Cell Biol 12:120–127.

von Heijne G (1986): A new method for predicting signal sequence cleavage sites. Nucleic Acids Res 14:4683–4690.

Wassarman PM (1983): Oogenesis: Synthetic events in the deveoping mammalian egg. In Hartmann JF (ed): "Mechanism and Control of Animal Fertilization." New York: Academic Press, pp 1–54.

Wassarman PM (1987a): The biology and chemistry of fertilization. Science 235:553–560.

Wassarman PM (1987b): Early events in mammalian fertilization. Annu Rev Cell Biol 3:109–142.

Wassarman PM (1988a): Zona pellucida glycoproteins. Annu Rev Biochem 57:415–442.

Wassarman PM (1988b): The mammalian ovum. In Knobil E, Neill JD (eds): "The Physiology of Reproduction," vol 1. New York: Raven Press, pp 69–102.

Wassarman PM (1989): Role of carbohydrates in receptor-mediated fertilization in mammals. In Bock G, Harnett S (eds): "Carbohydrate Recognition in Cellular Function," Ciba Foundation Symposium, no 145. New York: John Wiley & Sons, pp 135–155.

Wassarman PM (1990): Profile of a mammalian sperm receptor. Development 108:1–17.

Wassarman PM, ed (1991): "Elements of Mammalian Fertilization, vol 1, Basic Concepts; vol 2, Practical Applications." Boca Raton, FL: CRC Press.

Wassarman PM (1992): Cell surface carbohydrates and mammalian fertilization. In Fukuda M (ed): "Cell Surface Carbohydrates and Cell Development," Boca Raton, FL: CRC Press, pp 215–238.

Wassarman PM (1993): Mammalian fertilization: Sperm receptor genes and glycoproteins. In Wassarman PM (ed): "Advances in Developmental Biochemistry," vol 2. Greenwich, Conn: JAI Press, pp 155–195.

Wassarman PM, Mortillo S (1991): Structure of the mouse egg extracellular coat, the zona pellucida. In Jeon KW, Friedlander M (eds): "International Review of Cytology—Survey of Cell Biology," vol 130. New York: Academic Press, pp 85–110.

Wassarman PM, Bleil JD, Florman HM, Greve JM, Roller RJ, Salzmann GS, Samuels FG (1985): The mouse egg's receptor for sperm: What is it and how does it work? Cold Spring Harb Symp Quant Biol 50:11–19.

Yamamoto M, Ko LJ, Leonard MW, Beug H, Orkin SH, Engel JD (1990): Activity and tissue-specific expression of the transcription factor NF-E1 multigene family. Genes Dev 4:1650–1662.

Yanagimachi R (1988): Mammalian fertilization. In Knobil E, Neill JD (eds): "The Physiology of Reproduction," vol 1. New York: Raven Press, pp 135–185.

3. Building a Frog: The Role of *Xenopus Wnts* in Patterning the Embryonic Mesoderm

Jan L. Christian and Randall T. Moon

Department of Pharmacology, University of Washington School of Medicine, Seattle, Washington 98195

INTRODUCTION

How are cells informed of their position within the embryo, and how does this influence their fate? The idea that gradients of morphogens provide concentration-dependent positional values to embryonic cells, thereby determining their fate, remains central to many theories of morphogenesis (Wolpert, 1969). In invertebrates, morphogen gradients clearly play an important role. For example in *Drosophila* a gradient of bicoid protein, set up by diffusion during the early syncitial stage of development, provides the first step in determining cell fate along the anterior-posterior axis (Driever et al., 1990). While analogous systems may operate in higher organisms, recent evidence suggests that the vertebrate embryo is organized not so much by gradients of molecules which act in a concentration-dependent manner to instruct cells of their fate, but rather by the spatially restricted competence of cells to respond such signals. In particular, studies of the *Wnt* gene family in *Xenopus laevis* have focused attention on a new class of signaling agents which, rather than determining cell fate, locally restrict and/or modify the competence of cells to respond to more uniformly distributed inductive signals. This chapter presents an historical overview of the signals and inductive interactions which are important in patterning the *Xenopus* embryo, and integrates recent molecular data which have led to a revised model for mesodermal induction and patterning.

PATTERNING THE *XENOPUS* EMBRYO

In *Xenopus,* embryonic patterning is initiated during the first cell cycle by a rotation of cortical, or surface, cytoplasm relative to the internal cytoplasm of the egg. This so-called cortical rotation is believed to locally activate substances that are required for dorsal development on the prospective dorsal side of the embryo (Gerhart et al., 1989) (Fig. 1b). Shortly thereafter, the zygote begins a series of synchronous cleavages leading to the formation of the blastula

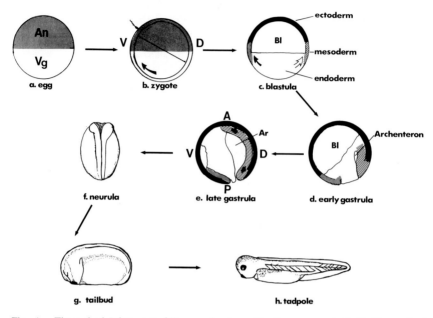

Fig. 1. The early development of *Xenopus laevis*. **a:** The *Xenopus* egg is divided into animal (An) and vegetal (Vg) hemispheres. **b:** Fertilization (sperm on left) initiates a cytoplasmic rotation (arrow), defining the future dorsal (D) and ventral (V) side of the embryo. **c:** During the early blastula stage, signals from vegetal cells (prospective endoderm, white) induce the formation of dorsal (hatched) and ventral (stippled) mesoderm from prospective ectodermal cells (black) of the marginal zone (Bl, blastocoel). **d–e:** During gastrulation, mesoderm invaginates, becoming sandwiched between ectoderm and endoderm, and the primitive gut, or archenteron (Ar) is formed. Dorsal mesoderm invaginates first, forming the dorsal lip of the blastopore. The earliest migrating cells take on anterior positional values, the latest take on posterior positional values, establishing the anterior-posterior (AP) axis. e: Dorsal mesoderm induces overlying ectoderm to form neural structures. **f–h:** Organogenesis encompasses the neurula, tailbud, and tadpole stages of embryogenesis.

(Fig. 1c). Two events of special consequence occur during this stage of development. The first is the initiation of zygotic gene transcription, which releases the embryo from its early reliance on maternal stocks of RNA and protein. This takes place approximately 7 hours after fertilization, at a point called the midblastula transition. The second event, the induction of the mesoderm, begins at about the 32-cell stage (Jones and Woodland, 1987). Prior to this time, the embryo consists of two primary germ layers: the pigmented animal hemisphere composed of prospective ectoderm, and the vegetal hemisphere consisting of prospective endoderm. The third primary germ layer, the mesoderm, differentiates from ectodermal cells residing in the equatorial (or marginal zone) region of the embryo in response to inductive signals provided by underlying endodermal cells (Nieuwkoop, 1969) (Fig. 1c). The mesoderm plays a central role in patterning the embryo, since positional information necessary for establishing the body plan is initially imprinted upon these cells. During gastrulation, the mesoderm invaginates around the lips of the blastopore, and this leads to the formation of the primitive gut, or archenteron (Fig. 1d–e). Near the end of gastrulation, dorsal mesodermal cells induce the overlying ectoderm to form neural tissue (Fig. 1e) which fuses at the dorsal midline and forms the future brain and spinal cord (Fig. 1f). Subsequently, organ rudiments differentiate and the organization of the anteroposterior and dorsal-ventral body axes becomes outwardly apparent (Fig. 1g–h).

In *Xenopus,* the body axes are organized through the action of at least two spatially and temporally distinct cell populations, the Nieuwkoop center and the Spemann organizer, each of which can induce a second body axis when transplanted into an appropriate host environment (see Gerhart et al., 1991 and references therein). The Nieuwkoop center is active from the early cleavage stages until at least the midblastula stage, and consists of dorsal cells which inherit cytoplasm activated by cortical rotation (see above). If cortical rotation is blocked by ultraviolet (UV) irradiation of the egg, development is limited to ventral extremes. Dorsal development can be rescued, however, by transplanting into these embryos single blastomeres from the Nieuwkoop center of nonirradiated cleavage-stage embryos (Gimlich, 1986). These cells can induce the overlying ectoderm to form dorsal mesoderm, whereas ventral and lateral cells induce only ventral mesoderm. Transplantation studies have shown that, at the 32-cell stage, the Nieuwkoop center (defined by the ability to induce a secondary embryo) includes both equatorial cells, which will self-differentiate into the gastrula stage, or Spemann's organizer, and vegetal blastomeres, which can induce overlying cells to form organizer tissue but are themselves fated to form ventral endodermal structures (Gimlich, 1986; Kaguera, 1990). Despite this evidence that vegetal blastomeres are sufficient to induce the formation of a Spemann organizer, which leads to the subsequent differentiation of dorsal axial structures, they are clearly not required for these functions in vivo, since normal dorsal development ensues following the removal of all four vegetal blastomeres from 32-cell embryos (Takasaki, 1987). Nevertheless, as will be discussed later, the ability of dorsal vegetal (endodermal) cells of cleavage-stage embryos to induce dorsal structures provides a convenient method for distinguishing between the activity of the Nieuwkoop center, and that of the Spemann organizer, which is restricted to a more equatorial, mesodermal cell population. In fact, cells of the Spemann organizer will self-differentiate as the most dorsal and anterior mesoderm of the body (the notochord and head mesoderm) (Smith and Slack, 1983). By the late blastula stage the mesoderm is specified as either extreme dorsal (organizer cells) or extreme ventral (ventral and lateral cells) (Stewart and Gerhart, 1990). Further patterning occurs under the influence of signals from the Spemann organizer. These signals induce lateral mesodermal cells to adopt a fate intermediate between extreme dorsal and extreme ventral (Dale and Slack, 1987).

The ability of these two organizing centers to self-differentiate, and to induce neighboring cells to change their pathway of differentiation, has led to a widespread search for endogenous inducing molecules secreted by these cells. Since signaling from the Nieuwkoop center leads to the formation of dorsal mesoderm, a useful screen for its potential components involves assaying soluble peptide growth factors for their ability to respecify the fate of ectoderm (isolated from the animal pole of blastula-stage embryos) as dorsal mesoderm. While such ectodermal explants will form epidermis in the absence of growth factors, the addition of activin can induce the differentiation of either dorsal or ventral mesoderm, depending on whether a high or low dose is applied (Green et al., 1990). This observation raised the intriguing possibility that a morphogenetic gradient of activin might account for both the dorsal mesoderm-inducing activity of the Nieuwkoop center and for the ventral mesoderm-inducing activity of ventral vegetal cells (Green and Smith, 1990). The coincident discovery that members of the fibroblast growth factor (FGF) family can also induce mesoderm, but primarily ventral mesoderm (Kimelman et al., 1987; Slack et al., 1987), led to the alternate hypothesis that the Nieuwkoop center and ventral vegetal-inducing activities consist of dorsally and ventrally segregated pools of activin and FGF respectively (Green et al., 1990). Unfortunately, several observations suggest that dorsal development is more complex than the picture provided by these simple, and thus appealing, hypotheses. First, studies to date have found no evidence for localization of FGFs, activins, or their receptors across the dorsoventral axis (Gillespie et al., 1989; Musci et al., 1990; Friesel and Dawid, 1991; Shiurba et al., 1991; Matthews et al., 1992). Second, activin and FGF can induce dorsal mesoderm to form only from ectoderm dissected from the dorsal, and not from the ventral, half of the embryo, an observation which suggests that the ectoderm has been prepatterned in its competence to respond to mesoderm-inducing agents (Ruiz i Altaba and Jessel, 1991; Sokol and Melton, 1991; Kimelman and Maas, 1992; Bolce et al., 1992). Finally, while endogenous agents active in the Nieuwkoop center can rescue dorsal development in embryos rendered axis-deficient by UV irradiation, and can induce the formation of a complete secondary embryo when introduced into ventral cells of cleavage-stage embryos (Gimlich, 1986; Kaguera, 1990), activin alone can do neither (Thomsen et al,. 1990; Sokol and Melton, 1991).

Xwnts AND EARLY MESODERMAL PATTERNING

In the midst of this confusion, several members of a third class of proteins, related to and including the protooncogene product Wnt-1, were found to have many of the properties expected of endogenous dorsal-inducing signals from the Nieuwkoop center (McMahon and Moon, 1989; Christian et al., 1991; Smith and Harland, 1991; Sokol et al., 1991; Wolda et al., 1992). Wnts are a recently described family of cysteine-rich, secreted glycoproteins, which are believed to function as autocrine and/or paracrine developmental signaling molecules (reviewed by McMahon, 1992; Nusse and Varmus, 1992). At least 12 *Xenopus Wnts* (*Xwnts*) have been identified, and these are expressed in spatially and temporally discrete, yet overlapping patterns during embryogenesis (reviewed by Moon, 1992). When synthetic RNA encoding Wnt-1, Xwnt-3A, or Xwnt-8 is injected into ventral cells of cleavage-stage embryos, a complete Siamese-twin embryo is formed (Fig. 2). Furthermore, when injected into selected blastomeres of UV-irradiated embryos, *Xwnt*-8 can restore development of a normal dorsal axis (Fig. 3). While it was not initially clear whether these *Wnts* were acting early (i.e., as a Nieuwkoop center), to induce dorsal mesoderm with Spemann organizer properties, or whether *Wnts* could directly mimic the Spemann organizer, thus bypassing the need for the early inductive signals, the issue was soon resolved. First, it was demonstrated that *Xwnt*-8, like the Nieuwkoop center, can induce a secondary axis even when the RNA is confined to vegetal blastomeres, which do not contribute progeny to dorsal axial structures (Smith and Harland, 1991). The further observation that *Xwnt*-8 cannot induce a secondary embryo when it is introduced into ventral cells at the time that the Spemann organizer is active confirms that it is not providing the same inductive signals as the gastrula stage organizer (Christian and Moon, 1993).

VENTRAL DORSAL

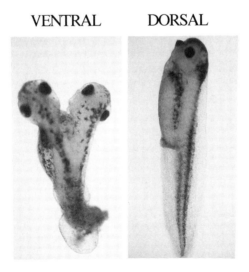

Fig. 2. Injection of *Xwnt*-8 RNA into cleavage stage embryos produces axial duplications. Injection of RNA encoding *Xwnt*-8 into ventral equatorial or vegetal blastomeres of 32-cell embryos produces a complete duplication of the body pattern. Injection into dorsal cells, which possess endogenous dorsal inducing factors, has no effect on the body pattern.

UV + XWNT-8　　　　UV LIGHT　　　　CONTROL

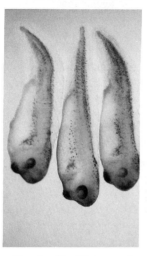  

Fig. 3. *Xwnt*-8 rescues dorsal development in UV-ventralized embryos. UV irradiation of the vegetal hemisphere of fertilized *Xenopus* eggs blocks dorsal development, producing a ventralized embryo (middle panel). Injection of RNA encoding *Xwnt*-8 into vegetal or equatorial blastomeres of ventralized cleavage stage embryos restores dorsal mesoderm formation and normal axial patterning (bottom panel).

Since ectopically expressed *Xwnt*-8 can restore the ability to form dorsal mesoderm to UV-ventralized embryos, it might be assumed that *Xwnt*-8 is itself a mesoderm-inducing agent. Surprisingly, this turns out not to be true. Injection of *Xwnt*-8 RNA into the animal pole of cleavage-stage embryos does not induce mesoderm formation either in the intact animal or in ectoderm explanted at the blastula stage (Christian et al., 1992). However, when ectodermal explants from embryos injected with *Xwnt*-8 are cultured in the presence of growth factors which would normally induce ventral mesoderm (such as FGF or low doses of activin), they now differentiate as embryoids containing dorsal mesoderm, and due to secondary inductions, neural tissue. Furthermore, although ectoderm isolated from UV-irradiated embryos, or from the ventral half of normal embryos, cannot form dorsal mesoderm even when exposed to high doses of activin, the addition of a *Xwnt*-8 signal enables these cells to differentiate as dorsal mesoderm in response to either FGF or activin (Christian et al., 1992; Sokol and Melton, 1992). Thus, dorsal mesoderm induction requires both a mesoderm-inducing signal and a modifying signal, the latter functioning to alter the competence of cells to respond to inducing agents. The ability of *Xwnt*-8 alone to reconstitute a Nieuwkoop center in UV-irradiated embryos implies that endogenous mesoderm-inducing signals are intact in these embryos. Together, these results suggest that the initial patterning of the mesoderm is not accomplished by dorsoventral differences in inducing signals, but by one or more mesoderm-inducing agents that are uniformly distributed across the dorsoventral axis, in combination with a dorsally localized modifying signal, that alters the response of ectodermal cells to induction (Christian et al., 1992; Kimelman et al., 1992). While activin and bFGF are excellent candidates for endogenous mesoderm-inducing agents, neither *Xwnt*-8, *Xwnt*-3A, nor *Xwnt*-1 are likely to be responsible for the endogenous modifying activity, since none of these have been detected prior to the midblastula transition, when the Nieuwkoop center is active. It may be that these ectopically expressed *Xwnts* are activating the receptor for another member of the *Xenopus Wnt* family that is the real dorsal modifier. This is plausible as several maternally expressed *Xwnts* have been described. While one of these (*Xwnt*-5A) shows no dorsal-inducing or dorsal-modifying activity (Moon et al., in press), a more recently identified maternal *Xwnt* (*Xwnt*-9) has at least partial axis-rescuing activity when injected into UV-ventralized embryos (M. Ku and D. Melton, unpublished data). Alternatively, the endogenous dorsal modifier may be an unrelated molecule that uses a signal transduction pathway that can also be stimulated by ectopically introduced *Xwnts*. This latter possibility has become more attractive with the discovery of a maternally expressed, secreted protein, noggin, that displays all of the dorsal-axis-inducing properties of *Xwnts* yet bears no sequence similarity (Smith and Harland, 1992). Although noggin is clearly capable of substituting for the Nieuwkoop center when ectopically introduced into embryos, several observations suggest that endogenous noggin may not perform the same role during normal development. First, the quality of noggin RNA which is present in the egg is less than the amount of synthetic noggin RNA which must be injected to induce dorsal structures. In addition, materal noggin transcripts are uniformly distributed in the early embryo, and UV treatment, which destroys the Nieuwkoop center, does not alter the quantity or distribution of the transcripts. None of these are insurmountable problems, however, since cortical rotation is believed to locally activate, rather than to locally redistribute, modifying components of the Nieuwkoop center. In fact, a functional Nieuwkoop center may consist of multiple components, some of which are required for the local activation of unrelated signaling molecules. Thus, it is possible that both a maternal *Xwnt*, as well as noggin, are required for establishment of the dorsal axis. Testing this possibility will require ablation of the respective signals, or blockade of the ability of cells to respond to these signals—difficult yet essential tasks for the future.

PATTERNING AFTER THE MIDBLASTULA TRANSITION

Although induction and patterning of the mesoderm is initiated by maternal gene products, it is far from over at the midblastula transition. A very simplified view of dorsal axis formation is that maternal gene products of the Nieuwkoop center lead to the expression of zygotic genes which create a functional Spemann organizer. For this reason, genes which are expressed shortly after the midblastula transition in cells of the Spemann organizer are of special interest. Three such examples include the putative transcription factors *goosecoid* (Cho et al., 1991), *Xlim*-1 (Taira et al., 1992), and *XFKH*-1 (Dirksen and Jamrich, 1992). Expression of each of these is induced by activin in dorsal ectodermal cells, in the absence of new protein synthesis, demonstrating that their expression is an early response to signals from the Nieuwkoop center. When RNA encoding *goosecoid* is injected into ventral marginal zone blastomeres, a complete secondary axis can occasionally be formed (Cho et al., 1991). Unlike some *Xwnts*, *goosecoid* cannot create a new Nieuwkoop center (Christian and Moon, 1993), but instead bypasses the requirement for these early inductive events, and initiates the subsequent steps required for generating a functional gastrula-stage organizer. The mechanism by which *goosecoid* does so almost certainly involves the transcriptional regulation of a complex hierarchy of genes, some of which directly generate the agents responsible for the dorsalizing and neural-inducing function of the Spemann organizer. Future studies aimed at uncovering organizer specific genes which are positively regulated by *goosecoid* may be helpful in identifying these elusive morphogens.

One putative downstream target of *goosecoid* is the endogenous *Xwnt*-8 gene. Normally, *Xwnt*-8 is expressed in ventral and lateral mesodermal cells shortly after the midblastula transition, with transcripts being specifically excluded from cells of the Spemann organizer field (Christian et al., 1991; Smith and Harland, 1991; Christian and Moon, 1993). When *goosecoid* RNA is injected into ventral marginal zone cells, expression of *Xwnt*-8 is specifically repressed in the ventral progeny of cells injected with *goosecoid* (Christian and Moon, 1993), demonstrating that the negative regulation of *Xwnt*-8 expression in organizer cells occurs downstream of the induction of *goosecoid*. During normal development, expression of *Xwnt*-8 is initiated within an hour or two of the time that *goosecoid* transcripts are first detected, and the spatial pattern of expression of these two genes in the marginal zone is mutually exclusive. Thus it is possible that *goosecoid* acts directly as a transcriptional repressor of *Xwnt*-8 in organizer cells, a possibility which is currrently being pursued.

As previously described, when *Xwnt*-8 is ectopically expressed in ventral blastomeres of cleavage-stage embryos, it functions as a dorsal modifier, leading to the formation of a secondary embryo. As with all ectopic expression studies, this gain of function phenotype must be interpreted cautiously since it tells us merely what potential roles *Xwnt*-8 can play when introduced into a given setting, but not necessarily what its role is during normal development. In fact, the timing of expression of endogenous *Xwnt*-8 (after the midblastula transition, at which time the Nieuwkoop center is no longer active), as well as its localization (in ventral and lateral cells), effectively precludes the idea that

Xwnt-8 might normally function as a dorsal inducer or modifier. Recent studies suggest that *Xwnt*-8 may instead function as an inhibitor of dorsal development, and illustrate how the same signal (*Xwnt*-8) can elicit dramatically different responses depending on its cellular context. The ventralizing role of endogenous *Xwnt*-8 was demonstrated by ectopically expressing *Xwnt*-8 in gastrula-stage embryos through the use of a plasmid expression construct, which is not transcribed until the midblastula transition (Christian and Moon, 1993). Since endogenous *Xwnt*-8 is normally expressed at this time, cellular responses to ectopically expressed *Xwnt*-8 should more closely mimic the normal response of cells to this gene product, and these experiments should facilitate the deduction of the true functions, rather than potential functions, of *Xwnt*-8. While expression of *Xwnt*-8 prior to the midblastula transition causes ventral blastomeres to differentiate with a dorsal fate, dorsal cells respond to *Xwnt*-8 after the midblastula transition by assuming a more ventral fate. Specifically, when *Xwnt*-8 is ectopically expressed in cells of the Spemann organizer field, prospective notochord cells are converted to muscle. In addition, head mesoderm is either defective or absent in these embryos, and this leads to a secondary

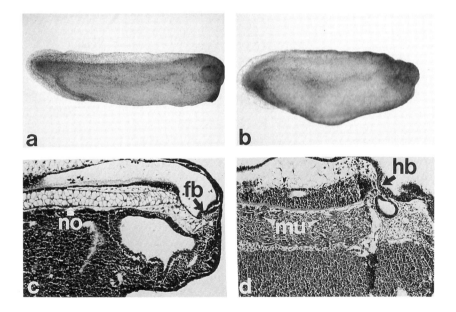

Fig. 4. Ectopic expression of *Xwnt*-8 in Spemann organizer cells results in loss of forebrain and notochord. **a:** *Xenopus* control embryos. **b:** *Xenopus* embryos made to ectopically express *Xwnt*-8 after the midblastula transition, in cells of the Spemann organizer field, lack anterior structures. **c:** Histologically, control embryos possess a normal formal forebrain (fb) and notochord (no). **d:** Treated embryos are missing the forebrain, but not the hindbrain (hb), and prospective notochord has differentiated as muscle (mu), a more ventral cell type.

loss of the forebrain (Fig. 4). These phenotypic effects are not due to the absence of dorsalizing signals, but rather to the diminished competence of organizer cells to respond to such signals. Thus, instead of differentiating as extreme dorsal mesoderm, organizer cells which ectopically express *Xwnt*-8 respond more like lateral mesodermal cells (which normally do express *Xwnt*-8) and differentiate as muscle. A ventralizing effect of *Xwnt*-8 can also be demonstrated in vitro, in that expression of *Xwnt*-8 after the midblastula stage in dorsal ectodermal explants inhibits activin-mediated induction of notochord. Interestingly, activin is a potent inducer of endogenous *Xwnt*-8 expression in ectodermal explants, but expression is restricted to cells derived from the ventral halves of such explants (Christian and Moon, 1993). The fact that activin can induce notochord and forebrain in dorsal, but not ventral, ectodermal cells (Sokol and Melton, 1991; Bolce et al., 1992) can most likely be attributed to the activin-induced expression of *Xwnt*-8 in ventral cells after the midblastula transition, which alters the response of these cells to dorsalizing signals. Together these data suggest that *Xwnt*-8 may function to attenuate the response of cells to endogenous dorsalizing signals. This limits extreme dorsal development to cells which do not express *Xwnt*-8, namely those derived from the organizer, and allows lateral cells, which receive both the *Xwnt*-8 signal and dorsalizing signals from the organizer, to differentiate as intermediate grades of mesoderm, such as muscle and lateral plate.

In addition to restricting dorsal development, *Xwnt*-8 may participate directly in the formation of ventral mesoderm. When *Xwnt*-8 is expressed in ectodermal explants after the midblastula transition, ventral mesoderm is formed autonomously, in the absence of exogenously added growth factors (Christian and Moon, 1993). Since mesoderm induction in vivo begins shortly after fertilization, it is clear that the initial steps of this process must be mediated by maternal factors, and *Xwnt*-8 is neither maternally expressed, nor can it induce ventral mesodermal differentiation when ectopically expressed prior to the midblastula transition (Christian et al., 1992). It is possible that endogenous *Xwnt*-8 can bypass these early steps, and can initiate the subsequent cascade of signals necessary for the differentiation of ventral mesoderm. Recent studies have implicated a number of additional zygotic gene products in the later steps of ventral mesoderm formation, and suggest that *Xwnt*-8 may require the cooperation of these or other proteins in this process. Examples of putative ventral mesodermalizing factors include bone morphogenetic protein-4 (BMP-4), which is a relative of the TGF-β family (Köster et al., 1991; Dale et al.,1992; Jones et al., 1992), as well as *Xbra*, the *Xenopus* homolog of mouse *brachyury* (Cunliffe and Smith, 1992). Further studies will be required to examine the hierarchy of these gene products, as well as the possible interactions between them, in the formation of ventral mesoderm.

A SYNERGISTIC MODEL FOR MESODERMAL PATTERNING

These recent molecular findings have been incorporated into a model for mesoderm formation (Fig. 5) that represents an expanded and revised version of our synergistic interpretation (Kimelman et al., 1992) of the well-known three-signal model (Smith and Slack, 1983; Slack, 1991). We suggest that the induction, and initial dorsoventral patterning, of the mesoderm requires two signals: a general mesoderm-inducing signal which is uniformly distributed across the dorsoventral axis, and a modifying activity (the Nieuwkoop center) which is restricted to cells on the dorsal side of the embryo. The modifying activity is established by the postfertilization cortical rotation. *Noggin* and/or a maternal *Xwnt* may be components of the modifying signal, while FGF and/or maternally expressed activin may function as inducers. At the midblastula transition, dorsal equatorial cells respond to inductive signals by activating genes such as *goosecoid*, which can initiate subsequent signalling from the Spemann organizer, and which can repress expression of *Xwnt*-8 in these cells. Ventral and lateral marginal zone cells respond to identical inductive signals by turning on genes required for ventral mesodermal differentiation, such as *Xwnt*-8. It is possible that *Xwnt*-8 does not provide a ventral-inducing signal itself, but instead makes marginal-zone cells competent to respond to more uniformly distributed inducing signals which might be provided, for example, by BPM-4. In addition, *Xwnt*-8 lowers the competence of cells to re-

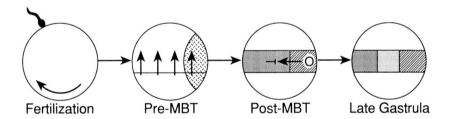

Fig. 5. A synergistic model for mesoderm induction and patterning. Cortical rotation (curved arrow) after fertilization establishes the Nieuwkoop center activity (cross-hatched) on the dorsal side of the pre-midblastula transition-stage (Pre-MBT) embryo. This activity prepatterns the competence of dorsal ectodermal cells to respond to mesoderm-inducing factors (vertical arrows) which are uniformly release from vegetal cells across the dorsal-ventral axis. After the midblastula transition (Post-MBT), cells patterned by the Nieuwkoop center are competent to form dorsal mesoderm (hatched area) with Spemann organizer (O) activity, while ventral and lateral cells are competent to form ventral mesoderm (dark stippling). The organizer (O) then emits a dorsalizing signal (horizontal arrow). Ventral and lateral cells express *Xwnt*-8 which, possibly in concert with BMP-4, attenuates the response of these cells (bar, Post-MBT) to the dorsalizing signal. While lateral cells differentiate as intermediate forms of mesoderm (light stippling, Late Gastrula), ventral cells, which are most distal to the organizer, continue to differentiate as ventral mesoderm (dark stippling, Late Gastrula). In the absence of *Xwnt*-8 signal, dorsal equatorial cells are fully competent to respond to dorsalizing signals and differentiate as extreme dorsal mesoderm (hatched area, Late Gastrula).

spond to dorsalizing signals emitted by the Spemann organizer. As a result, organizer cells, which do not express *Xwnt*-8, respond to dorsalizing signals by differentiating as extreme dorsal mesoderm, while adjacent lateral cells, which express *Xwnt*-8, respond to the same signal by forming intermediate grades of mesoderm. Ventral cells, which both express *Xwnt*-8 and are most distal to the dorsalizing signal, differentiate as extreme ventral mesoderm. In this way, a complete dorsal to ventral spectrum of mesoderm is formed, and is later translated into the final body plan.

These studies highlight an important, but often overlooked, concept in developmental biology: that pattern is established not only through a cascade of inducing signals, which instruct cells to follow a given developmental pathway, but also through spatial and temporal differences in the competence of cells to respond to such signals. This recurring theme may enable a limited number of signaling molecules and signal transduction pathways to determine a broad range of cell fates during morphogenesis. The challenge for the future lies not only in identifying the biochemical nature of inducing agents, but also in deciphering the nature, and the mechanisms of action, of agents which regulate the local competence of responding cells.

REFERENCES

Bolce ME, Hemmati-Brivanlou A, Kushner PD, Harland RM (1992): Ventral mesoderm of *Xenopus* forms neural tissue, including hindbrain, in response to activin. Development 115:681–688.

Cho KWY, Blumberg B, Steinbesser H, DeRobertis EM (1991): Molecular nature of Spemann's organizer: The role of the *Xenopus* homeobox gene *goosecoid*. Cell 67:1111–1120.

Christian JL, Moon RT (1993): Interactions between the *Xwnt*-8 and Spemann organizer signaling pathways generate dorsoventral pattern in the embryonic mesoderm of *Xenopus*. Genes Dev 7:13–28.

Christian JL, McMahon JA, McMahon AP, Moon RT (1991): *Xwnt*-8, a *Xenopus Wnt*-1/*int*-1-related gene responsive to mesoderm inducing growth factors, may play a role in ventral mesodermal patterning during embryogenesis. Development 111:1045–1055.

Christian JL, Olson DJ, Moon RT (1992): *Xwnt*-8 modifies the character of mesoderm induced by bFGF in isolated *Xenopus* ectoderm. EMBO J 11:33–41.

Cunliffe V, Smith JC (1992): Ectopic mesoderm formation in *Xenopus* embryos caused by widespread expression of a *Brachyury* homologue. Nature 358:427–430.

Dale L, Slack JMW (1987): Regional specification within the mesoderm of early embryos of *Xenopus laevis*. Development 100:279–295.

Dale L, Howes G, Price BMJ, Smith JC (1992): Bone morphogenetic protein-4: A ventralizing factor in early *Xenopus* development. Development 115:573–585.

Dirksen ML, Jamrich M (1992): A novel, activin-inducible, blastopore lip-specific gene of *Xenopus laevis* contains a *fork head* DNA-binding domain. Genes Dev 6:599–618.

Driever W, Siegel V, Nüsslein-Volhard C (1990): Autonomous determination of anterior structures in the early *Drosophila* embryo by the *bicoid* morphogen. Development 109:811–820.

Friesel R, Dawid IB (1991): cDNA cloning and developmental expression of fibroblast growth factor receptors from *Xenopus laevis*. Mol Cell Biol 11:2481–2488.

Gerhart J, Danilchik M, Doniach T, Roberts S, Rowning B, Stewart R (1989): Cortical rotation of the *Xenopus* egg: Consequences for the anteriposterior pattern of embryonic dorsal development. Development 107(Suppl):37–51.

Gerhart JC, Stewart R, Doniach T (1991): Organizing the *Xenopus* organizer. In Keller R, Clark W Jr, Griffin F (eds): "Gastrulation: Movements, Patterns and Molecules." New York: Plenum, pp 57–77.

Gillespie LL, Paterno GD, Slack JM (1989): Analysis of competence: Receptors for fibroblast growth factor in early *Xenopus* embryos. Development 106:203–208.

Gimlich RL (1986): Acquisition of developmental autonomy in the equatorial region of the *Xenopus* embryo. Dev Biol 115:340–352.

Green JBA, Howes G, Symes K, Cooke J, Smith JC (1990): The biological effects of XTC-MIF: Quantitative comparison with *Xenopus* bFGF. Development 108:173–183.

Green JBA, Smith JC (1990): Graded changes in dose of a *Xenopus* activin A homologue elicit stepwise transitions in embryonic cell fate. Nature 347:391–394.

Jones EA, Woodland HR (1987): The development of animal cap cells in *Xenopus*: A measure of the start of animal cap competence to form mesoderm. Development 101:557–563.

Jones CM, Lyons KM, Lapan PM, Wright CVE, Hogan BLM (1992): DVR-4 (bone morphogenetic protein-4) as a posterior-ventralizing factor in early *Xenopus* mesoderm induction. Development 115:639–647.

Kaguera H (1990): Spatial distribution of the capacity to initiate a secondary embryo in the 32-cell embryo of *Xenopus laevis*. Dev Biol 142:432–438.

Kimelman D, Maas A (1992): Induction of dorsal and ventral mesoderm by ectopically expressed *Xenopus* basic fibroblast growth factor. Development 114:261–269.

Kimelman D, Abraham J, Haaparanta T, Palisi TM, Kirschner M (1987): The presence of fibroblast growth factor in the frog egg: Its role as a natural mesoderm inducer. Science 242:1053–1056.

Kimelman D, Christian JL, Moon RT (1992): Synergistic principles of development: Overlapping systems in *Xenopus* mesoderm induction. Development 116:1–9.

Köster M, Plessow S, Clement JH, Lorenz A, Tiedemann H, Knöchel W (1991): Bone morphogenetic protein-4 (BMP-4), a member of the TGF- family, in early embryos of *Xenopus laevis*: Analysis of mesoderm inducing activity. Mech Dev 33:191–200.

Matthews LS, Vale WW, Kintner CR (1992): Cloning of a second type of activin receptor and functional characterization in *Xenopus* embryos. Science 255:1702–1705.

McMahon AP, Moon RT (1989): Ectopic expression of the proto-oncogene *int*-1 in *Xenopus* embryos leads to duplication of the embryonic axis. Cell 58:1075–1084.

McMahon AP (1992): The *Wnt* superfamily of developmental regulators. Trends Genet 8:1–5.

Moon RT (1992): In pursuit of the functions of the *Wnt* family of developmental regulators: Insights from *Xenopus laevis*. Bioessays 15:91–97.

Moon RT, Campbell RM, Christian JL, McGrew LL, Shih J, Fraser S: *Xwnt*-5A: A maternal unit which affects morphogenetic movements after overexpression in embryos of *Xenopus laevis*. Development (in press).

Musci TJ, Amaya E, Kirschner MW (1990): Regulation of the fibroblast growth factor receptor in early *Xenopus* embryos. Proc Natl Acad Sci 87:8365–8369.

Nieuwkoop PD (1969): The formation of the mesoderm in urodelean amphibians, I. The induction by the endoderm. W Roux' Arch Ent Org 100:599–638.

Nusse R, Varmus HE (1992): *Wnt* genes. Cell 69:1073–1087.

Ruiz i Altaba A, Jessell T (1991): Retinoic acid modifies mesodermal patterning in early *Xenopus* embryos. Genes Dev 5:175–187.

Shiurba RA, Jing N, Sakakura T, Godsave SF (1991): Nuclear transition of fibroblast growth factor during *Xenopus* mesoderm induction. Development 113:487–493.

Slack, JMW (1991): "From Egg to Embryo." Cambridge/Cambridge University Press.

Slack JMW, Darlington BG, Heath JK, Godsave SF (1987): Mesoderm induction in early *Xenopus* embryos by heparin binding growth factors. Nature 326:197–200.

Smith W, Harland RH (1991): Injected *Xwnt*-8 acts early in *Xenopus* embryos to promote formation of a vegetal dorsalizing center. Cell 67:753–765.

Smith W, Harland RH (1992): Expression cloning of *noggin*, a new dorsalizing factor localized to Spemann's organizer in *Xenopus* embryos. Cell (in press).

Smith JC, Slack JMW (1983): Dorsalization and neural induction: properties of the organizer in *Xenopus laevis*. J Embryol Exp Morph 78:299–317.

Sokol S, Melton DA (1991): Pre-existent pattern in *Xenopus* animal pole cells revealed by induction with activin. Nature 351:409–411.

Sokol S, Melton DA (1992): Cooperation of *Wnt* and activin in the induction of dorsal mesoderm in *Xenopus* animal pole cells. Dev Biol 154:348–355.

Sokol S, Christian JL, Moon RT, Melton DA (1991): Injected *Wnt* RNA induces a complete body axis in *Xenopus* embryos. Cell 67:753–765.

Stewart R, Gerhart JC (1990): The anterior extent of dorsal development of the *Xenopus* embryonic axis depends on the quantity of organizer in the late blastula. Development 109:363–372.

Taira M, Jamrich M, Good PJ, Dawid IB (1992): The LIM domain-containing homeobox *Xlim*-1 is expressed specifically in the organizer region of *Xenopus* gastrula embryos. Genes Dev 6:356–366.

Takasaki H (1987): Fates and roles of the presumptive organizer region in the 32-cell embryo in normal development of *Xenopus laevis*. Dev Growth Differ 29:141–152.

Thomsen G, Woolf T, Whitman M, Sokol S, Melton DA (1990): Activins are expressed in *Xenopus* embryogenesis and can induce axial mesoderm and anterior structures. Cell 63:485–493.

Wolda SL, Moody CJ, Moon RT (1992): Overlapping expression of *Xwnt*-3A and *Xwnt*-1 in neural tissue of *Xenopus laevis* embryos. Dev Biol 155:46–57.

Wolpert, L (1969): Positional information and the spatial pattern of cellular formation. J Theor Biol 25:1–47.

4. Components of Intercellular Adhesive Junctions and Their Roles in Morphogenesis

Pamela Cowin and Anthony M. C. Brown

Departments of Cell Biology and Dermatology, New York University Medical Center, New York, New York 10016 (P.C.); Department of Cell Biology and Anatomy, Cornell University Medical College, New York, New York 10021 (A.M.C.B.)

INTRODUCTION

Desmosomes and adherens junctions are major sites of cell adhesion and constitute membrane anchorage points for intermediate filaments and elements of the actin cytoskeleton, respectively (Farquhar and Palade, 1963). The textbook image of such junction–filament complexes has been one of stable mechanical networks that form the "buttons, belts, and suspenders" of a tissue. In contrast, research over the last decade has shown that several junctional components play a far more instructive role in determining cell shape and tissue form. For example, the adhesive components of intercellular junctions, which are transmembrane glycoproteins of the cadherin superfamily, play a morphoregulatory role during embryogenesis that embodies both a cellular recognition function and a capacity to alter cell phenotype (Nose et al., 1988). Cadherins can induce striking changes in cellular morphology by causing a reorganization of the subcortical and filamentous cytoskeleton, and in some cases this involves second messenger signaling pathways (Ozawa et al., 1989; McNeill et al., 1990; Doherty et al., 1991). In order to effect cell–cell adhesion, cadherins must become physically associated with specific subcortical proteins, including members of the plakoglobin family (Cowin et al., 1986; Ozawa et al., 1989; McCrea et al., 1991). One model arising from the existing data is that proteins of the plakoglobin family may provide important links in the transduction or implementation of morphogenetic signals by regulating the recruitment of cadherins into junctional sites. The regulated assembly of these junctions in turn leads to reorganization of the actin and intermediate filament cytoskeletal networks, thereby altering cell shape and tissue architecture. Regulatory aspects of subcortical junctional proteins will be the principal focus of this review.

Both desmosomes and adherens junctions are formed early during vertebrate embryogenesis. Individual junctions are first detected at the two-cell stage and subsequently increase in frequency just prior to gastrulation, at a time when intercellular attachments are required to withstand the forces of epiboly (Lentz and Trinkaus, 1971; Pereda and Coppo, 1987). In adult tissues both type of junction occur in a wide variety of cell types but are particularly abundant in tissues subject to mechanical stress, such as epidermis and cardiac muscle (Cowin et al., 1985; Cowin and Garrod, 1983; Cowin et al., 1984a).

The two types of junction have similar ultrastructural morphology. Both are associated with a widened intercellular space and a pair of electron-dense mats, called plaques, which underlie and support the membrane. However, the intercellular space of desmosomes is characteristically bisected by a midline plate, and the junctions can be further distinguished by the type of filament that associates with their plaques (Farquhar and Palade, 1963).

COMPONENTS OF ADHERENS JUNCTIONS

The adhesion molecules of cell–cell adherens junctions were discovered through screening for antibodies that inhibited either calcium-dependent adhesion of various cells in culture or early developmental processes such as compaction of the eight-cell-stage mouse morula (Damsky et al., 1983; Gallin et al., 1986a; Hyafil et al., 1980; Ogou et al., 1983; Vestweber and Kemler, 1984). These proteins were the first described members of the cadherin superfamily of calcium-dependent homophilic adhesion molecules (Gallin et al., 1986b; Nagafuchi et al., 1987; Ringwald et al., 1987). At least three closely related members of the cadherin family, E-cadherin (uvomorulin), L-CAM, and N-cadherin (A-CAM) have each been immunolocalized to regions of cell–cell contact and are found concentrated at intercellular adherens-type junctions (Boller et al., 1985; Hirano et al., 1987; Mege et al., 1988).

Colocalization studies have also identified a number of intracellular proteins of adherens junctions. These include plakoglobin ($M_r\sim 83,000$), vinculin ($M_r\sim 130,000$), and radixin ($M_r\sim 82,000$), as well as protein kinases of the src family (Cowin et al., 1986; Geiger and Ginsberg, 1991; McCrea et al., 1991; Tsukita et al., 1991; Tsukita and Tsukita, 1989). In addition, immunoprecipitations with E-cadherin antibodies coprecipitate three cadherin-associated cytoplasmic proteins termed α-, β-, and γ-catenin ($M_r \sim 120,000$, 92,000, and 83,000, respectively) (Nagafuchi and Takeichi, 1988; Ozawa et al., 1989; Ozawa et al., 1990b; McCrea and Gumbiner, 1991). While the identity of γ-catenin is presently unknown, the sequences of α- and β-catenin show them to be closely related to vinculin and plakoglobin, respectively (Herrenknecht et al., 1991; McCrea et al., 1991; Nagafuchi et al., 1991) (Tables I, II).

TABLE I.
Components of Intercellular Adherens Junctions

Adhesive glycoproteins	Subcortical components	Minor components
Cadherins (E, N and L-CAM)	Plakoglobin	Radixin
	Vinculin	Zyxin
	Catenins (α, β, γ)	220kD protein
		c-src
		c-yes

COMPONENTS OF DESMOSOMES

Subcellular fractions enriched in desmosomes can be readily obtained from citric acid extractions of the upper layers of epidermis and tongue (Gorbsky and Steinberg, 1981; Skerrow and Matoltsky, 1974). As a consequence, a great deal is known about desmosomes from these sources, whereas our knowledge of desmosomal constituents from other tissues has mostly been inferred from immunological studies (Cowin and Garrod, 1983; Cowin et al., 1984a). In general, the major components of desmosomal plaques appear to be highly conserved throughout different vertebrate species and tissues, whereas the transmembrane glycoproteins show considerable immunological variation (Cowin and Garrod, 1983; Cowin et al., 1984a; Suhrbier and Garrod, 1986). Gene cloning studies are beginning to reveal the molecular basis of this variation and have shown it to be generated both by a multigene family and by alternative splicing (Collins et al., 1991; Goodwin et al., 1990; Koch et al., 1992; Koch et al., 1990; Mechanic et al., 1991; Parker et al., 1991; Wheeler et al., 1991; Buxton et al., 1993).

Six major polypeptide bands can be resolved from epidermal desmosomes on SDS-PAGE. Three of these bands correspond to desmoplakins I and II ($M_r \sim 215,000$ and $230,000$), and plakoglobin ($M_r \sim 83,000$), all of which have now been localized to the cytoplasmic plaques of desmosomes by immunological

TABLE II.
Components of Desmosomes

Adhesive glycoproteins		Subcortical components	Minor components
Desmogleins		Plakoglobin	22kD glycoprotein
Dsg 1	(Desmoglein I)	Desmoplakins I and II	140kD glycoprotein
Dsg 2	(Pemphigus Vulgaris antigen)		Band 6
Dsg 3	(HDGC)		Plectin
Desmocollins			Desmocalmin
Dsc 1a, 1b	(Desmocollins I and II)		Lamin B-like protein
Dsc 2a, 2b			Desmoyokin
Dsc 3a, 3b			

studies (Cowin and Garrod, 1983; Cowin et al., 1985a,b; Cowin et al., 1986). The desmoplakins share sequence similarity in their carboxyl-termini with the 1B rod domain of intermediate filament proteins and have been proposed to mediate filament attachment to the desmosome (Green et al., 1990; Stappenbeck and Green, 1992). This same function has also been assigned to four other minor components of the desmosome: desmocalmin, plectin, desmosomal band 6, and a 140-kd protein immunochemically related to lamin B (Cautaud et al., 1990; Kapprell et al., 1988; Tsukita and Tsukita, 1985; Wiche et al., 1984). It is notable that plakoglobin is the only junctional component present both at desmosomes and at adherens junctions and, in addition to its membrane-bound forms, is also found in a cytosolic pool as a soluble dimer (Cowin et al., 1986). The potential function of plakoglobin and other proteins related to it will be discussed later.

The remaining three protein bands contain the transmembrane calcium-binding glycoproteins desmoglein I ($M_r \sim 165,000$) and desmocollins I and II ($M_r \sim 130,000$ and $115,000$). The adhesive role of these molecules has been established by studies showing that antibodies directed against their ectodomains inhibit desmosome formation and disrupt cell adhesion. For example, monovalent fragments of antibodies to desmocollins inhibit desmosomes formation by Modin-Derby canine kidney (MDCK) cells in culture (Cowin et al., 1984b). Moreover, in a striking clinical disorder called pemphigus foliaceus or fogo selvagem (Wild Fire), patients have circulating pathogenic autoantibodies that recognize desmoglein I (Eyre and Stanley, 1987). These antibodies are deposited within the epidermis and cause extensive blistering by disrupting the intercellular adhesion of suprabasal keratinocytes. We and others have cloned and sequenced desmoglein I and desmocollins I and II and shown that they have extensive sequence similarity in their ectodomain to cadherins, yet form distinct subtypes within this superfamily (Collins et al., 1991; Goodwin et al., 1990; Koch et al., 1990; Koch et al., 1992; Mechanic et al., 1991; Parker et al., 1991; Wheeler et al., 1991).

THE CADHERIN SUPERFAMILY

The cadherin superfamily has burgeoned in recent years and now contains over 20 members. All show similarity in the sequence of their ectodomain (Fig. 1). Vertebrate cadherins can be categorized into three classes, however, on the basis of their distinct cytoplasmic regions. For the purpose of this review, those cadherins with cytoplasmic sequences that show a high degree of homology to those of E-cadherin will be referred to as class I cadherins; those which resemble desmoglein and desmocollin will be designated class II and III, respectively.

With the exception of T-cadherin (Ranscht and Bronner, 1991), the members of all three classes of cadherins are single-pass transmembrane proteins,

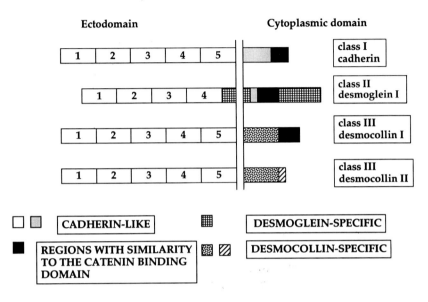

Fig. 1. The major structural differences between the three families of proteins that compose the vertebrate cadherin superfamily. All three families share similarity in the sequence and organization of their ectodomain into four repeats (1–4). More variation in sequence is found in the fifth region, particularly among the desmogleins. The class I cadherins are highly conserved in their cytoplasmic sequences, particularly in the region known to bind catenins. Class II cadherins contain a sequence with some similarity to the catenin-binding domain in the center of their long cytoplasmic tail. Class III cadherins are produced as two forms by alternative splicing, which introduces a region similar to the class I cadherin carboxy terminus in desmocollin I.

which are synthesized as precursors and proteolytically cleaved to the mature form in the trans-Golgi network (Pasdar et al., 1991; Penn et al., 1987a,b; Shore and Nelson, 1991). The prosegments show little sequence similarity and are thought to prevent adhesive interactions of cadherins within intracellular compartments during biosynthesis. In vertebrates, the ectodomain of mature cadherin polypeptides can be aligned into four repeated sequences of 100 to 120 amino acids each (Hatta et al., 1988). A fifth region that is highly variable in length and sequence, particularly among the class II desmogleins, precedes the transmembrane segment (Amagai et al., 1991; Hatta et al., 1988; Mechanic et al., 1991). Each of the four repeats contains putative calcium-binding pockets (Ringwald et al., 1987), and site-directed mutagenesis has confirmed the calcium-binding properties of one such motif (Ozawa et al., 1990a). The binding of cations to this site confers on the cadherin conformational changes that are a prerequisite for cell adhesion. The few cadherins so far described in invertebrates have a much larger ectodomain than the vertebrate forms. For example the product of the *Drosophila* gene *fat* contains as many as 34 tandem cadherin repeats followed by 4 epidermal growth factor (EGF)-like repeats (Mahoney et al., 1991).

The cytoplasmic domains of all class I cadherins show a striking degree of homology, about 80% of the residues being identical (Hatta et al., 1988). In contrast, the desmosomal cadherins show little homology with class I cadherins in their cytoplasmic domains, except in the region implicated in catenin binding (see below). These differences presumably reflect the requirement of these molecules to interact with distinct cytoskeletal components (Koch et al., 1990; Mechanic et al., 1991; Parker et al., 1991; Wheeler et al., 1991).

CADHERINS AND MORPHOGENESIS

Cell adhesion mediated by cadherins clearly plays a fundamental role in development, by allowing the formation of cohesive cell sheets. Such cohesion also appears to be a prerequisite for the formation of tight and gap junctions (Gumbiner et al., 1988). Apart from these purely adhesive functions, class I cadherins play at least two other roles in morphogenesis. These involve both cell recognition and a capacity to reorganize the cytoskeleton and alter cell phenotype.

Cadherins and Cellular Recognition

The role of cadherins in cell recognition has been elegantly demonstrated by experiments showing that fibroblasts transfected with class I cadherins acquire the capacity to recognize and selectively adhere to cells expressing the same cadherin through calcium-dependent, homophilic interactions (Nose et al., 1988). Moreover, these studies have shown that cells expressing two different cadherins will physically sort out from each other within mixed aggregates (Nose et al., 1988). The numerous members of the cadherin family provide a wide repertoire of specificity for the many intricate cellular recognition and aggregation events that occur during embryogenesis.

Studies of the expression patterns of class I cadherins in adult and embryonic tissue have shown them to be differentially expressed in a manner consistent with such a morphoregulatory role (Takeichi, 1988; Takeichi, 1990; Takeichi, 1991; Takeichi et al., 1990). In general, individual cell types express more than one cadherin species, and their complement of cadherins may vary in a spatially and temporally regulated manner. Such changes often correlate with morphogenetic events, particularly those involving separation of one cell sheet from another. For example, cells of the presumptive neural tube lose expression of E-cadherin and begin to express N-cadherin just prior to their separation from the overlying epidermis that continues to express E-cadherin (Takeichi, 1991). The separation of these two cell layers can be impaired if N-cadherin is ectopically expressed in the epidermis, illustrating the crucial importance of differential cadherin expression in this morphogenetic event (Detrick et al., 1990; Fujimori et al., 1990).

Evidence for a similar role in cellular recognition by desmosomal glycoproteins is as yet preliminary, since the full diversity of this family remains to be determined. Already, however, three desmoglein-like proteins and four desmocollin-type isoforms have been described (Amagai et al., 1991; Goodwin et al., 1990; Mechanic et al., 1991; Koch et al., 1991, 1992; Wheeler et al., 1991; Parker et al., 1991), and immunological studies suggest the presence of many more related isoforms (Cowin and Garrod, 1983; Cowin et al., 1984a; Suhrbier and Garrod, 1986). Desmoglein I (Dsg1) and desmocollins I and II (Dscla, Dsclb) are restricted to the suprabasal layers of epidermis and tongue. Another desmoglein-like protein, pemphigus vulgaris antigen (Dsg3) is restricted to the lower region of the suprabasal layers of stratified squamous tissues, while a third desmoglein named HDGC (Dsg2) and a second pair of desmocollins (Dsc2a, Dsc2b) are found in a range of simple epithelia, as well as the basal cells of stratified tissue (Theis et al., 1993; Arnemann et al., 1993; Koch et al., 1992). The precise regulation of expression of these isoforms suggests that like the class I molecules, desmosomal cadherins may contribute to selective cell recognition during morphogenesis.

The sequences important for the homophilic specificity of cell adhesion mediated by class I cadherin lie in the amino-terminal 113 residues of the mature protein. A crucial region has been pinpointed by site-directed mutagenesis to the residues surrounding a highly conserved HAV tripeptide, found at position 78 in the first extracellular repeat (Nose et al., 1990). Point mutations in amino acids flanking the HAV motif in E-cadherin do not affect the strength of adhesive interactions but do significantly alter the recognition properties (Nose et al., 1990). An HAV motif is also found in influenza strain A hemagglutinins and fibroblast growth factor receptors, where mutational studies indicate that the tripeptide is required for protein function (Blaschuk et al., 1990; Byers et al., 1992). The HAV sequences are absent from desmosomal and *Drosophila* cadherins.

Cadherins and Growth Control

While most of our understanding of adhesive junctions thus far has been derived from biochemical and cell biological studies in vertebrate systems, additional insights have recently emerged from *Drosophila* genetics. A search for cadherins in *Drosophila* has so far yielded two new members of the superfamily, the products of the genes *fat* and *dachsous*, of which the former is the better characterized (Horsch et al., 1991; Mahoney et al., 1991). Loss of function mutations in the *fat* locus result in massive overgrowth of cells in the imaginal discs, indicating that *fat* has the properties of a tumor suppressor gene (Mahoney et al., 1991). Although these proteins have several structural features that are distinct from vertebrate cadherins, they are clearly members of the cadherin superfamily, and their mutant phenotypes indicate that they can act to restrain cell proliferation. Whether vertebrate cadherins might participate in contact-mediated inhibition of cell growth remains to be determined, but in this regard it is notable that synthetic peptides containing the conserved sequence motif HAV and surrounding residues of E-cadherin are able to induce mitogenic effects on MDCK cells, in addition to disrupting cell adhesion (Liaw et al., 1990).

Cadherins and Cellular Differentiation

A number of studies demonstrate that cadherins can bring about dramatic and characteristic changes in cell shape and phenotype. For example, mouse L-cells transfected by cDNA encoding E-cadherin displayed a remarkable change in their morphology and adhesive phenotype; the typical bipolar appearance of these fibroblasts gave way to a more flattened morphology and the cells grouped together in colonies (Nose et al., 1988). Concomitant with these changes, the cells underwent a major reorganization of their ankyrin and fodrin subcortical cytoskeleton and a polarization of Na,K-ATPase to the basolateral cell borders (McNeill et al., 1990). Thus, cells of mesenchymal origin could be induced to assume epithelial characteristics by ectopic expression of a cadherin found in many epithelial cell types. Consistent with this effect are studies in which impaired expression or function of E-cadherin caused epithelial cells to adopt more fibroblastic properties, including a more invasive phenotype, which could be corrected to some degree by transfection of E-cadherin cDNAs (Frixen et al., 1991; Vleminckx et al., 1991). E-Cadherin levels have also been inversely correlated with the invasive properties of numerous tumor cell lines (Behrens et al., 1989; Behrens et al., 1991; Birchmeier et al., 1991).

Another striking change in cellular phenotype that is mediated by cadherins has been described in the pheochromocytoma cell line PC12. These cells, which express endogenous N-cadherin, can be induced to assume a neuronal morphology by contact with NIH3T3 cells transfected with N-cadherin cDNA (Doherty et al., 1991; Doherty et al., 1992). The extension of neurites occurs in the absence of transcription but can be inhibited by pertussis toxin and calcium channel antagonists. These findings suggest that second messenger signaling pathways are required for this effect. Subsequent studies have shown that these changes can be mimicked if the second messengers are activated directly (Saffell et al., 1992). These data imply that, in this instance, neuronal cell differentiation is effected via a signaling pathway that can be activated by cadherin homophilic interaction, but which is not necessarily dependent on cell–cell adhesion. On the basis of these observations, the authors have argued that cadherins are instructive molecules (Doherty et al., 1991).

REGULATION OF CADHERIN-MEDIATED ADHESION BY THE CYTOPLASMIC DOMAIN AND ITS ASSOCIATED PROTEINS

The changes in cell shape and behavior that result from E-cadherin expression have been shown by deletion analysis to depend upon sequences in the highly conserved cytoplasmic tail, as well as on the homophilic interactions of the ectodomain. Mutant forms of E-cadherin deleted in the carboxy-terminal 69 amino acids are unable to bring about cell–cell adhesion even though they reach the plasma membrane and retain the normal conformation and binding properties of the ectodomain (Nagafuchi and Takeichi, 1988; Ozawa et al., 1989). Without this cytoplasmic domain, cadherins fail to cluster into junctional sites and are unable to reorganize the cytoskeleton (Ozawa et al., 1989). The importance of this cytoplasmic region has been confirmed by studies of N-cadherin mutant proteins expressed in *Xenopus* embryos (Kintner, 1992). The same stretch of 69 amino acids, as well as a small region close to the cytoplasmic face of the membrane, were found independently to produce dominant negative effects on the function of the wild-type protein and to disrupt cell adhesion. The above observations have focused attention on the proteins which interact with these cytoplasmic regions. The carboxy-terminal region is known to be the binding site for the catenins (Ozawa and Kemler, 1992; Ozawa et al., 1990a,b).

α-Catenin

The importance of α-catenin in junction formation has been demonstrated by a recent study employing the lung carcinoma cell line PC9 (Hirano et al., 1992). These cells express E-cadherin, but have no detectable α-catenin, and normally grow as loose aggregates of cells in suspension. Transfection of these cells with a cDNA encoding α-N-catenin, an isoform of α-catenin found in brain, promoted strong intercellular adhesion and caused the cells to form highly ordered polarized spherical cysts in culture (Hirano et al., 1992). Thus, α-catenins are required for E-cadherins to engage in cell adhesion and multicellular organization. Interestingly, the synthesis and/or stability of α-catenin is increased by ectopic E-cadherin expression in fibroblasts, implying that the two molecules are engaged in reciprocal regulation (Nagafuchi et al., 1991).

The sequence of α-catenin shows homology with three distinct regions including the self-association domain of vinculin, a major component of the cytoplasmic plaques of adherens junctions and focal contacts (Herrenknecht et al., 1991; Nagafuchi et al., 1991). This suggests that cadherin-catenin complexes cluster through oligomerization of α-catenin, either with itself or with vinculin. Alternatively, α-catenin and vinculin may be components in a chain of physically associated molecules that serve to tether the cadherins to actin microfilaments. In this case the clustering of cadherins in the membrane would occur by a mechanism analogous to actin-mediated capping of other receptors (Herrenknecht et al., 1991; Nagafuchi et al., 1991; Ozawa and Kemler, 1992; Ozawa et al., 1990a, b).

THE PLAKOGLOBIN FAMILY

In addition to α-catenin, three other cytoplasmic proteins have been found physically associated with members of the cadherin superfamily in coprecipitation experiments (McCrea et al., 1991; Ozawa et al., 1989). These are γ-catenin, about which little is currently known, β-catenin, and plakoglobin (Cowin et al., 1986; Ozawa et al., 1989). Plakoglobin and β-catenin are closely related to each other (63% identity) and to the product of the *Drosophila* segment polarity gene *armadillo* (63%–70% identity) (Franke et al., 1989; McCrea et al., 1991; Peifer and Weischaus, 1990). The plakoglobin family also includes a more distantly related protein named p120, a putative substrate for tyrosine phosphorylation by pp60^{src}.

Plakoglobin

Plakoglobin was first described as an 83-kd component of isolated desmosomes and was originally termed band 5 (Skerrow and Matoltsky, 1974). Some years ago, we carried out immunological studies that localized plakoglobin to both the adherens junction and the desmosome (Cowin et al., 1986; Kapprell et al., 1985), and recent data suggest that it associates with several different transmembrane junctional components. First, a strong association between plakoglobin and both desmoglein I and pemphigus vulgaris antigen (PVA), has been described on the basis of coprecipitation data (Eyre and Stanley, 1987). Secondly, plakoglobin colocalizes with a chimeric fusion protein containing the carboxy-terminal 65 amino acids of desmocollin I, although direct binding to this domain has not as yet been demonstrated (Troyanovsky et al., 1993). Thirdly, plakoglobin is a weakly associated constituent of the cadherin-catenin complexes precipitated by E- or N-cadherin antibodies (Ozawa et al., 1989; Piepenhagen and Nelson, 1992; DeMarais and Moon, 1992). Although the cytoplasmic sequences of the different classes of cadherin are generally divergent, desmocollin I, desmoglein, PVA, and class I cadherins all contain a region of similar sequence that in class I cadherins has been shown to interact with β-catenin (see below). We have recently demonstrated that plakoglobin binds to the related sequence of desmoglein 1 (Mathur, L. Goodwin, P. Cowin, unpublished results). By virtue of its association with all three classes of cadherins, plakoglobin is an attractive candidate for regulation of the adhesive function of both major forms of intercellular adhesive junction in a coordinate manner.

β-Catenin

β-Catenin has been co-immunoprecipitated with both E- and N-cadherin, and most probably associates with all class I cadherins (Knudsen and Wheelock, 1992; Ozawa et al., 1989). This protein remains associated under relatively stringent conditions suggesting that it may bind directly to the carboxyl termini of the cadherins (McCrea and Gumbiner, 1991; Ozawa and Kemler, 1992). While the association of both α- and β-catenin with cadherins is apparently a prerequisite for cell adhesion, other observations clearly indicate that formation of this complex is not sufficient by itself to trigger junction formation. First, both α- and β-catenin are found associated with cadherin mutants deleted or chimeric in their ectodomain (Ozawa et al., 1989). Secondly, both of these catenins become associated during the early steps of cadherin biosynthesis rather than at the time of junction assembly (Ozawa and Kemler, 1992). β-Catenin becomes bound when the cadherin precursor is in the endoplasmic reticulum (ER) and α-catenin joins the complex following proteolytic processing of the cadherin precursor in the medial stacks of the Golgi network. It is therefore likely that other regulatory events or protein interactions must intervene to initiate junction formation.

Regulation by Phosphorylation

For many years it has been known that adherens junctions are disrupted in cells transformed by $pp60^{v\text{-}src}$ (Volberg et al., 1991; Warren and Nelson, 1987). Moreover, the zonula adherens, the major form of adherens junction in epithelial tissues, contains a concentration of several kinases of the *src* family, for example *c-yes* and *c-src* (Maher et al., 1985; Tsukita et al., 1991). This suggests that phosphorylation of the junctional components themselves may occur and could play a key role in regulating junctional stability. Consistent with this view are data showing that kinase inhibitors prevent the disassembly of intercellular junctions that normally occurs when extracellular calcium is removed (Citi, 1992), as well as reports that desmocollins are preferentially phosphorylated on serines when junctions are rapidly forming and breaking down (Parrish et al., 1990).

Two recent studies have correlated the degree of phosphorylation of members of the plakoglobin family with the impaired adhesive behavior seen in cells transformed by *v-src* (Matsuyoshi et al., 1992; Reynolds et al., 1992). The first study examined 3Y1 fibroblast derivatives transformed by *src* and *fos*. These cells had equivalent levels of cadherin and catenin expression to the nontransformed parental line, yet showed impaired ability to aggregate and to organize their cadherins into regions of cell–cell contact (Matsuyoshi et al., 1992). The aggregation properties were restored by treatment of these cells with a tyrosine kinase inhibitor, whereas an inhibitor of phosphatase adversely affected aggregation of the parental cell line. This suggests that cadherin function was regulated by phosphorylation. Analysis of the cadherin-catenin complex showed that the effects on aggregation were mirrored by the degree of phosphorylation of β-catenin and to a lesser extent the cadherins (Matsuyoshi et al., 1992).

The second study focused on p120, a membrane-associated phosphotyrosine-containing protein in *src* transformed cells, which has a modest but significant resemblance to plakoglobin in its sequence (Reynolds et al., 1992). In contrast to several other putative *src* substrates, phosphorylation of p120 was not induced by a mutant of *src* that retains kinase activity but is transformation-defective (Reynolds et al., 1992; Reynolds et al., 1989). Although it is not yet known whether p120 is associated with junctions, these observations suggest that phosphorylation of p120 may be crucial for the transforming effects of *src*.

Further clues from *Drosophila*

As mentioned above, another member of the plakoglobin family has been described in *Drosophila,* as the product of the segment polarity gene *armadillo* (Peifer and Weischaus, 1990). Segment polarity genes are required for the correct establishment of cell patterning within each body segment of the embryo, a pattern initially manifested in the morphology of the cuticle overlying the epidermis (Ingham, 1991). Mutations in the *armadillo* gene can result in specific defects in this pattern, and give a phenotype similar to that of mutations in another segment polarity gene, *wingless* (Nusslein-Volhard and Wieschaus, 1980). In wild-type embryos, the level of Armadillo protein is elevated in a broad stripe of cells within each segment (Riggleman et al., 1990). Within these stripes there seems to be an increase in both cytoplasmic and membrane-localized Armadillo protein (M. Peiffer and E. Weischaus, personal communication), and the modulation of Armadillo in these cells is apparently an early response to signals from the *wingless* gene (Riggleman et al., 1990). The subcellular distribution of Armadillo suggests that this protein may have more in common with plakoglobin than with β-catenin: since a significant proportion of plakoglobin in vertebrate cells appears to be present in a cytosolic pool in addition to membrane-bound forms (Cowin et al., 1986).

The vertebrate homolog of *Drosophila wingless* is the protooncogene *Wnt*-1, a gene which normally functions in early central nervous system development in the mouse and which, like *wingless*, is known to encode a secreted protein (see Nusse and Varmus, 1992 for review). Modulation of

Armadillo protein in the *Drosophila* embryo is a response to *wingless*-dependent signals, and it is possible that components of this signaling pathway have been conserved in certain vertebrate cells. The rat pheochromocytoma cell line PC12 displays increased cellular adhesion as a result of expressing an exogenous Wnt-1 gene (G. Shackleford and H. Varmus, personal communication; R. Bradley, P. Cowin, and A. Brown, manuscript in preparation). We have recently observed elevated plakoglobin levels in these cells, particularly in the membrane-associated fraction (R. Bradley, P. Cowin, and A. Brown, unpublished data). The amount of E-cadherin detectable at the cell membrane is also increased, and it is tempting to speculate that the phenotypic changes induced by *Wnt-1* in this cell line may result, at least in part, from the modulation of cadherin function. Similarly, the *wingless*-dependent changes in Armadillo protein described in *Drosophila* suggest that regulation of adhesive junctions might play an essential role in the interpretation of positional information that is necessary for correct morphogenesis of the body segments. In principal, such junctions could be required simply to hold cells together and facilitate the passage of other signals. Alternatively, the nonuniform distribution of junctions within a segment might itself provide patterning cues by effecting a graded reorganization of the local cytoskeletal network.

CONCLUSIONS

The adhesive components of both desmosomes and intercellular adherens junctions are now known to be members of the cadherin superfamily. Their functions in cell adhesion and morphogenesis are governed not only by differential expression but also by regulation of their membrane distribution. The subcortical and adhesive components of these junctions are engaged in a dialogue of mutual recruitment and stabilization. The association of members of the plakoglobin and vinculin protein families with cadherins, as well as their posttranslational modification, may be important points of regulation of junction assembly and adhesive function. A provocative area of research emerging at present involves the growing number of membrane-bound phosphatases and kinases which either have structural similarity to junctional components or cellular distribution patterns that suggest an intimate association with them (Gu et al., 1991; Yang and Tonks, 1991). A fuller understanding of the interactions of junctional elements with such molecules promises to be an exciting area of future research.

ACKNOWLEDGMENTS

Work in our laboratories is supported by the National Institutes of Health (GM47429 and CA47207), The American Cancer Society (P.C.), and The March of Dimes (A.B.). The authors gratefully acknowledge the Pew Charitable Trusts for scholarships that significantly fostered this collaboration (P.C. and A.B.).

REFERENCES

Amagai M, Klaus KV, Stanley JR (1991): Autoantibodies against a novel epithelial cadherin in pemphigus vulgaris, a disease of cell adhesion. Cell 67:869–877.

Arnemann J, Sullivan KH, Magee AI, King IA, Buxton RS (1993): Stratification-related expression of isoforms of the desmosomal cadherins in human epidermis. J Cell Sci 104:741–750.

Behrens J, Mareel MM, Van RFM, Birchmeier W (1989): Dissecting tumor cell invasion: Epithelial cells acquire invasive properties after the loss of uvomorulin-mediated cell-cell adhesion. J Cell Biol 108:2345–2447.

Behrens J, Weidner KM, Frixen UH, Schipper JH, Sachs M, Arakaki N, Daikuhara Y, Birchmeier W (1991): The role of E-cadherin and scatter factor in tumor invasion and cell motility. Experientia Suppl 59:109–126.

Birchmeier W, Behrens J, Weidner KN, Frixen UH, Schipper J (1991): Dominant and recessive genes involved in tumor cell invasion. Curr Opin Cell Biol 3:832–840.

Blaschuk OW, Pouliot Y, Holland PC (1990): Identification of a conserved region common to cadherins and influenza strain A hemagglutinins. J Mol Biol 211:679–682.

Boller K, Vestweber D, Kemmler R (1985): Cell adhesion molecule uvomorulin is localised in the intermediate junctions of adult intestinal epithelial cells. J Cell Biol 100:327–332.

Buxton RS, Cowin P, Franke WW, Garrod DR, Green KJ, King IA, Koch PJ, Magee AI, Rees DA, Stanley JR, Steinberg MS (1993): Nomenclature of the desmosomal cadherins. J Cell Biol 121:481–483.

Byers S, Amaya E, Munro S, Blaschuk O (1992): Fibroblast growth factor receptor contain a conserved HAV region common to cadherins and influenza strain A hemagglutinins: A role in protein-protein interactions. Dev Biol 152:411–416.

Cautaud A, Ludosky MA, Courvalin JC, Cautaud J (1990): A protein antigenically related to nuclear lamin B mediates the association of intermediate filaments with desmosomes. J Cell Biol 111:581–588.

Citi S (1992): Protein kinase inhibitors prevent junction dissociation induced by low extracellular calcium in MDCK cells. J Cell Biol 117:169–178.

Collins JE, Legan PK, Kenny TP, MacGarvie J, Holton JL, Garrod DR (1991): Cloning and sequence analysis of desmosomal glycoproteins 2 and 3 (desmocollins): cadherin-like desmosomal adhesion molecules with heterogeneous cytoplasmic domains. J Cell Biol 113:381–391.

Cowin P, Garrod D (1983): Antibodies to epithelial desmosomes show wide tissue and species cross-reactivity. Nature 302:148–150.

Cowin P, Mattey D, Garrod DR (1984a): Distribution of desmosomal components in tissues of vertebrates, studied by fluorescent antibody staining. J Cell Sci 66:119–132.

Cowin P, Mattey D, Garrod DR (1984b): Identification of desmosomal surface components (desmocollins) and inhibition of desmosome formation by specific FAB'. J Cell Sci 70:41–60.

Cowin P, Franke WW, Grund C, Kapprell H-P, Kartenbeck J (1985a): The desmosome intermediate filament complex. In Edelman GM, Thiery J-P (eds). "The Cell in Contact." New York: John Wiley & Sons, pp 427–461.

Cowin P, Kapprell HP, Franke WW (1985b): The complement of desmosomal plaque proteins in different cell types. J Cell Biol 101:1442–1454.

Cowin P, Kapprell H-P, Franke WW, Tamkun J, Hynes RO (1986): Plakoglobin: A protein common to different kinds of intercellular adhering junctions. Cell 46:1063–1073.

Damsky CH, Richa J, Solter D, Knudsen K, Buck CA (1983): Identification and purification of a cell surface glycoprotein mediating intercellular adhesion in embryonic and adult tissue. Cell 34:455–466.

DeMarais AA, Moon RA (1992): The *armadillo* homologs beta-catenin and plakoglobin are differentially expressed during early development of *Xenopus laevis*. Dev Biol 153:337–343.

Detrick RJ, Dickey D, Kintner CR (1990): The effects of N-cadherin misexpression on morphogenesis in Xenopus embryos. Neuron 4:493–506.
Doherty P, Ashton SV, Moore SE, Walsh FS (1991): Morphoregulatory activities of NCAM and N-cadherin can be accounted for by G protein-dependent activation of L- and N-type neuronal Ca^{2+} channels. Cell 67:21–33.
Doherty P, Ashton SV, Skaper SD, Leon A, Walsh FS (1992): Ganglioside modulation of neural cell adhesion molecule and N-cadherin-dependent neurite outgrowth. J Cell Biol 117:1093–1099.
Eyre RW, Stanley JR (1987): Human autoantibodies against a desmosomal protein complex with a calcium-sensitive epitope are characteristic of pemphigus foliaceus patients. J Exp Med 165:1719–1724.
Farquhar MG, Palade GE (1963): Junctional complexes in various epithelia. J Cell Biol 17:375–412.
Franke WW, Goldschmidt MD, Zimblemann R, Mueller HM, Schiller DL, Cowin P (1989): Molecular cloning and amino acid sequence of plakoglobin, common junctional plaque protein. Proc Natl Acad Sci USA 86:4027–4031.
Frixen UH, Behrens J, Sachs M, Eberle G, Voss B, Warda A, Lochner D, Birchmeier W (1991): E-cadherin-mediated cell-cell adhesion prevents invasiveness of human carcinoma cells. J Cell Biol 113:173–185.
Fujimori T, Miyatani S, Takeichi M (1990): Ectopic expression of N-cadherin perturbs histogenesis in Xenopus embryos. Development 110:97–104.
Gallin WJ, Chuong CM, Finkel LH, Edelman GM (1986a): Antibodies to liver cell adhesion molecule perturb inductive interactions and alter feather pattern and structure. Proc Natl Acad Sci 83:8235–8239.
Gallin W, Sorkin BC, Edelman GM, Cunningham BA (1986b): Sequence analysis of a cDNA clone encoding the liver cell adhesion molecule, L-CAM. Proc Natl Acad Sci 74:2808–2812.
Geiger B, Ginsberg D (1991): The cytoplasmic domain of adherens-type junctions. Cell Motil Cytoskeleton 20:1–6.
Goodwin L, Hill JE, Raynor K, Raszi L, Manabe M, Cowin P (1990): Desmoglein shows extensive homology to the cadherin family of cell adhesion molecules. Biochem Biophys Res Commun 173:1224–1230.
Gorbsky G, Steinberg MS (1981): Isolation of the intercellular glycoproteins of desmosomes. J Cell Biol 90:243–248.
Green KJ, Parry DA, Steinert PM, Virata ML, Wagner RM, Angst BD, Nilles LA (1990): Structure of the human desmoplakins. Implications for function in the desmosomal plaque. J Biol Chem 265:2603–2612.
Gu M, York JD, Warshawsky I, Majerus PW (1991): Identification, cloning and expression of a cytosolic megakaryocyte protein tyrosine-phosphatase with sequence homology to cytoskeletal protein 4.1. Proc Natl Acad Sci USA 88:5867–5871.
Gumbiner B, Stevenson B, Grimaldi A (1988): The role of the cell adhesion molecule uvomorulin in the formation and maintenance of the epithelial junctional complex. J Cell Biol 107:1575–1587.
Hatta K, Nose A, Nagafuchi A, Takeichi M (1988): Cloning and expression of cDNA encoding a neural calcium-dependent cell adhesion molecule: Its identity in the cadherin gene family. J Cell Biol 106:873–881.
Herrenknecht K, Ozawa M, Eckerskorn C, Lottspeich F, Lenter M, Kemler R (1991): The uvomorulin-anchorage protein alpha catenin is a vinculin homologue. Proc Natl Acad Sci 88:9156–9160.
Hirano S, Nose A, Hatta K, Kawakami A, Takeichi M (1987): Calcium-dependent cell-cell adhesion molecules (cadherins): Subclass specificities and possible involvement of actin bundles. J Cell Biol 105:2501–2510.

Hirano S, Kimoto N, Shimoyama Y, Hirohashi S, Takeichi M (1992): Identification of a neural alpha-catenin as a key regulator of cadherin function and multicellular organization. Cell 70:293–301.
Horsch M, Goodman CS (1991): Cell and substrate adhesion molecules in Drosophila. Ann Rev Cell Biol 7:505–557.
Hyafil F, Morello D, Babinet C, Jacob F (1980): A cell surface glycoprotein involved in the compaction of embryonal carcinoma cells and cleavage stage embryos. Cell 21:927–934.
Ingham P (1991): Segment polarity genes and cell patterning within the Drosophial body segment. Curr Biol 1:261–267.
Kapprell HP, Cowin P, Franke WW, Ponstingl H, Opferkuch HJ (1985): Biochemical characterization of desmosomal proteins isolated from bovine muzzle epidermis: Amino acid and carbohydrate composition. Eur J Cell Biol 36:217–229.
Kapprell HP, Owaribe K, Franke WW (1988): Identification of a basic protein of Mr 75,000 as an accessory desmosomal plaque protein in stratified and complex epithelia. J Cell Biol 106:1679–1691.
Kintner C (1992): Regulation of embryonic cell adhesion by the cadherin cytoplasmic domain. Cell 69:225–236.
Knudsen KA, Wheelock MJ (1992): Plakoglobin, or an 83-kD homologue distinct from beta-catenin, interacts with E-cadherin and N-cadherin. J Cell Biol 118:671–679.
Koch PJ, Walsh MJ, Schmelz M, Goldschmidt MD, Zimbelmann R, Franke WW (1990): Identification of desmoglein, a constitutive desmosomal glycoprotein, as a member of the cadherin family of cell adhesion molecules. Eur J Cell Biol 53:1–12.
Koch P, Goldschmidt MD, Walsh MJ, Zimbleman R, Franke WW (1991): Complete amino acid sequence of the epidermal desmoglein precursor polypeptide and identification of a second type of desmoglein gene. Eur J Cell Biol 55:200–208.
Koch PJ, Goldschmidt MD, Zimbelmann R, Troyanovsky R, Franke WW (1992): Complexity and expression patterns of the desmosomal cadherins. Proc Natl Acad Sci USA 89:353–357.
Lentz TL, Trinkaus JP (1971): Differentiation of the junctional complex of surface cells in the developing Fundulus blastoderm. J Cell Biol 48:455–472.
Liaw CW, Tomaselli KJ, Cannon C, Davis D, Bryant K, Rubin L (1990): Disruption of MDCK and bovine endothelial cell tight junctions with cadherin synthetic peptides. J Cell Biol 111:408a.
Maher P, Pasquale EB, Wang JY, Singer SJ (1985): Phosphotyrosine-containing proteins are concentrated in focal adhesions and intercellular junctions in normal cells. Proc Natl Acad Sci USA 82:6576–6580.
Mahoney PA, Weber U, Onofrechuk P, Biessmann H, Bryant PJ, Goodman CS (1991): The fat tumor suppressor gene in Drosophila encodes a novel member of the cadherin gene superfamily. Cell 67:853–868.
Matsuyoshi N, Hamaguchi M, Taniguchi S, Nagafuchi A, Tsukita S, Takeichi M (1992): Cadherin-mediated cell-cell adhesion is perturbed by v-src tyrosine phosphorylation in metastatic fibroblasts. J Cell Biol 118:703–714.
McCrea PD, Gumbiner BM (1991): Purification of a 92-kDa cytoplasmic protein tightly associated with the cell-cell adhesion molecule E-cadherin (uvomorulin). Characterization and extractibility of the protein complex from the cell cytostructure. J Biol Chem 266:4514–4520.
McCrea PD, Turck CW, Gumbiner B (1991): A homolog of the armadillo protein in Drosophila (plakoglobin) associated with E-cadherin. Science 254:1359–1361.
McNeill H, Ozawa M, Kemler R, Nelson WJ (1990): Novel function of the cell adhesion molecule uvomorulin as an inducer of cell surface polarity. Cell 62:309–316.
McNeill H, Ryan TA, Smith SJ, Nelson WJ (1993): Spatial and temporal dissection of immediate and early events following cadherin-mediated epithelial cell adhesion. J Cell Biol 120:1217–1226.
Mechanic S, Raynor K, Hill JE, Cowin P (1991): Desmocollins form a novel subset of the cadherin family of cell adhesion molecules. Proc Natl Acad Sci USA 88:4476–4480.

Mege RM, Matsuzaki F, Gallin WJ, Goldberg JI, Cunningham BA, Edelman GM (1988): Construction of epithelioid sheets by transfection of mouse sarcoma cells with cDNAs for chicken cell adhesion molecules. Proc Natl Acad Sci USA 85:7274–7278.

Nagafuchi A, Takeichi M (1988): Cell binding function of E-cadherin is regulated by the cytoplasmic domain. EMBO J 7:3679–3684.

Nagafuchi A, Shirayoshi Y, Okazaki K, Yasuda K, Takeichi M (1987): Transformation of cell adhesion properties by exogenously introduced E-cadherin cDNA. Nature 329:341–343.

Nagafuchi A, Takeichi M, Tsukita S (1991): The 102 kd cadherin-associated protein: Similarity to vinculin and posttranscriptional regulation of expression. Cell 65:849–857.

Nose A, Nagafuchi A, Takeichi M (1988): Expressed recombinant cadherins mediate cell sorting in model systems. Cell 54:993–1001.

Nose A, Tsuji K, Takeichi M (1990): Localization of specificity determining sites in cadherin cell adhesion molecules. Cell 61:147–155.

Nusse R, Varmus H (1992): Wnt genes. Cell 69:1073–1087.

Nusslein-Volhard C, Wieschaus E (1980): Mutations affecting segment number and polarity in Drosophila. Nature 287:795–801.

Ogou S, Yoshida Noro C, Takeichi M (1983): Calcium-dependent cell-cell adhesion molecules common to hepatocytes and teratocarcinoma stem cells. J Cell Biol 97:944–948.

Ozawa M, Kemler R (1992): Molecular organization of the uvomorulin-catenin complex. J Cell Biol 116:989–996.

Ozawa M, Baribault H, Kemler R (1989): The cytoplasmic domain of the cell adhesion molecule uvomorulin associates with three independent proteins structurally related in different species. EMBO J 8:1711–1717.

Ozawa M, Engel J, Kemler R (1990a): Single amino acid substitutions in one Ca^{2+} binding site of uvomorulin abolish the adhesive function. Cell 63:1033–1038.

Ozawa M, Ringwald M, Kemler R (1990b): Uvomorulin-catenin complex formation is regulated by a specific domain in the cytoplasmic region of the cell adhesion molecule. Proc Natl Acad Sci USA 87:4246–4250.

Palade GE (1985): Differentiated microdomains in cellular membranes: Current status. In Edelman G (ed). "The Cell in Contact." New York: John Wiley & Sons; pp 9–27.

Parker AE, Wheeler GN, Arnemann J, Pidsley SC, Ataliotis P, Thomas CL, Rees DA, Magee AI, Buxton RS (1991): Desmosomal glycoproteins II and III. Cadherin-like junctional molecules generated by alternative splicing. J Biol Chem 266:10438–10445.

Parrish EP, Marston JE, Mattey DL, Measures HR, Venning R, Garrod DR (1990): Size heterogeneity, phosphorylation and transmembrane organisation of desmosomal glycoproteins 2 and 3 (desmocollins) in MDCK cells. J Cell Sci 96:239–248.

Pasdar M, Krzeminski KA, Nelson WJ (1991): Regulation of Desmosome assembly in MDCK cess: Coordination of membrane core and cytoplasmic plaque domain assembly at the plasma membrane. J Cell Biol 113:645–657.

Peifer M, Weischaus E (1990): The segment polarity gene armadillo encodes a functionally modular protein that is the Drosophila homolog of human plakoglobin. Cell 63:1167–1178.

Penn EJ, Burdett ID, Hobson C, Magee AI, Rees DA (1987a): Structure and assembly of desmosome junctions: biosynthesis and turnover of the major desmosomes components of Madin-Darby canine kidney cells in low calcium medium. J Cell Biol 105:2327–2334.

Penn EJ, Hobson C, Rees DA, Magee AI (1987b): Structure and assembly of desmosome junctions: Biosynthesis, processing, and transport of the major protein and glycoprotein components in cultured epithelial cells. J Cell Biol 105:57–68.

Pereda J, Coppo M (1987): Ultrastructure of a two-cell human embryo. Anat Embryol (Berl) 177:91–96.

Piepenhagen P, Nelson WJ (1993): Gamma catenin and the phosphoprotein plakoglobin are distinct E-cadherin-associated proteins in Madin-Darby canine kidney epithelial cells. J Cell Sci 104:751–762.

Ranscht B, Bronner FM (1991): T-cadherin expression alternates with migrating neural crest cells in the trunk of the avian embryo. Development 111:15–22.

Reynolds AB, Rosel DJ, Kanner SB, Parsons JT (1989): Transformation-specific tyrosine phosphorylation of a novel cellular protein in chicken cells expressing oncogenic variants of the avian cellular *src* gene. Mol Cell Biol 9:629–638.

Reynolds AB, Herbert L, Cleveland JL, Berg ST, Gaut JR (1992): p120, a novel substrate of protein tyrosine kinase receptors and of p60 vsrc, is related to cadherin-binding factors Beta catenin, plakoglobin and armadillo. Oncogene 7:2439–2445.

Riggleman R, Schedl P, Wieschaus E (1990): Spatial expression of the Drosophila segment polarity gene *armadillo* is post-transcriptionally regulated by *wingless*. Cell 63:549–560.

Ringwald M, Schuh R, Vestweber D, Eistetter H, Lottspeich F, Engle J, Dolz R, Jahnig F, Epplen J, Mayer S et al., (1987): The structure of cell adhesion molecule uvomorulin. Insights into the molecular mechanism of calcium-ion dependant cell adhesion. EMBO J 6:3647–3653.

Saffell JL, Walsh F, Doherty P (1992): Direct Activation of second messenger pathways mimics cell adhesion molecule-dependent neurite outgrowth. J Cell Biol 118:663–670.

Shore EM, Nelson WJ (1991): Biosynthesis of the cell adhesion molecule uvomorulin (E-cadherin) in Madin-Darby canine kidney epithelial cells. J Biol Chem 266:19672–19680.

Skerrow CJ, Matoltsky AG (1974): Isolation of epidermal desmosomes. J Cell Biol 63:515–523.

Stappenbeck T, Green KJ (1992): The desmoplakin carboxyl terminus coaligns with and specifically disrupts intermediate filament networks when expressed in cultured cells. J Cell Biol 116:1197–1209.

Suhrbier A, Garrod D (1986): An investigation of the molecular components of desmosomes in epithelia cells of five vertebrates. J Cell Sci 81:223–242.

Takeichi M (1988): The cadherins: Cell-cell adhesion molecules controlling animal morphogenesis. Development 102:639–655.

Takeichi M (1990): Cadherins: A molecule family important in selective cell-cell adhesion. Annu Rev Biochem 59:237–252.

Takeichi (1991): Cadherin cell adhesion receptors as a morphogenetic regulator. Science 251:1451–1455.

Takeichi M, Inuzuka H, Shimamura K, Fujimori T, Nagafuchi A (1990): Cadherin subclasses: Differential expression and their roles in neural morphogenesis. Cold Spring Harb Symp Quant Biol 55:319–325.

Theis D, Koch PJ, Franke WW (1993): Differential synthesis of type 1 and type 2 desmocollin mRNAs in human stratified epithelia. Int J Dev Biol 37:101–110.

Troyanovsky SM, Eshkind LG, Troyanovsky RB, Leube RE, Franke WW (1993): Contributions of cytoplasmic domains of desmosomal cadherins to desmosome assembly and intermediate filament anchorage. Cell 72:561–574.

Tsukita S, Tsukita S (1985): Desmocalmin: A calmodulin-binding high molecular weight protein isolated from desmosomes. J Cell Biol 101:2070–2080.

Tsukita S, Tsukita S (1989): Isolation of cell-to-cell adherens junctions from rat liver. J Cell Biol 108:31–41.

Tsukita S, Oishi K, Akiyama T, Yamanashi Y, Yamamoto T (1991): Specific proto-oncogene tyrosine kinases of a src family are enriched in cell-to-cell adherens junctions where the level of tyrosine phosphorylation is elevated. J Cell Biol 113:867–879.

Vestweber D, Kemler R (1984): Rabbit antiserum against a purified surface glycoprotein decompacts mouse preimplantation embryos and reacts with specific adult tissues. Cell Res 152:169–178.

Vleminckx K, Vakaet LJ, Mareel M, Fiers W, Van RF (1991): Genetic manipulation of E-cadherin expression by epithelial tumor cells reveals an invasion suppressor role. Cell 66:107–119.

Volberg T, Geiger B, Dror R, Zick Y (1991): Modulation of intercellular adherens-type junction and tyrosine phosphorylation of their components in RSV-transformed cultured chick lens cells. Cell Regul 2:105–120.

Wacker IA, Rickard JE, DeMey JR, Kreis TE (1992): Accumulation of a microtubule binding protein, pp170, at desmosomal plaques. J Cell Biol 117:813–824.

Warren SL, Nelson WJ (1987): Nonmitogenic morphoregulatory action of pp60v-src on multicellular epithelial structures. Mol Cell Biol 7:1326–1337.

Wheeler GN, Parker AE, Thomas CL, Ataliotis P, Poynter D, Arnemann J, Rutman AJ, Pidsley SC, Watt FM, Rees DA et al. (1991): Desmosomal glycoprotein DGI, a component of intercellular desmosome junctions, is related to the cadherin family of cell adhesion molecules. Proc Natl Acad Sci USA 88:4796–4800.

Wiche G, Krepler R, Artlieb U, Pytela R, Aberer W (1984): Identification of plectin in different human cell types and immunolocalization at epithelial basal cell surface membranes. Exp Cell Res 155:43–49.

Yang Q, Tonks N (1991): Isolation of a cDNA clone encoding a human protein-tyrosine phosphatase with homology to the cytoskeletal-associated proteins band 4.1, ezrin, and talin. Proc Natl Acad Sci USA 88:5949–5953.

Molecular Basis of Morphogenesis, pages 69–77
© 1993 Wiley-Liss, Inc.

5. Eyes, Ears, and Homeobox Genes in Zebrafish Embryos

Monte Westerfield, Marie-Andrée Akimenko, Marc Ekker, and Andreas Püschel

Institute of Neuroscience, University of Oregon, Eugene, Oregon 97403

INTRODUCTION

Major advances in our understanding of pattern formation during vertebrate development have been provided by analyses of gene families, most notably the homeobox-containing genes. Homeobox sequences encode domains which are thought to function as transcription factors that may regulate developmental processes (Akam, 1987; Ingham, 1988; Scott et al., 1989). Analyses of the encoded sequences and studies of the expression patterns of homeobox genes have suggested that they can be divided into several families (Scott et al., 1989) some of which include the *Hox* (Duboule and Dollé, 1989; Graham et al., 1989), *paired box* or *Pax* (Kessel and Gruss, 1990), *muscle segment homeobox* or *Msh* (Holland, 1991; Akimenko et al., 1991), and *distal-less* or *Dlx* (Porteus et al., 1991; Price et al., 1991; Robinson et al., 1991; Ekker et al., 1992) gene families. The *Hox* genes, which are organized in tightly linked clusters, are thought to function in the specification of positional cues or cellular identities along the anteroposterior axis, from the hindbrain through the trunk (see review by Kessel and Gruss, 1990). Cells at specific axial positions express particular combinations of *Hox* genes, both in the nervous system and peripheral structures derived from the neural crest, and in axial structures derived from the sclerotome. This review summarizes recent studies of zebrafish that have identified *Pax*, *Msh*, and *Dlx* genes which are expressed in particular combinations in cells of the developing eye and ear and which may function to specify positional information and cell identities in these organs.

HOMEOBOX GENES IN THE DEVELOPING ZEBRAFISH EYE

The expression of two *Pax* genes demarcates regions in the developing eye of zebrafish embryos towards the end of gastrulation. An oval-shaped domain of cells in the prospective prosencephalon expresses *pax-6* (Püschel et al.,

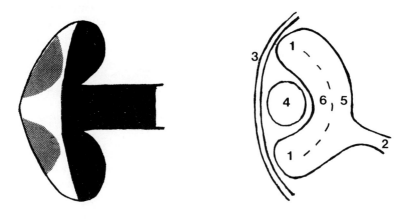

	Hox-7.1	Pax-2	Pax-6	Hox-8.1	Dlx-1
1. Ciliary margin	+	?	?	–	–
2. Optic stalk	–	+	–	–	–
3. Overlying ectoderm	–	–	+	+	–
4. Lens	–	–	+	+	–
5. Outer retinal layers	–	+	–	+	+
6. Inner retinal layers	–	–	+	+	+

Fig. 1. A homeobox gene code for the eye. **Top left:** Early expression of *pax-2* (light shading) and *pax-6* (dark shading) divide the eye primordium into distinct parts. Dorsal view of the developing head, anterior to the left. (Adapted from Püschel et al., 1992a,b, with permission of the publisher.) **Top right:** Diagram of a vertebrate eye. The numbers refer to various structures within the eye as listed in the table at the bottom. **Bottom:** Expression of each gene that has been detected in the various structures of the developing eye is indicated. The data come from Monaghan et al., 1991, for *Hox-7.1* and *Hox-8.1*; Krauss et al., 1991a,c and Püschel et al., 1992a for *pax-2*; Krauss et al., 1991c, and Püschel et al., 1992b for *pax-6*; and Dollé et al., 1992 for *Dlx-1*.

1992b; also called *pax[zf-a]*, Krauss et al., 1991a,c) shortly after epiboly is completed and before the tailbud is visible. It contains a stripe of cells at its posterior margin which express higher levels of *pax-6* than cells farther anterior. A few hours later, the *pax-6* expressing cells in this region are located in the optic vesicles, but not in the presumptive optic stalk (Fig. 1, top). Expression in the optic vesicle is highest in the lateral and posterior regions which will invaginate to form the optic cup. The anteroposterior and dorsoventral boundaries of expression are identical in the prosencephalon and the optic vesicle. In addition, the overlying ectoderm, which is known in other species to interact with the optic vesicles to form the lens, expresses *pax-6*. By the end of the first day of development, when most of the somites have formed, the expression of *pax-6* in the eye includes both the optic cup and the lens (Fig. 1, bottom). Later, during formation of the retina, cells in the bipolar and ganglion cell layers continue to express *pax-6*.

Cells in the developing zebrafish eye express a second *Pax* gene, *pax-2* (Püschel et al., 1992a; also termed *pax[zf-b]*, Krauss et al., 1991b). Initial *pax-2* expression in the eye appears a few hours later than that of *pax-6* and, unlike *pax-6*, includes the region that will later form the optic stalk (Fig. 1, top). At this stage, the cells expressing the *pax-2* and *pax-6* genes occupy distinct non-overlapping regions of the eye. During the next few days of development, expression of *pax-2* persists in the stalk and in the ventral medial portion of the optic cup, and in contrast to *pax-6*, cells in the developing lens never express *pax-2* (Fig. 1, bottom).

Spatially restricted subsets of cells in the developing eye express at least three other homeobox containing genes. Transcripts from two members of the *Msh* gene family, *Hox-7.1* and *Hox-8.1* in mice (Monaghan et al., 1991), appear in the inner layer of the optic cup. Cells of the overlying ectoderm, the lens, and the neural retina express *Hox-8.1*; whereas cells of the presumptive iris and ciliary body express *Hox-7.1*. The fifth gene expressed by the early eye is *Dlx-1*, a member of the *distal-less* gene family (Dollé et al., 1992). As with *Hox-8.1*, cells of the developing neural retina express *Dlx-1*, but unlike *Hox-8.1*, the lens and overlying ectoderm are negative for *Dlx-1*. Thus, distinct regions of the developing eye are specified in terms of their patterns of gene expression before they exhibit overt signs of morphological differentiation (Fig. 1, bottom). These homeobox genes may function in specifying various parts of the eye, as suggested by analysis of mutations in the human (Ton et al., 1991) and mouse (Hill et al., 1991) homologues of *pax-6*, which produce an absence of the iris or reduced or missing eyes.

HOMEOBOX GENES IN THE DEVELOPING ZEBRAFISH EAR

The inner ear of all jawed vertebrates arises from the epithelium of the otic vesicle and contains three semicircular canals, otoliths, and sets of sensory neurons, all positioned precisely within the cranium to detect head orientation and movement. We have found that precursors of these various parts of the ear in zebrafish embryos express specific combinations of at least four different homeobox genes, including two members of the *Msh* gene family, *msh-C* and *msh-D* (Ekker et al., 1992), one *distal-less* gene, *dlx-3* (Ekker et al., 1992), and one *Pax* gene, *pax-2* (Püschel et al., 1992a).

In zebrafish, the otic vesicle forms by a condensation of cells, which becomes visible as the otic placode (Waterman and Bell, 1984) during early somite stages. Several hours earlier, *dlx-3* transcripts appear at high levels in a loosely packed cluster of cells in the region where the placode will form. These cells are morphologically indistinguishable from neighboring cells at this stage. Expression is uniform within this region, a situation which suggests that most, if not all cells express *dlx-3*. A couple of hours later, the otic vesicle is beginning to form from the placode and *dlx-3* expression is more restricted. The dorsal and posterior regions of the vesicle contain higher levels of *dlx-3* transcripts and, within this region, the transcripts appear in cells of the otic epithelium.

During formation of the otic vesicle, the expression domains of *msh-D* and *pax-2* in the ear are established. Cells in the dorsal aspect, along the entire anterior/posterior extent of the vesicle, contain *msh-D* transcripts (Fig. 2, top) and cells on the medial surface of the vesicle express *pax-2*. The *dlx-3* transcripts are now more restricted to cells in the dorsal and medioposterior aspects of the epithelium than at earlier stages (Fig. 2, top). Thus, the regions of *dlx-3*, *pax-2*, and *msh-D* expression are aligned with the axes of the vesicle, dorsal-ventral for *msh-D*, medial-lateral for *pax-2*, and anterior-posterior for *dlx-3*. The semicircular canals later extend along these same axes (Fig. 2, bottom).

During the second day of development, the semicircular canals (Figure 2, bottom) begin to form as projections of the epithelium into the cavity, first from the lateral and then from the anterior, posterior, and finally the ventral surfaces of the vesicle (Waterman and Bell, 1984). The otic maculae, which contain the sensory hair cells, form between these projections (Fig. 2, bottom). The precursors of the macular cells, including the hair cells, specifically express both the *msh-C* and *msh-D* genes as they begin to differentiate. The expression of the *msh* genes by sensory neurons in the maculae could specify the particular fates of these cell types.

The initial induction of the otic placode is thought to occur in response to signals from the chordamesoderm (Waddington, 1937). Later, additional signals from the hindbrain neuroectoderm are required (Waddington, 1937; Van de Water and Conley, 1982). By the stage when *dlx-3* transcripts are localized in the nascent otic placode, the placode is not yet morphologically distinguished, and induction of the vesicle by the hindbrain neuroectoderm has probably not yet begun. Thus, the *dlx-3* gene product is present and could function during these inductive interactions. The *dlx-3* gene encodes a product which presumably acts as a transcription factor localized within the nucleus, making it more likely that it plays a regulatory, rather than a direct role in the induction process. For example, *dlx-3* expression may render cells responsive to the inductive signals.

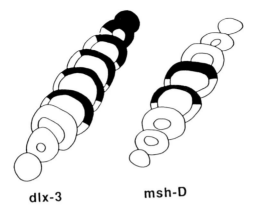

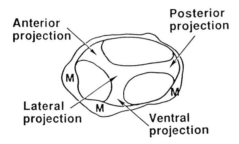

Fig. 2. Summary of expression of *dlx-3* and *msh-D* in the otic vesicle and formation of the semicircular canals. **Top:** Reconstructions of sagittal sections, from lateral to medial, through the otic vesicle at the end of the first day of development hybridized with *dlx-3* (left) or *msh-D* (right) probes. The cells expressing the genes are indicated by black shading. **Bottom:** Camera lucida drawing of the otic vesicle in a 2.5-day-old embryo. The sensory maculae (M) and the four projections of the epithelium which form the walls of the semicircular canals are labeled. Anterior to the left and dorsal to the top. (Reprinted from Ekker et al., 1992, with permission of the publisher.)

Expression of *msh-D*, *pax-2*, and *dlx-3* in particular regions of the otic vesicle could specify the morphogenesis and subsequent orientation of the semicircular canals. Our results demonstrate that the restricted expression domains of these three genes are aligned along the same axes which the extending semicircular canals later follow. Thus, their expression may help specify position within the epithelium. Positional information is presumably required to specify which epithelial cells will form the canals and in which direction they will project.

The restricted expression patterns of these four homeobox genes appear after the otic vesicle has been induced. Thus, their restriction may be a consequence of the induction by factors like retinoic acid (Represa et al., 1990) or int-2 (Represssa et al., 1990). Recent studies have shown that the *int-2* gene product, which is structurally related to members of the fibroblast growth factor family (Moore et al., 1986), is an inductive signal for formation of the otic vesicle (Represssa et al., 1991). Blocking expression of *int-2* inhibits formation of the otic vesicle, while application of exogenous int-2 protein induces vesicle formation. The *int-2* gene product is transiently expressed by the hindbrain (Wilkinson et al., 1989) and is thought to be a diffusible molecule which can act on the adjacent placodal cells. Expression of the *int-2* gene may be regulated by other early genes, like *Krox-20* or the *Hox* genes that are expressed in the hindbrain (Hunt and Krumlauf, 1991; Hunt et al., 1991a,b; Wilkinson and Krumlauf, 1990).

Consistent with this idea, Lufkin et al. (1991) and Chisaka et al. (1992) have demonstrated that disruption of the *Hox-1.6* gene produces defects in the inner ear. The inner ears of *Hox-1.6* mutants form only inflated thin-walled sacs containing a few sensory cells. These defects could reflect inappropriate expression of the homeobox genes we have described, due either to very early alterations in positional specification of the ectodermal placode or to a failure of trophic interactions with the hindbrain or mesoderm (Hogan and Wright, 1992).

CONSERVATION OF HOMEOBOX GENE REGULATION

These studies of developing eyes and ears, as well as previous studies of the hindbrain and spinal cord, suggest that homeobox genes may have similar functions: to specify positional information in many parts of the developing embryo. Moreover, comparisons of the patterns of embryonic development in vertebrates as diverse as fish and mice suggest that they use similar genes and, hence, similar genetic programs at some stages of development. By isolating from zebrafish potential developmental control genes that have homologues in mammals, we could compare directly their structures, expression patterns, and regulation in divergent vertebrates.

Among the genes we examined, the *Pax* genes have the most highly conserved structures. The *pax-2* (Krauss et al., 1991b) and the *pax-6* (Püschel et al., 1992b) genes encode proteins that are more than 95% identical to their mouse homologues. For the msh proteins (Ekker et al., 1992), on the other hand, there is little sequence similarity between the two zebrafish products or among them and the mouse Hox-7 and Hox-8 proteins outside of the homeodomain, with the exception of a decapeptide sequence near the amino terminal which is well conserved among all of the vertebrate msh proteins described thus far. Within the homeodomains, there is at least 95% identity among the four proteins.

In two cases, for *pax-2* and *pax-6,* we could make direct comparisons between the expression patterns of the homologous genes in zebrafish in mice. We found that the tissue specificity and time course of expression of these genes relative to developmental processes is almost identical during the early development of these two organisms (Püschel et al., 1992a,b). The most significant differences were in the time courses of genes that are expressed transiently. For example, some cells in zebrafish express *pax-2* for a slightly longer developmental time than do the corresponding cells in mice (Püschel et al., 1992a).

The evolutionary conservation of the structures and expression patterns of these genes suggests that their regulation may also have been conserved. To examine this idea, we developed a transient expression system (using mosaically transgenic zebrafish) which allows rapid analysis of transgene expression. We examined the activities of two mammalian *Hox* gene promoters, mouse *Hox-1.1* and human *HOX-3.3* (Westerfield et al., 1992). We found that these Hox promoters are activated in specific regions and tissues of developing zebrafish embryos and that this specificity depends upon the same regulatory elements within the promoters that specify the spatial expression of these genes in mice. Our results suggest that, like their structures and expression patterns, the promoter activities of homeobox genes have been remarkably conserved from fish to mammals.

CONCLUSIONS

The patterns of expression of homeobox genes in developing eyes and ears support the notion that, as in the hindbrain and spinal cord, positional information and cell fates are specified by the expression of particular combinations of these genes. Many of these genes have identifiable homologues in both fish and mammals. Preliminary comparisons demonstrate that the structures, promoter activities, and expression patterns of at least some of these genes have been conserved. Thus, homeobox genes probably subserve developmental functions that have been well conserved during evolution of the vertebrate lineage. Moreover, these results suggest that different vertebrate species apparently share a number of developmental mechanisms, control genes and, thus, genetic programs of development. Identification of additional developmental control genes in more than one species will presumably further elucidate the common and unique mechanisms of development in diverse organisms.

REFERENCES

Akam M (1987): The molecular basis of metameric pattern in the *Drosophila* embryo. Development 101:1–22.

Akimenko M-A, Ekker M, Westerfield M (1991): Characterization of three zebrafish genes related to *Hox-7*. In Hinchliffe JR, Hurle J, Summerbell D (eds): "Developmental Patterning of the Vertebrate Limb," NATO ASI Series. New York: Plenum Press, pp 61–63.

Chisaka O, Musci T, Capecchi M (1992): Developmental defects of the ear, cranial nerves and hindbrain resulting from targeted disruption of the mouse homeobox gene *Hox-1.6*. Nature 355:516–520.

Dollé P, Price M, Deboule D (1992): Expression of the murine *Dlx-1* homeobox gene during facial, ocular and limb development. Differentiation 49:93–99.

Duboule D, Dollé P (1989): The structural and functional organization of the murine HOX gene family resembles that of *Drosophila* homeotic genes. EMBO J 8:1497–1505.

Ekker M, Akimenko M-A, Bremiller R, Westerfield M (1992): Expression of three related homeobox genes in zebrafish embryonic vestibular organ. Neuron 9:27–35.

Graham A, Papalopulu N, Drumlauf R (1989): The murine and *Drosophila* homeobox gene complexes have common features of organization and expression. Cell 57:687–378.

Hill R, Favor J, Hogan B, Ton C, Saunders G, Hanson I, Prosser J, Jordan T, Hastle N, Heyningen V (1991): Mouse *Small* eye results from mutations in a paired-like homeobox-containing gene. Nature 54:522–526.

Hogan B, Wright C (1992): The making of the ear. Nature 355:494–495.

Holland, PWH (1991): Cloning and evolutionary analysis of msh-like homeobox genes from mouse, zebrafish and ascidian. Gene 98:253–257.

Hunt P, Krumlauf R (1991): Deciphering the *Hox* code: Clues to patterning branchial regions of the head. Cell 66:1075–1078.

Hunt P, Gulisano M, Cook M, Sham M-H, Faiella A, Wilkinson D, Boncinelli E, Krumlauf R (1991a): A distinct *Hox* code for the branchial region of the vertebrate head. Nature 353:861–864.

Hunt P, Whiting J, Muchamore I, Marshall H, Krumlauf R (1991b): Homeobox genes and models for patterning the hindbrain and branchial arches. Development Suppl 1:187–196.

Ingham PW (1988): The molecular genetics of embryonic pattern formation in *Drosophila*. Nature 335:25–34.

Kessel M, Gruss P (1990): Murine developmental control genes. Science 249:374–379.

Krauss S, Johanson T, Korzh V, Fjose A (1991a): Expression pattern of zebrafish *pax* genes suggest a role in early brain regionalization. Nature 353:267–270.

Krauss S, Johanson T, Korzh V, Fjose A (1991b): Expression of the zebrafish paired box gene *pax[zf-b]* during early neurogenesis. Development 113:1193–1206.

Krauss S, Johanson T, Korzh V, Moens U, Ericson J, Fjose A (1991c): Zebrafish *pax[zf-a]*: a paired box-containing gene expressed in the neural tube. EMBO J 10:3609–3619.

Lufkin T, Dierich A, LeMeur M, Mark M, Chambon P (1991): Disruption of the *Hox 1.6* homeobox gene results in defects in a region corresponding to its rostral domain of expression. Cell 66:1105–1119.

Monaghan AP, Davidson DR, Sime C, Graham E, Baldock R, Bhattacharya SS, Hill RE (1991): The *msh*-like homeobox genes define domains in the vertebrate eye. Development 112: 1053–1061.

Porteus MH, Bulfone A, Ciaranello RD, Rubenstein JLR (1991): Isolation and characterization of a novel cDNA clone encoding a homeodomain that is developmentally regulated in the ventral forebrain. Neuron 7:221–229.

Price M, Lemaistre M, Pischetola M, Di Lauro R, Duboule D (1991): A mouse gene related to distal-less shows a restricted expression in the developing forebrain. Nature 351:748–751.

Püschel A, Dressler G, Westerfield M (1992a): Comparative analysis of *Pax-2* protein distributions during neurulation in mice and zebrafish. Mech Dev 38:197–208.

Püschel AW, Gruss P, Westerfield M (1992b): Sequence and expression pattern of *pax-6* are highly conserved between zebrafish and mice. Development 114:643–651.

Repressa J, Sanchez A, Miner C, Lewis J (1990): Retinoic acid modulation of the early development of the inner ear is associated with the control of c-fos expression. Development 110:1081–1090.

Repressa J, León Y, Miner C, Giraldez F (1991): The int-2 proto-oncogene is responsible for induction of the inner ear. Nature 353:561–563.

Robinson G, Wray S, Mahon K (1991): Spatially restricted expression of a member of a new family of murine *distal-less* homeobox genes in the developing forebrain. New Biol 3:1183–1194.

Scott MP, Tamkun JW, Hartzell GW (1989): The structure and function of the homeodomain. Biochim Biophys Acta 989:25–48.

Ton C, Hirronen H, Miwa H, Weil M, Monaghan P, Jorden T, Heyningen V, Hastle N, Meijers-Heijboer H, Drechsler M, Royer-Pokora B, Collins F, Swaroop D, Strong C, Saunders G (1991): Positional cloning and characterization of a paired-box and homeobox-containing gene from the aniridia region. Cell 67:1559–1074.

Van de Water T, Conley E (1982): Neural inductive message to the developing mammalian inner ear: Contact mediated versus extra-cellular matrix interactions. Anat Rec 202:195A.

Waddington CH (1937): The determination of the auditory placode in the chick. J Exp Biol 14:232–239.

Waterman RE, Bell DH (1984): Epithelial fusion during early semicircular canal formation in the embryonic zebrafish, *Brachydanio rerio*. Anat Rec 210:101–114.

Westerfield M, Wegner J, Jegalian B, DeRobertis E, Püschel A (1992): Specific activation of mammalian *Hox* promoters in mosaic transgenic zebrafish. Genes Dev 6:591–598.

Wilkinson DG, Krumlauf R (1990): Molecular approaches to the segmentation of the hindbrain. Trends Neurosci 13:335–339.

Wilkinson DG, Bailes JA, Champion JE, McMahon AP (1987): A molecular analysis of mouse development from 8 to 10 days post coitum detects changes only in embryonic globin expression. Development 99:493–500.

Wilkinson DG, Bhatt S, Cook M, Boncinelli E, Krumlaut R (1989): Segmental expression of *Hox-2* homeobox-containing genes in the developing mouse hindbrain. Nature 341:405–409.

6. Cardiac Morphogenesis in the Zebrafish, Patterning the Heart Tube Along the Anteroposterior Axis

Didier Y. R. Stainier and Mark C. Fishman

Cardiovascular Research Center, Massachusetts General Hospital, Charlestown, Massachusetts 02129

INTRODUCTION

The development of the heart, the earliest organ to form during vertebrate morphogenesis, has fascinated scientists and philosophers for centuries because of its vital function and also because of its spontaneously rhythmical contractions. Currently, the physiological properties of the adult heart are being extensively characterized, but our understanding of cardiac morphogenesis is mainly limited to descriptive studies in various species. We are using the zebrafish embryo in an attempt to define the molecular forces that fashion the heart tube. Although heart development has traditionally been studied in the chick and the frog, the zebrafish offers several distinct advantages. The transparency of its embryo, its rapid development (the heart is functional within a day after fertilization), the prominence of the heart, and the potential for a genetic analysis make the zebrafish an excellent vertebrate model system to study cardiac morphogenesis.

ZEBRAFISH HEART DEVELOPMENT

The teleost heart consists of four chambers in series: the sinus venosus, the atrium, the ventricle, and the bulbus arteriosus, with the most prominent being the atrium and the ventricle. Two concentric cellular tubes form the heart with the inner, endocardial, tube being separated from the outer, myocardial, tube by extracellular matrix referred to as cardiac jelly. Zebrafish embryonic development starts with a period of rapid cleavages; then the blastoderm thins and spreads to cover the yolk cell in a process termed epiboly. During gastrulation, which starts at 50% epiboly, deep cells involute and move anteriorly (Fig. 1). As cells converge toward one side of the gastrula, a local thickening appears along the margin, the embryonic shield (Fig. 1, arrow), which marks the dorsal side of the embryo (D). Cardiac progenitor cells involute

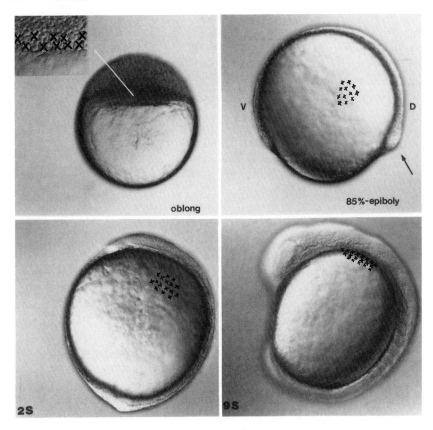

Fig. 1. Schematic representation of the position and migration of cardiac progenitor cells at early developmental stages. Single cells located at, or near, the margin of blastula embryos were labeled with rhodamine-dextran and their progeny followed until 24 hours after fertilization to determine their contribution to the beating heart tube. Cells from the marginal region at 90° longitude contribute to the heart tube but also to the outflow tract and possibly to pharyngeal endoderm. From the 2000-cell stage to the 9-somite stage, cardiac progenitors divide once or twice. Timetable of development at 28.5°C (Westerfield, 1989): oblong, 3.75 hours; 50% epiboly (onset of gastrulation), 5.2 hours; 85% epiboly, 8.5 hours; 2-somite, 10.6 hours; 9-somite, 13.5 hours. V, ventral; D, dorsal. (Reprinted from Stainier and Fishman, 1992, with permission from the publisher.)

early, shortly after the onset of gastrulation (Warga and Kimmel, 1990; D.Y.R. Stainier, R.K. Lee, and M.C. Fishman, manuscript in preparation). Lineage analysis shows that at the midblastula stage, cells located at the blastoderm margin midway between the future dorsal and ventral axis (90° longitude) contribute to the heart tube. After involuting, cardiac progenitors migrate to the embryonic axis (Fig. 1), where they first arrive at the 8-somite stage and coalesce to form the heart tube.

The early heart tube lies in an anteroposterior (A-P) orientation under the otic capsule, adjacent to the yolk sac. By the 18-somite stage, prior to the onset of contractions, the heart tube, which really consists of two tubes on either side of the midline, can be distinguished from surrounding tissues by staining for tropomyosin. At an early stage, as shown in Figure 2A, the presumptive atrium sits anteriorly and the presumptive ventricle posteriorly. This relative positioning is revealed by microscopic observation of the developing embryo, cell lineage analysis of heart-forming cells, and immunohistochemistry with chamber-specific monoclonal antibodies (Stainier and Fishman, 1992). During subsequent development, the head lifts dorsally, the distance between the eyes and the otic capsules progressively diminishes, and the yolk gets depleted. As described by Senior (1909) in the shad, the heart tube appears to remain attached at its arterial end to the embryo, while the venous end accompanies the retreating yolk. The net result is that the heart tube swings, reversing its orientation, so that, at later stages, venous return is to the posterior end of the tube and arterial outflow from its anterior end (Fig. 2B,C). Thus, unlike most other organs, in which relative axial positioning remains aligned throughout development, the zebrafish heart tube resides transiently in a reversed orientation. By the 26-somite stage, the ventricle, which is located slightly rostral to the otic vesicle, has started beating; and by the end of the 26-somite stage (less than 30 minutes later at 28.5°C), the atrium, which resides slightly caudal to the midbrain-hindbrain boundary (MHB), has also started beating. Immunohistochemical analysis using chamber-specific monoclonal antibodies confirms this arterial end to venous end progression of differentiation (Stainier and Fishman, 1992), which is also observed in the avian embryo (DeHaan, 1965; Litvin et al., 1992).

By 36 hours, the heart tube has looped leaving the atrium left-sided and bringing the ventricle medially (Fig. 2B). Starting at the end of the second day (Fig. 2C), endocardial cells in specific locales along the heart tube migrate into the cardiac jelly and undergo a mesenchymal transformation to form the valves. Because of the transparency of the embryo and the readily accessible location of the heart, this process can be followed live in the intact zebrafish embryo. By the fifth day, the heart has assumed its adult configuration with the atrium sitting on top of the ventricle (Fig. 2D).

HEART TUBE POLARITY AND RETINOIC ACID–INDUCED CARDIAC TRUNCATIONS

As part of the evolution from primitive chordates to vertebrates, the heart tube achieved polarity (Bourne, 1980). Polarity is established during embryonic life and is revealed by the gradient of differentiation along the early heart tube, and by the differential expression of myosin heavy-chain isotypes which delineates the presumptive cardiac chambers (Gonzalez-Sanchez and Bader, 1984; Evans et al., 1988; Stainier and Fishman, 1992). It is further evidenced by the gradient of intrinsic beat rates of different parts of the embryonic heart

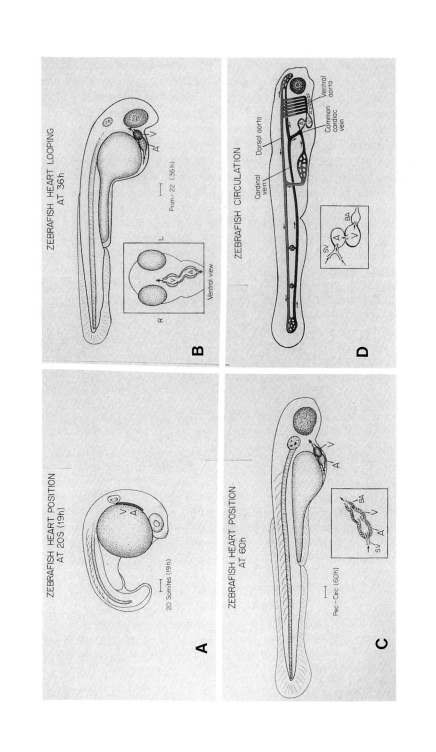

isolated in culture: tissues from the inflow, venous, part of the chick heart beat more rapidly than the outflow regions (DeHaan, 1963).

Although the mechanisms by which internal organs such as the heart acquire polarity are not known, the use of retinoic acid (RA) has provided some insight into the mechanisms of axial patterning in vertebrate embryos. In *Xenopus* embryos, RA treatment leads to the progressive deletion of anterior structures (Durston et al., 1989; Sive et al., 1990; Ruiz i Altaba and Jessell, 1991). In mouse embryos, it induces homeotic transformations of vertebrae (Kessel and Gruss, 1991). In zebrafish, RA treatment at high doses during early development causes the truncation of the A-P axis (Stainier and Fishman, 1992). Low doses of RA applied at times from 50% epiboly (which is the onset of gastrulation) to the 9-somite stage, cause the deletion of parts of the heart tube (Fig. 3B,C). This effect is dose-dependent, as well as stage-dependent. Beginning at gastrulation, progressively higher doses are needed to cause the same extent of deletion (Stainier and Fishman, 1992).

We identified monoclonal antibodies (mAbs) that distinguish between the different cardiac chambers (Stainier and Fishman, 1992); mAb S46 specifically stains the atrium and MF20 (Bader et al., 1982) reacts with all cardiac chambers. The double-labeling of RA-treated embryos with these antibodies reveals that RA causes a dose-dependent truncation of the heart tube that progresses from the arterial end to the venous end (Fig. 4A–G). The deletion of parts of chambers precedes their complete ablation, suggesting that RA reveals a continuous gradient along the entire heart tube, one that exists in at least partial independence from chamber assignment. Hence, the primitive heart tube may be subject to axial patterning mechanisms similar to those that enforce the A-P axis for the overall body plan.

Our data do not indicate the fate of the missing cardiac cells, that is, whether they die or are transformed. At higher doses, RA treatment can lead to the death of certain anterior and posterior cells. It is feasible to visualize cell death directly in zebrafish embryos; under a dissecting microscope, the normally transparent areas appear opaque, and under Nomarski optics, the cells appear refractile. Following a high-dose RA treatment, a region in the caudal midbrain-rostral hindbrain area clearly exhibits numerous round and

---

Fig. 2. Zebrafish heart development. **A:** At the 20-somite stage (19 h), the embryonic heart tube, revealed by mAB CH1 staining (for tropomyosin), lies under the embryonic axis, adjacent to the yolk sac. Scale bar, 100 μm. **B:** The heart starts beating at the 25-somite stage (21.5 h) and by 25 hours, circulation is evident. The initial looping of the heart tube is complete by 36 hours. Scale bar, 100 μm. **C:** At 50 hours, circulation is strong in the segmental vessels and appears in the pectoral fin rudiment. By 60 hours, the valves are formed and functional. SV, sinus venosus; BA, bulbus arteriosus. Scale bar, 100 μm. **D:** Schematic representation of the zebrafish circulation. Branchial arches are numbered I through VI.

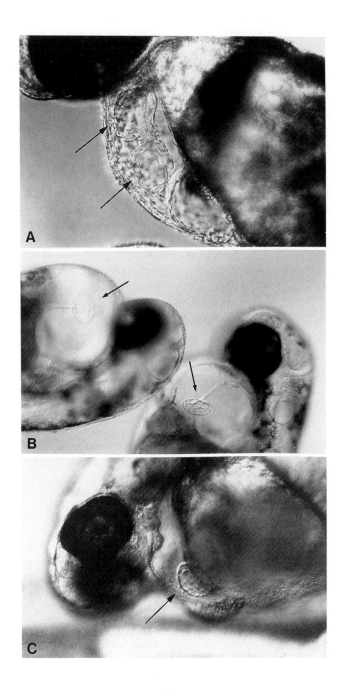

refractile cells that appear to have loosened up from the surrounding tissues (Fig. 5). This area appears to correspond to the engrailed positive (4D9 positive) region recently reported to be missing following RA treatment of zebrafish embryos (Holder and Hill, 1991). The dying cells appear by the 15-somite stage and are resorbed by 48 hours after fertilization. Embryos treated with doses of RA that cause cardiac truncations do not exhibit such patterns of cell death in the region of the developing heart tube.

CARDIAC CELL DIFFERENTIATION AND SPECIFICATION

Low concentrations of RA applied as early as the onset of gastrulation truncate the heart tube from its arterial end, without visibly affecting the rest of the embryo (Stainier and Fishman, 1992). Although exogenously supplied RA may be retained in the embryo for some time (Durston et al., 1989), the specificity of this effect, and the high sensitivity and gradual resistance of the heart tube to RA suggest that RA interferes directly with cardiac polarity, and does so shortly after exposure. Therefore, it appears that cardiac progenitors acquire their axial coordinate during early gastrulation, a stage when heart mesoderm seems to acquire its fate in *Xenopus* (Sater and Jacobson, 1989) and in the avian embryo (Gonzalez-Sanchez and Bader, 1990).

In support of this model, lineage analysis of single heart progenitors at late blastula stages shows the restriction of labeled cells within one cardiac chamber (D.Y.R. Stainier, R.K. Lee, and M.C. Fishman, manuscript in preparation). Figure 6 superimposes two stages of zebrafish development, the late blastula stage and the early heart tube stage. Two cardiac progenitors are shown as they involute and migrate to contribute to different parts of the heart tube; the cell closest to the margin involutes first, extends further rostrally, and populates the atrium; the other cell contributes to the ventricle. The local concentration of a morphogen (possibly RA itself) might specify the fate of these

---

Fig. 3. Retinoic acid, at low doses, deletes parts of the heart tube. **A:** Control embryo at 2 days (*sih* mutant embryo is shown here to illustrate the finer structure and position of the heart). Arrows point to the two main chambers, the atrium and the ventricle. **B,C:** Embryos treated with retinoic acid. Arrows point to the remaining heart structure which can consist of one (C) or more (B) chambers. (B) Two-day embryos that had been treated at the 9-somite stage with 0.4 μM RA for 1 hour. (C) Two-day embryo that had been treated at 50% epiboly with 0.04 μM RA for 1 hour. Embryos treated at these doses during the window from 50% epiboly to the 9-somite stage look normal in general appearance. The effects seem to be cardiac-specific, and it is likely that edema is a direct result of cardiac malfunction. Even the most anterior structure, the hatching gland cells (which sit medially on the sac surrounding the heart), are always present in normal numbers, although they are sometimes mildly affected in their final migration and/or distribution. The MHB looks normal in these embryos. This specificity is not seen when treatment is applied at earlier times, that is, during the blastula stage, as RA also affects anterior structures other than the heart. (Reprinted from Stainier and Fishman, 1992, with permission of the publisher.)

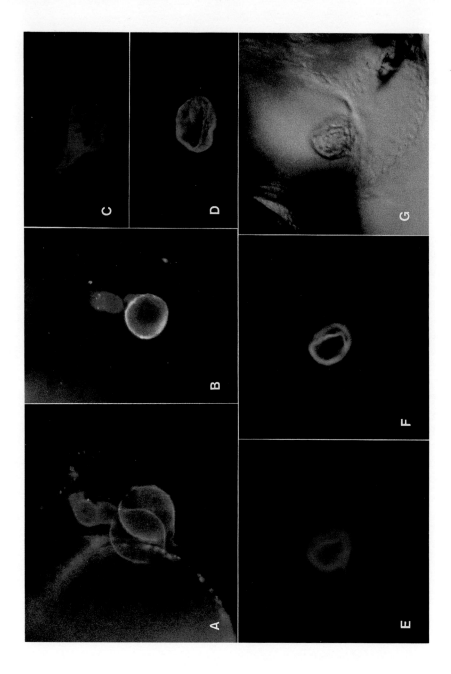

progenitors at, or around, gastrulation. For example, cells in a low concentration of morphogen would tend to assume ventricular fates and those exposed to higher concentrations would assume atrial fates. If the morphogen concentration is raised or perturbed by transiently bathing the embryo in a low dose of RA, the A-P character of the precardiac mesoderm will be modified: presumptive ventricular cells will assume a more venous fate, and the heart tube will develop truncated from its arterial end. How the remaining heart's size might compensate for such changes is not addressed by this type of study. At this point, it is still unclear whether cardiac commitment and axial specification happen simultaneously or successively, although single-cell analysis of isolated chick cardiogenic cells grown in vitro (Gonzalez-Sanchez and Bader, 1990) suggests that they are successive, and that ventricular differentiation is the "default mode" of differentiation. Subsequent interaction with the pharyngeal endoderm seems to be necessary for the maturation of cardiac cells into beating cardiomyocytes, as illustrated by work on the *cardiac lethal* (cl) mutation found in the axolotl, *Ambystoma mexicanum* (Davis and Lemanski, 1987).

GENETIC ANALYSIS OF ZEBRAFISH HEART DEVELOPMENT

Mutational analysis, combined with an approachable embryological system, offers the opportunity to identify molecular decisions involved in cardiogenesis, to place them in epistatic relationships, and, ultimately, to clone the relevant genes. The mouse has several advantages as a genetic system, but the inability to observe and manipulate early stages of development easily, its small litter size and its high maintenance cost make it impractical to use for a large-scale screen for mutations affecting heart formation and function. The zebrafish combines excellent embryology with genetics [see Rossant and Hop-

---

Fig. 4. Retinoic acid progressively deletes the heart tube from the arterial end to the venous end. Embryos were treated with a pulse of RA for 1 hour and stained at 2 days with mAbs S46 (which labels the atrium and the sinus venosus) and MF20 (which labels the entire heart). Note that as the dose is increased, there is progressive loss of MF20-labeled structures, with full retention of S46 labeling until the highest doses. Parts of a chamber are deleted before that entire chamber disappears. **A:** Embryo treated at 50% epiboly with 0.02 μM RA showing the presence of the atrium and most of the ventricle. (The sample was shifted slightly between photographic exposures in order to separate the red and green staining.) **B:** Embryo treated at 100% epiboly with 0.2 μM RA. (The sample was not shifted so that double-labeled tissue appears yellow.) **C,D:** Embryo treated at the 9-somite stage with 0.4 μM RA, showing the presence of the atrium and some ventricular tissue. (C) MF20 staining; (D) S46 staining. **E,F,G:** Embryo treated at the 9-somite stage with 0.4 μM RA, showing the presence of the atrium only. (E) MF20 staining; (F) S46 staining; (G) Nomarski image. (Reprinted from Stainier and Fishman, 1992, with permission of the publisher.)

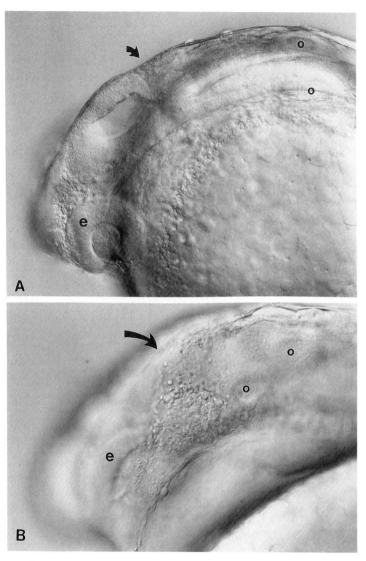

Fig. 5. Retinoic acid treatment leads to cell death in anterior tissues. **A:** Nomarski micrograph of the anterior region of an untreated 25-somite–stage embryo. The arrow points to the midbrain-hindbrain boundary. e, eye; o, otic vesicle. **B:** Nomarski micrograph of the anterior region of a 25-somite–stage embryo treated at 50% epiboly with a pulse of 1 μM RA for 1 hour. The area between the eyes and the otic vesicles is reduced in length and contains numerous round and refractile cells that appear to have loosened up from the surrounding tissues (arrow). [See also Grunwald et al. (1988) for the optical description of cell degeneration in the CNS of *ned-1* mutant embryos]. The tails of these embryos also exhibit some dying cells.

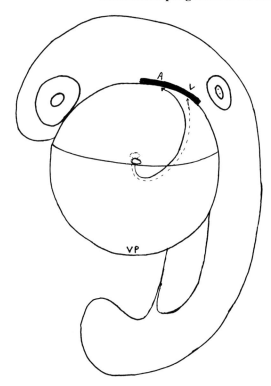

Fig. 6. Acquisition of anteroposterior polarity by the zebrafish heart tube. Two stages of development are superimposed in this drawing: the late blastula stage and the early heart tube stage. Cardiac progenitors involute and migrate to contribute to different parts of the heart tube: the cell closest to the margin (solid line) involutes first, migrates further rostrally, and populates the atrium; the other cell (dashed line) contributes to the ventricle. (This relative positioning of the cardiac progenitors is deduced from general cell movement during gastrulation but has to be confirmed directly by double-labeling lineage tracing). The local concentration of a morphogen (possibly RA itself) would specify the fate of these progenitors, at or around gastrulation, with the atrial progenitor being exposed to the higher concentration. The perturbation of this morphogenetic gradient by bathing the embryo in low doses of RA modifies the A-P character of the precardiac mesoderm, shifting presumptive ventricular cells to a more venous (atrial) fate. This treatment results in the progressive truncation of the embryonic heart tube from the arterial, or ventricular, end.

kins (1992) for a detailed comparative analysis of the mouse and the zebrafish as model systems for a mutational analysis of vertebrate embryonic development]. More specifically, for a cardiovascular mutational screen, the zebrafish heart is prominent and is functional within a day, the circulation is easily scored, and the embryo is small enough that it can live up to a week at 28.5°C without a functioning heart, relying on diffusion for oxygen and nutrients. This allows us to screen visually and rapidly, at low magnification under a dissecting microscope, for mutations affecting heart development.

Two γ-ray–induced mutations that affect heart development were isolated in Oregon by Walker and Kimmel, and they have been particularly useful in defining the constraints of the system. The *fibrils unbundled* (*fub*) mutation affects both skeletal and cardiac muscle function [(Felsenfeld et al., 1990). The mutant embryos are paralyzed and have only a weak heartbeat: the atrium beats rhythmically but is enlarged, while the ventricle is mostly silent. Electron microscopy of the cardiac muscle reveals myofibrillar disarray (D.Y.R. Stainier and M.C. Fishman, unpublished results) as is also observed in the skeletal muscle. The *silent heart* (*sih*) mutation is cardiac-specific and affects the heartbeat in a most dramatic fashion. The mutant heart, though morphologically normal, never starts beating (D.Y.R. Stainier and M.C. Fishman, manuscript in preparation). *Fub* and *sih* mutant embryos, which do not have a functional heart, develop edema of the pericardium and of the yolk sac but they live for at least a week.

CONCLUSION

The zebrafish embryo offers several distinct advantages for studying early vertebrate development, and, more specifically, cardiac morphogenesis. Its excellent embryology combined with its promising genetics should permit a genetic characterization of heart development. The acquisition of anteroposterior polarity by the heart tube is central to the normal development of its form and function. Retinoic acid, at low doses, causes the progressive truncation of the zebrafish heart tube from the arterial end to the venous end. The analysis of the heart tube's sensitivity to RA and its timing suggest that polarity is established during, or shortly after, initial commitment to the cardiac lineage. Axial coordinates would thereby become available as a template for subsequent cardiac morphogenesis.

ACKNOWLEGMENTS

This work is supported by a grant from the Helen Hay Whitney Foundation (to D.Y.R.S.) and by grants from the NIH (to M.C.F.) and Bristol-Myers Squibb (to M.C.F.).

REFERENCES

Bader D, Masaki T, Fischman DA (1982): Immunochemical analysis of myosin heavy chain during avian myogenesis in vivo and in vitro. J Cell Biol 95:763–770.

Bourne GH (1980): "Hearts and Heart-like Organs." New York: Academic Press.

Davis LA, Lemanski LF (1987): Induction and myofibrillogenesis in cardiac lethal mutant axolotl hearts rescued by RNA derived from normal endoderm. Development 99:145–154.

DeHaan RL (1963): Regional organization of pre-pacemaker cells in the cardiac primordia of the early chick embryo. J Embryol Exp Morphol 11:65–76.

DeHaan RL (1965): Morphogenesis of the vertebrate heart. In DeHaan RL, Ursprung H (eds): "Organogenesis." Holt, Rinehart and Winston, pp 377–420.

Durston AJ, Timmermans JPM, Hage WJ, Hendriks HFJ, deVries NJ, Heideveld M, Nieuwkoop PD (1989): Retinoic acid causes an anteroposterior transformation in the developing central nervous system. Nature 340:140–144.

Evans D, Miller JB, Stockdale FE (1988): Developmental patterns of expression and coexpression of myosin heavy chains in atria and ventricles of the avian heart. Dev Biol 127:376–383.

Felsenfeld AL, Walker C, Westerfield M, Kimmel C, Streisinger G (1990): Mutations affecting skeletal muscle myofibril structure in the zebrafish. Development 108:443–459.

Gonzalez-Sanchez A, Bader D (1984): Immunochemical analysis of myosin heavy chains in the developing chicken heart. Dev Biol 103:151–158.

Gonzalez-Sanchez A, Bader D (1990): In vitro analysis of cardiac progenitor cell differentiation. Dev Biol 139:197–209.

Grunwald DJ, Kimmel CB, Westerfield M, Walker C, Streisinger G (1988): A neural degeneration mutation that spares primary neurons in the zebrafish. Dev Biol 126:115–128.

Holder N, Hill J (1991): Retinoic acid modifies development of the midbrain-hindbrain border and affects cranial ganglion formation in zebrafish embryos. Development 113:1159–1170.

Kessel M, Gruss P (1991): Homeotic transformations of murine vertebrae and concomitant alteration of *Hox* codes induced by retinoic acid. Cell 67:89–104.

Litvin J, Montgomery M, Gonzalez-Sanchez A, Bisaha JG, Bader D (1992): Commitment and differentiation of cardiac myocytes. Trends Cardiovasc Med 2:27–32.

Rossant J, Hopkins N (1992): Of fin and fur: mutational analysis of vertebrate embryonic development. Genes Dev 6:1–13.

Ruiz i Altaba A, Jessell TM (1991): Retinoic acid modifies mesodermal patterning in early *Xenopus* embryos. Genes Dev 5:175–187.

Sater AK, Jacobson AG (1989): The specification of heart mesoderm occurs during gastrulation in *Xenopus laevis*. Development 105:821–830.

Senior HD (1909): The development of the heart in shad. Am J Anat 9:212–276.

Sive HL, Draper BW, Harland RM, Weintraub H (1990): Identification of a retinoic acid-sensitive period during primary axis formation in *Xenopus laevis*. Genes Dev 4:932–942.

Stainier DYR, Fishman MC (1992): Patterning the zebrafish heart tube: Acquisition of anteroposterior polarity. Dev Biol 153:91–101.

Warga RM, Kimmel CB (1990): Cell movements during epiboly and gastrulation in zebrafish. Development 108:569–580.

Westerfield, M. (1989). "The Zebrafish Book." Eugene: University of Oregon Press.

7. Genetic Hierarchy Controlling Flower Development

Detlef Weigel and Elliot M. Meyerowitz

California Institute of Technology, Division of Biology 156-29, Pasadena, California 91125

INTRODUCTION

Genetic hierarchies have emerged as a common theme in the study of animal development. Recent experiments indicate that similar genetic mechanisms control an important aspect of plant development, the formation of flowers.

As with other developmental processes, flower development can be broken down into a series of steps in temporal succession. When embryogenesis is completed, the plant has two meristems, a root meristem, which gives rise to the root system, and a shoot meristem, which gives rise to the aerial portions of the plant. Upon floral induction, the vegetative shoot meristem is thought to be reorganized into an inflorescence meristem, from which the different parts of the inflorescence derive. The inflorescence meristem generates floral meristems, which produce floral organ primorida. The inflorescence meristem may either retain its identity during the whole life cycle of the plant, in which case it can produce a potentially indeterminate number of floral meristems, or the inflorescence meristem itself can be converted into a floral meristem, in which case the inflorescence terminates with a flower. The floral organ primordia produced by the floral meristem adopt different fates according to their position within the developing flower, and they finally differentiate into the different organ types, such as sepals, petals, stamens, and carpels.

Recently, several groups have started to conduct systematic mutageneses to identify mutations that interfere with normal flower development (reviewed by Schwarz-Sommer et al., 1990; Coen and Meyerowitz, 1992). Many of the isolated mutations affect only one step of flower development, which allows for ordering them into a temporal hierarchy of action. Accumulating genetic and molecular evidence suggests that this temporal hierarchy corresponds to a regulatory hierarchy, in which earlier-acting genes control the expression of later-acting ones.

About 30 loci that are required for wild-type flower development in the common lab weed, *Arabidopsis*, have been described (Table I), and at least five of

TABLE I.
Arabidopsis Genes Affecting Flower Development

Locus Name	Molecular Biology	Reference
Genes affecting flowering transition		
LATE FLOWERING:		
FCA		1
FD		1
FE		1
FHA		1
FPA		1
FT		1
FVE		1
FWA		1
FY		1
CONSTANS (CO = FG)		1
GIGANTEA (GI)		1
TERMINAL FLOWER (TFL)		2
Meristem identity genes		
LEAFY (LFY)	Novel protein	3
APETALA1 (AP1)	MADS-box	4, 5
TERMINAL FLOWER (TFL)		2
CAULIFLOWER (CAL)		5
Cadastral genes		
SUPERMAN (SUP) = FLO10		6
AGAMOUS (AG)	MADS-box	7, 8, 9
APETALA2 (AP2)		7, 8, 10
Homeotic genes		
AG		
AP2		
APETALA3 (AP3)	MADS-box	7, 8, 11
PISTILLATA (PI)	MADS-box	7, 8, 12
Organ number		
AG		
AP2		
SUP		
CLAVATA1, 2, 3 (CLV1, 2, 3)		13
FL-82		14
PERIANTHIA		15
Other		
PINFORMED		16
FL-54		17
FL-84		17
FL-165		14
UNUSUAL FLORAL ORGANS (UFO)		18

References: (1) Koornneef et al., 1991; (2) Shannon and Meeks-Wagner, 1991; Alvarez et al., 1992; (3) Schultz and Haughn, 1991; Weigel et al., 1992; Huala and Sussex, 1992; (4) Irish and Sussex, 1990; Mandel et al., 1992; (5) Bowman, 1991; (6) Schultz et al., 1991; Bowman et al., 1992; (7) Bowman et al., 1989; (8) Bowman et al., 1991; (9) Yanofsky et al., 1990; (10) Kunst et al., 1989; (11) Jack et al., 1992; (12) K. Goto and E.M.M., unpublished observations; (13) Koornneef et al., 1983, S. Clark and E.M.M., unpublished observations; (14) Okada et al., 1989; (15) M. Running and E.M.M., unpublished observations; (16) Goto et al., 1991; (17) Komaki et al., 1988; (18) J. Levin and E.M.M., unpublished observations.

these have been cloned. In this article, we will review some of the data corroborating the notion that many of these genes act in a regulatory hierarchy.

WILD-TYPE FLOWER DEVELOPMENT

The *Arabidopsis* seedling starts out with a vegetative shoot meristem, which produces a limited number of vegetative leaves. Since there is little internode elongation between the vegetative leaves, they form a basal rosette. Upon floral induction, the shoot meristem begins to behave as an inflorescence meristem, which first produces a small number of cauline leaves with associated secondary inflorescence meristems, before it goes on to produce floral meristems, from which the floral organs derive. In contrast to many other plant species, flowers in *Arabidopsis* are not normally subtended by leaflike bracts. Whereas leaves and flowers arise in a spiral phyllotaxis on the shoot meristem, floral organs are produced in rings, or whorls, by the floral meristem. Four types of floral organs constitute the flower, with sepals in the first or outermost whorl, petals in the second whorl, stamens in the third whorl, and carpels forming the gynoecium in the fourth or innermost whorl (Fig. 1A).

GENES AFFECTING FLORAL TRANSITION

The first class of mutations alters the timing of the transition from vegetative to reproductive development, that is, the time until a plant starts to flower. *Arabidopsis* is a facultative long-day plant, in which the time before a plant flowers, called flowering time (FT) by Koornneef et al. (1991), is shortened by exposure to long days (Napp-Zinn, 1985). Apart from long days, vernalization (exposure of vegetative plants to cold) shortens FT in many ecotypes. The different environmental stimuli are thought to result in the production of a factor that mediates floral induction. It has been demonstrated for many plant species that differences in daylength are perceived in the leaves, and that the leaves produce a signal that will ultimately act on the vegetative shoot meristem and convert it into an inflorescence meristem (Bernier, 1988). The perception of the floral induction signal in the shoot meristem is referred to as floral evocation.

Eleven loci, known as *LATE FLOWERING GENES*, have been described, and mutations in these delay flowering without affecting other aspects of the plant life cycle (Table I). An elegant genetic study by Koornneef and coworkers (1991) has shown that these genes appear to act in at least three different pathways mediating floral induction and/or evocation. Since the three pathways function independently, all single and double mutants retain the ability to flower. A triple mutant, in which all three pathways are defective, has not yet been constructed. Double mutants between the different late flowering mu-

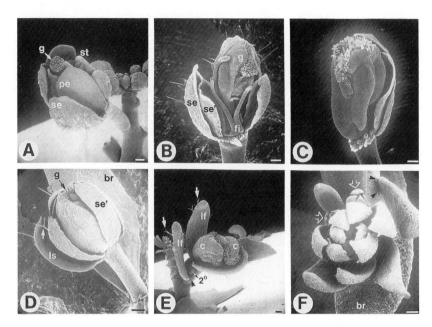

Fig. 1. Phenotypes of wild type and mutant flowers. Scanning electron micrographs of mature or nearly mature flowers. **A:** Wild type. The four floral organ types, sepals (se), petals (pe), stamens (st), and carpels comprising the central gynoecium (g) are visible. **B:** *ap3-3*. Petals are transformed into extra sepals (se'). Stamens are replaced by filamentous organs (fi) or extra carpels which contribute to the enlarged gynoecium (g). **C:** *sup-1 ; ap3-1*. The double mutant phenotype closely resembles the *ap3* single mutant phenotype. **D:** *lfy-6*. The flower is subtended by a cauline leaflike bract (br). One of the outer organs is a leaflike sepal (ls). Another sepal-like organ (se') bears a stellate trichome typical for leaves (arrow). **E:** *ap2-2 ; lfy-6*. The outer organs are more leaflike (lf) than in *lfy* single mutants. Note stellate trichomes (arrows) on the leaflike organs, and stipules (arrowhead) at their base. Secondary flowers (2°) due to the *lfy* mutation are visible in the axil of the outer organs. The central organs are carpels (c) as in *ap2-2* single mutants. **F:** *ap1-1 ; lfy-6*. All organs are leaflike, and are very similar to the bract (br) subtending the flower. Note stipules (arrowheads) at the base of the leaflike organs, and secondary flowers in their axils (arrows). Bars represent 100 μm.

tants fall into two classes: In the first class, double mutants flower at the same time as the later-flowering parent. In the second class, double mutants flower later than the later-flowering parent. In double mutants of the first class, both mutations appear to affect the same pathway and, therefore, they do not show a stronger effect than the later parent. In double mutants of the second class, different pathways are thought to be affected and, therefore, a more severe effect on the delay of flowering than in either single parent is observed.

Mutations in the *TERMINAL FLOWER* (*TFL*) gene have an early flowering phenotype. In addition, *tfl* mutations affect the inflorescence meristem (Shannon and Meeks-Wagner, 1991; Alvarez et al., 1992; see below). Epistatic relations between *tfl* and *late flowering* mutations have not yet been reported.

MERISTEM IDENTITY GENES

The establishment of the inflorescence meristem is the prerequisite for the action of the next class of genes, which control the identity of inflorescence and floral meristems. The best understood of the meristem identity genes in *Arabidopsis* is the *LEAFY* (*LFY*) gene, mutations in which cause a complete transformation of early-arising flowers and a partial transformation of late-arising flowers into secondary inflorescence shoots (Schultz and Haughn, 1991; Weigel et al., 1992; Huala and Sussex, 1992). The partial transformation of late-arising flowers is evident from the development of leaflike organs and secondary flowers on the primary flowers, the development of cauline leaf-like bracts subtending the flowers (Fig. 1D), and a phyllotaxis of floral organs that is often intermediate between the whorled mode of wild-type flowers and the spiral mode of inflorescences (Fig. 2A). From the phenotype it has been concluded that the *LFY* wild-type product is required to promote floral as opposed to inflorescence meristem identity.

The *LFY* gene has been cloned, and its expression pattern in both wild-type and mutant plants has been studied by in situ hybridization to tissue sections. In the wild type, *LFY* is expressed in floral anlagen and floral primordia (Fig. 3A), but not in the inflorescence meristem proper, so that *LFY* may act autonomously in the floral primordia. Activation of *LFY* expression is the earliest known molecular consequence of floral induction in *Arabidopsis*, since high levels of *LFY* RNA accumulation are specific to floral primordia. Only

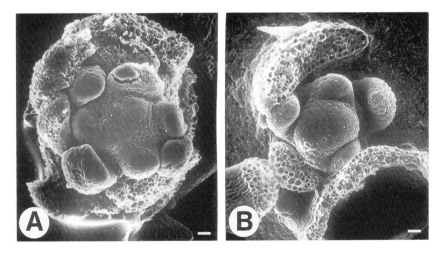

Fig. 2. Phyllotaxis in **A:** *lfy-6*. **B:** *lfy-6 ag-2*. Scanning electron micrographs of early floral primordia with the outer sepal primordia manually removed. Note that the organs do not emerge in rings or whorls, but in a spiral fashion. The spiral phyllotaxis is not always perfect, as in (A). Bars represent 10 μm.

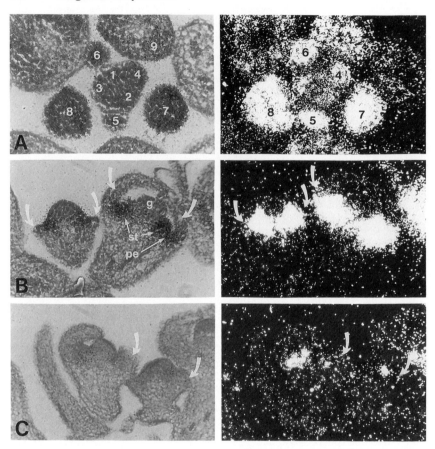

Fig. 3. RNA distribution of floral control genes in developing flowers. RNA accumulation was visualized by in situ hybridization to tissue sections. The left panels show phase contrast, and the right panels dark-field views. **A:** Expression of *LFY* in wild type. A transverse section through an inflorescence apex is shown. The floral anlagen and primordia surrounding the inflorescence meristem proper are labeled according to their age, with "1" being the youngest. **B:** Expression of *AP3* in the wild type. On the left floral primordium, sepal primordia have emerged (arrows). Expression is detected interior to the sepal primordia in the anlagen of the petals and stamens, but not in the anlage of the central gynoecium. This pattern is maintained through later stages, when petal (pe) and stamen (st) primordia have become visible. **C:** Expression of *AP3* in *lfy-6*. Two floral primordia of similar age as in panel (B) are shown. The level of *AP3* RNA is much reduced.

very weak *LFY* expression is detected in cauline leaf primordia, which arise in positions homologous to the positions of flowers on the shoot meristem (Weigel et al., 1992). Several lines of circumstantial evidence indicate that *LFY* is indeed regulated by factors mediating floral induction, including the products of the *LATE FLOWERING* genes. Several factors, which delay floral induction, such as short days, low temperature, or *late flowering* mutations, also delay the switch from the production of cauline leaves to flowers in the inflorescence. This correlation is compatible with a model in which the same factors control the initiation of the primary inflorescence meristem and the initiation of flowers, which we think to be equivalent to the activation of *LFY*. Significantly, all situations that delay floral induction also enhance the *lfy* mutant phenotype (Weigel et al., 1992; Huala and Sussex, 1992; D.W. and E.M.M., unpublished observations). In addition, suboptimal floral induction in the related genus *Sinapis* followed by growth under poor light causes inflorescence reversion, which is often associated with the production of flowers that are subtended by bracts and have partial inflorescence character (Bagnard et al., 1972). This phenotype might be due to insufficient maintenance of *LFY* expression (G. Bernier, personal communication).

Even apparently null mutations in the *LFY* gene do not effect a complete transformation of all flowers into secondary inflorescences, indicating that determination of floral meristem identity requires factors in addition to *LFY*. The *lfy* mutant phenotype, i.e., the conversion of floral into inflorescence meristems, is strongly enhanced by mutations in the *APETALA1* (*AP1*) gene, indicating that *LFY* and *AP1* act synergistically in promoting floral over inflorescence meristem identity (Fig. 1F) (Weigel et al., 1992; Huala and Sussex, 1992). The same factors that enhance the *lfy* phenotype, low temperature and short days, also enhance the *ap1* phenotype, the production of additional floral meristems by the primary floral meristem (Irish and Sussex, 1990; Huala and Sussex, 1992; J. Bowman and J. Alvarez, personal communication). This finding suggests that *LFY* and *AP1* expression are activated by a similar mechanism. Early expression of *LFY* is not affected in *ap1* mutants and confirms that the *ap1* phenotype is not mediated by *LFY* (Weigel et al., 1992). *AP1* has been cloned, and it has been shown to be expressed at a similarly early stage as *LFY* in the developing floral primordia (Mandel, et al., 1992).

Recently, a mutation at a third locus, *CAULIFLOWER* (*CAL*), has been demonstrated to affect floral meristem identity. A *cal* mutation enhances the *ap1* mutant phenotype such that all of the primary floral meristems go through several rounds of producing supernumerary floral meristems (Bowman, 1991). The *cal ap1* double mutant appears to affect floral meristem identity at a slightly later stage than *lfy* mutations, since *cal ap1; lfy* triple mutants exhibit a phenotype very similar to *ap1; lfy* double mutants (D.W., J. Bowman, and E.M.M., unpublished observations).

Finally, mutations at the *TFL* locus cause a phenotype that is somewhat opposite to the phenotype of the other three meristem identity mutants. The wild-type inflorescence of *Arabidopsis* is indeterminate, indicating that the inflorescence meristem retains its identity throughout the life span of the plant.

In contrast, *tfl* mutant inflorescence meristems produce only a limited number of flowers, before giving rise to terminal flowers, indicating a transformation of the inflorescence meristem into a floral meristem (Shannon and Meeks-Wagner, 1991; Alvarez et al., 1992). As expected, *LFY* RNA is ectopically expressed in the transformed inflorescence meristems (Weigel et al., 1992). *LFY* is, however, not the only factor mediating the transformation of the inflorescence meristem in *tfl* mutants, since *tfl; lfy* double mutant inflorescences are still determinate (D.W. and E.M.M., unpublished observations). Not only is the *tfl* mutant phenotype somewhat opposite to the *ap1* and *lfy* mutant phenotypes, but it is also affected by short days and low temperature in the opposite way. Whereas *lfy* and *ap1* mutant phenotypes are enhanced by short days and low temperature, the *tfl* mutant phenotype, which is associated with ectopic expression of *LFY*, and probably also of *AP1*, is relieved under the same conditions (Shannon and Meeks-Wagner, 1991; Alvarez et al., 1992). These results corroborate the notion that floral inductive signals might have a rather direct effect on the expression of floral meristem identity genes such as *LFY* and *AP1*.

CADASTRAL GENES

Organ identity in the developing flower is determined by the homeotic genes, of which at least three are expressed in a region-specific fashion (see below). This pattern contrasts with the uniform expression of the floral meristem identity genes *LFY* and *AP1* in the early floral primordium. The available data do not support a model in which *LFY* and *AP1* directly control the region-specific expression of homeotic genes. Rather, it appears that another class of genes, the cadastral genes, have an important role in specifying the boundaries of homeotic gene expression.

In the wild type, the *APETALA3* (*AP3*) gene is expressed in the prospective second and third whorls, from which petals and stamens arise, but not in the center of the flower, from which the gynoecium develops (Fig. 3B) (Jack et al., 1992). In *superman* (*sup* = *flo10*) mutants, the expression domain of *AP3* expands toward the center of the floral primordium (Bowman et al., 1992). The ectopic *AP3* expression is associated with the development of additional whorls of stamens at the expense of the central gynoecium. That the *sup* phenotype is causally related to ectopic *AP3* expression has been demonstrated by constructing *sup; ap3* double mutants, which have essentially the same phenotype as *ap3* single mutants (Fig. 1B,C) (Schultz et al., 1991; Bowman et al., 1992).

Whereas the *SUP* gene appears to function only as a cadastral gene, two other genes, *AGAMOUS* (*AG*) and *APETALA2* (*AP2*), have both cadastral and homeotic functions, and are discussed in the following section.

HOMEOTIC GENES

The four canonical homeotic genes, *AP2, AG, AP3,* and *PISTILLATA* (*PI*), fall into three classes, with each class being required in two adjacent whorls of organs (Bowman et al., 1989, 1991; Kunst et al., 1989; Hill and Lord, 1989; Jack et al., 1992). The observed phenotypes of single, double, and triple mutants are consistent with a model in which the combination of homeotic gene activities in a given whorl determines organ identity (Bowman et al., 1991). *AP2* activity alone determines sepal identity, *AP2* plus *AP3* and *PI* activity determines petal identity, *AG* plus *AP3* and *PI* activity determines stamen identity, and *AG* activity alone determines carpel identity. In support of this model, it has been found that *AP3* and *PI* RNAs are expressed in the prospective second and third whorls of the early floral primordium (Fig. 3B) (Jack et al., 1992; K. Goto and E.M.M., unpublished observations), and that *AG* RNA is expressed in the prospective third and fourth whorls (Drews et al., 1991). The expression domains of *AP3, PI,* and *AG* are maintained until quite late in flower development, in contrast to the expression of the meristem identity gene *LFY*, whose expression pattern is more dynamic.

In addition to determining organ identity, *AP2* and *AG* control each other's (genetic) expression domains (Bowman et al., 1989, 1991). For AG it has been shown that the RNA expands into the first and second whorls in *ap2* mutants (Drews et al., 1991). *AG* and *AP2* have therefore dual roles, as cadastral and as homeotic genes.

The regulatory interactions between homeotic genes are not confined to cadastral functions, as has been demonstrated by Jack et al. (1992) for *PI*, which is required for the maintenance of *AP3* expression. Furthermore, in *ap2* mutant flowers, the initial domain of *AP3* RNA is smaller, and the level of *AP3* RNA accumulation is lower than in the wild type. This effect is apparently mediated by ectopic *AG* activity in *ap2* mutants, since *AP3* expression is restored in *ag ap2* double mutants (Jack et al., 1992). Thus, although *AP3* expression in the wild type is apparently initiated independently of other homeotic genes, perturbing the expression domains of other homeotic genes interferes with normal *AP3* activation.

Which factors then activate homeotic gene expression? Since expression of floral homeotic genes is confined to flowers, the floral meristem identity genes are good candidates for activators of the homeotic genes. Both genetic and molecular experiments have confirmed this assumption. The phenotypes of double mutants between *ap1* and homeotic mutations are essentially additive, indicating that *AP1* on its own does not have a major role in activating floral homeotic genes (Irish and Sussex, 1990; Bowman, 1991). A different picture has emerged from the analysis of double mutants involving *lfy* mutations. Neither the supernumerary secondary inflorescences, nor the spiral phyllotaxis of *lfy* mutant flowers are suppressed by homeotic mutations, confirming that *LFY* acts before the homeotic genes (Fig. 2B) (Schultz and Haughn, 1991; Weigel et al., 1992; Huala and Sussex, 1992). Strong *lfy* mutant flowers consist predominantly of sepal-like and carpel-like organs (Fig. 1D). Similarly, *ap3* and *pi* mutant flowers consist only of sepals and carpels (Fig. 1B). *ap3; lfy* and *pi lfy* double mutant flowers exhibit essentially the *lfy* single mutant phenotype, suggesting that *AP3* and *PI* activation is dependent on *LFY* activity (Weigel et al., 1992; Huala and Sussex, 1992). In *ag* mutant flowers, carpels are homeotically transformed into sepals, and one might, therefore, expect that carpels are absent from *ag; lfy* double mutants. Surprisingly, carpels do develop in *ag; lfy* double mutant flowers (Weigel et al., 1992; Huala and Sussex, 1992), a fact which indicates that the organ identity function of *AG* is not required in *lfy* mutants. Alternatively, the absence of carpels in *ag* mutant flowers might result from ectopic *LFY* activity, which is eliminated in *ag; lfy* double mutants. The fact that carpels develop in other double and triple mutant combinations involving *ag* mutations (Bowman et al., 1989, 1991) supports the latter interpretation. *AG* also controls the determinacy of the floral meristem, and this aspect of *AG* function is independent of *LFY* activity, since double mutant flowers exhibit the indeterminacy of *ag* single mutants. The strongest effect on organ identity in *lfy* mutant flowers is caused by either weak or strong *ap2* mutations. As in *ap2* single mutants, sepals are replaced by carpels or leaves in *ap2; lfy* double mutants (Fig. 1E) (Weigel et al., 1992; Huala and Sussex, 1992). Strong *ap2* alleles also cause a reduction in floral organ number due to ectopic *AG* expression (Bowman et al., 1991). This effect is apparently related to *AG*'s function in floral meristem determinacy, and it is also observed in *ap2; lfy* double mutants, consistent with the finding that the determinacy function of *AG* is active in *lfy* mutants (Weigel et al., 1992). Surprisingly, an intermediate *ap2* allele, which produces sepal/carpel chimeric organs in the first whorl, does not appear to have a major effect on organ identity in a strong *lfy* mutant background (Schultz and Haughn, 1991).

The strongest prediction regarding the effect of *LFY* on homeotic gene expression could be made for the *AP3* and *PI* genes whose genetic activity is apparently very reduced in *lfy* mutants. In situ hybridization experiments confirm that the level of *AP3* RNA is indeed much lower in a strong *lfy* mutant than in the wild type (Fig. 3B,C) (D.W., T. Jack, and E.M.M., unpublished observations). The pattern of *AP3* expression, however, is still similar to the wild-type pattern, indicating that *LFY* acts mainly as an activator of *AP3* and not as a regulator of its spatial domain of expression. It will be interesting to learn whether AP3 expression is even further reduced, or completely eliminated, in *ap1; lfy* double mutants.

EVOLUTIONARY CONSERVATION OF FLORAL CONTROL GENES

Although the overall appearance of flowers in different species varies dramatically, the basic organization of flowers is very conserved. There are four primary floral organ types—sepals, petals, stamens, and carpels—and they are almost without exception found in the same arrangement, with sepals being outermost, and carpels being innermost (Gifford and Foster, 1988). It is thus likely that homologous genes control organ identity, and perhaps other aspects of flower development, in many species of flowering plants. With the recent comparison of floral control genes in *Arabidopsis* and *Antirrhinum*, evidence for the conservation of the basic mechanisms controlling flower development has emerged. Like *Arabidopsis, Antirrhinum* is easily amenable to genetic analysis, and several mutations affecting flower development have been isolated (e.g., Klemm, 1927; Stubbe, 1966; Sommer et al., 1990; Carpenter and Coen, 1990; Schwarz-Sommer et al. 1990; Huijser et al., 1992). Many mutations in *Arabidopsis* and *Antirrhinum* cause identical, or similar, phenotypes suggesting that they affect homologous genes. In two cases, this has been confirmed by molecular analysis.

ap3 and *deficiens* (*def*) mutations both cause homeotic transformations of petals into sepals and of stamens into carpels (Klemm, 1927; Bowman et al., 1989; Sommer et al., 1990; Jack et al., 1992). Both genes belong to the MADS-box family of transcription factors, and the encoded proteins have 58% sequence identity (Sommer et al., 1990; Jack et al., 1992). A number of other MADS-box genes has been isolated from both *Arabidopsis* and *Antirrhinum* (Schwarz-Sommer et al., 1990; Ma et al., 1991; Huijser et al., 1992; A. Mandel and M. Yanofsky, personal communication). Sequence comparison among all MADS-box genes (E.M.M., unpublished observations) revealed that *AP3* and *DEF* are more similar to each other than to any MADS-box gene from the same species, corroborating the notion that *AP3* and *DEF* are cognate homologues (Jack et al., 1992). The homology extends to very similar expression patterns. Both genes are expressed at high levels in the anlagen and primordia of petals and stamens, although *DEF* expression appears to be initiated slightly later than *AP3* expression (Schwarz-Sommer et al., 1992; Jack et al., 1992).

The *LFY* and *FLORICAULA* (*FLO*) proteins are even more similar than *AP3* and *DEF*, with 70% of the amino acid sequence being identical (Coen et al., 1990; Weigel et al., 1992). The phenotypes of null alleles, however, are different. Whereas in *flo* mutants all flowers are replaced by inflorescence shoots, only the early-arising flowers are completely transformed into inflorescence shoots in *lfy* mutants, indicating that the *lfy* phenotype is weaker than the *flo* phenotype (Carpenter and Coen, 1990; Schultz and Haughn, 1991; Weigel et al., 1992; Huala and Sussex, 1992). The earliest expression of *LFY* and *FLO* is detected in the anlagen and primordia of flowers, and uniform expression is maintained until the first organ primordia emerge (Coen et al., 1990; Weigel et al., 1992). Thus, although the gene products and the expression patterns are very similar, the function of both genes appears to be somewhat different.

The comparison of *AP3/DEF* and *LFY/FLO* indicates that the basic mechanisms of flower development are probably conserved between *Arabidopsis*

Nicotiana	L H C L D E E A S N A L R R A F K E R G E N V G A
Arabidopsis	. . . . S
Antirrhinum	. . . . . . a . . . .
Solanum	. . . . . . D . . . .
Zea	. . . . V . . . Y . a . . . .
Gingko	. . . . . s . q . . h . . . i Y . . . .

Fig. 4. Conservation of *LFY* related sequences in other species. The *LFY* homolog in *Antirrhinum* is the *FLO* gene, which has been isolated and sequenced by Coen et al. (1990). The tobacco (*Nicotiana*) homolog of *LFY* has been cloned and sequenced by M. Bonnlander and D.R. Meeks-Wagner (personal communication). Short pieces of *LFY*-related sequences have been isolated from nightshade (*Solanum*), maize (*Zea*), and *Gingko* by polymerase chain reaction (D.W. and E.M.M., unpublished observations). Dots indicate identical amino acids, uppercase letters conservative exchanges, and lower-case letters nonconservative exchanges.

and *Antirrhinum*, although some variations exist. Since *Arabidopsis* and *Antirrhinum* belong to two different, only distantly related subclasses within the class of dicotyledonous plants (Cronquist, 1981), it is likely that similar mechanisms operate in most or all species of flowering plants. Preliminary evidence indicates that *LFY*-related sequences are present in the genome of tobacco and nightshade, as well as in maize, a monocotyledonous species (Fig. 4). *LFY*-related sequences could even be detected in a nonflowering seed plant, *Gingko* (Fig. 4), although *LFY* expression and function are flower-specific in *Arabidopsis*. No other homolog of *LFY* has been detected in *Arabidopsis* (Weigel et al., 1992), suggesting that the *Gingko* gene is the cognate homologue of *Arabidopsis LFY*.

CONCLUSIONS

Genetic and molecular experiments indicate that a genetic hierarchy controls flower development in *Arabidopsis* (Fig. 5), and probably in other flowering plants as well. In *Arabidopsis*, it has been demonstrated that floral meristem identity genes and cadastral genes control the expression of homeotic genes. The activity of floral meristem identity genes itself is likely to be under the control of genes mediating floral induction. In addition to the control of later-acting genes by earlier-acting genes, cross-regulatory interactions between genes activated at the same stage have been found. Further experiments are required to determine whether any of these interactions are direct, for example, through transcriptional activation or

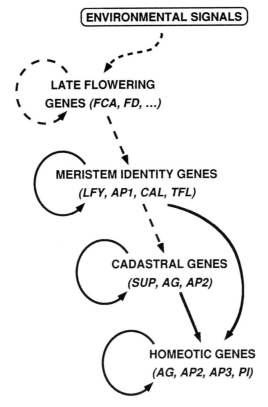

Fig. 5. Summary of regulatory interactions between floral control genes in *Arabidopsis*. Solid lines indicate interactions observed at the level of RNA expression, dashed lines indicate genetic interactions.

repression. Finally, the functional comparison of floral control genes and their cognate homologues between flowering and nonflowering plants may help us to understand how flowers evolved.

ACKNOWLEDGMENTS

We thank J. Bowman and J. Alvarez, G. Bernier, K. Goto, T. Jack, A. Mandel and M. Yanofsky, and D.R. Meeks-Wagner for communicating unpublished results, and E. Huala and I. Sussex for a preprint of their manuscript on *LFY*. We thank T. Jack for the *ap3-3* micrograph, L. Brockman for maize DNA, and L. Sieburth for *Gingko* DNA. D.W. has been supported by an EMBO Long-Term Fellowship, and by a Senior Fellowship from the American Cancer Society, California Division. Our work is supported by National

Science Foundation grant DCB 8403439, by NIH grant GM45697, and by US Department of Energy Division of Energy Biosciences grant DE-FG03-88ER13873 to E.M.M.

REFERENCES

Alvarez J, Guli CL, Yu X-H, Smyth DR (1992): *terminal flower*: A gene affecting inflorescence development in *Arabidopsis thaliana*. Plant J 2:103–116.

Bagnard C, Bernier G, Arnal C (1972): Étude physiologique et histologique de la réversion de l'inflorescence chez *Sinapis alba* L. Physiol Vég 10:237–254.

Bernier G (1988): The control of floral evocation and morphogenesis. Annu Rev Plant Physiol Plant Mol Biol 39:175–219.

Bowman JL (1991): Molecular genetics of flower development in *Arabidopsis thaliana*. California Institute of Technology, Pasadena, CA: Ph.D. Thesis.

Bowman JL, Smyth DR. Meyerowitz EM (1989): Genes directing flower development in *Arabidopsis*. Plant Cell 1:37–52.

Bowman JL, Smyth DR, Meyerowitz EM (1991): Genetic interactions among floral homeotic genes of *Arabidopsis*. Development 112:1–20.

Bowman JL, Sakai H, Jack T, Weigel D, Mayer U, Meyerowitz EM (1992): *SUPERMAN*, a regulator of floral homeotic genes in *Arabidopsis*. Development 114:599–615.

Carpenter R, Coen ES (1990): Floral homeotic mutations produced by transposon-mutagenesis in *Antirrhinum majus*. Genes Dev 4:1483–1493.

Coen ES, Meyerowitz EM (1991): The war of the whorls: Genetic interactions controlling flower development. Nature 353:31–37.

Coen ES, Romero JM, Doyle S, Elliott R, Murphy G, Carpenter R (1990): *floricaula*: A homeotic gene required for flower development in Antirrhinum majus. Cell 63:1311–1322.

Cronquist A (1981): "An Integrated System of Classification of Flowering Plants," New York: Columbia University Press.

Drews GN, Bowman JL, Meyerowitz EM (1991): Negative regulation of the *Arabidopsis* homeotic gene *AGAMOUS* by the *APETALA2* product. Cell 65:991–1002.

Gifford EM, Foster AS (1988): "Morphology and Evolution of Vascular Plants," 3rd ed. New York: Freeman.

Goto N, Katoh N, Kranz AR (1991): Morphogenesis of floral organs in *Arabidopsis*: Predominant carpel formation of the pin-formed mutant. Jpn J Genet 66:551–567.

Hill JP, Lord EM (1989): Floral development in *Arabidopsis thaliana*: Comparison of the wild-type and the homeotic *pistillata* mutant. Can J Bot 67:2922–2936.

Huala E, Sussex IM (1992): *LEAFY* interacts with floral homeotic genes to regulate *Arabidopsis* floral development. Plant Cell 4:901–913.

Huijser P, Klein J, Lönnig W-E, Meijer H, Saedler H, Sommer H (1992): Bracteomania, an inflorescence anomaly, is caused by the loss of function of the MADS-box gene *squamosa* in *Antirrhinum majus*. EMBO J 11:1239–1249.

Irish VS, Sussex IM (1990): Function of the *apetala-1* gene during *Arabidopsis* floral development. Plant Cell 2:741–751.

Jack T, Brockman LL, Meyerowitz EM (1992): The homeotic gene *APETALA3* of Arabidopsis thaliana encodes a MADS-box and is expressed in petals and stamens. Cell 68:683–697.

Klemm M (1927): Vergleichende morphologische und entwicklungsgeschichtliche Untersuchungen einer Reihe multipler Allelomorphe bei *Antirrhinum majus*. Bot Archiv 20:423–474.

Komaki MK, Okada K, Nishino E, Shimura Y (1988): Isolation and characterization of novel mutants of *Arabidopsis thaliana* defective in flower development. Development 104:195–203.

Koornneef M, van Eden J, Hanhart CJ, Stam P, Braaksma FJ, Feenstra WJ (1983): Linkage map of *Arabidopsis thaliana*. J Hered 74:265–272.

Koornneef M, Hanhart CJ, van der Veen JH (1991): A genetic and physiological analysis of late flowering mutants in *Arabidopsis thaliana*. Mol Gen Genet 229:57–66.

Kunst L, Klenz JE, Martinez-Zapater J, Haughn GW (1989): *AP2* Gene determines the identity of perianth organs in flowers of *Arabidopsis thaliana*. Plant Cell 1:1195–1208.

Ma H, Yanofsky MF, Meyerowitz EM (1991): *AGL1-AGL6*, an *Arabidopsis* gene family with similarity to floral homeotic and transcription factor genes. Genes Dev 5:484–495.

Mandel MA, Gustafson-Brown C, Savidge B, Yanofsky MF (1992): Molecular characterization of the *Arabidopsis* floral homeotic gene *APETALA1*. Nature 360:273–277.

Napp-Zinn K (1985) *Arabidopsis thaliana*. In Halevy HA (ed): "CRC Handbook of Flowering." Boca Raton, FL: CRC Press, pp 492–503.

Okada K, Komaki MK, Shimura Y (1989): Mutational analysis of pistil structure and development of *Arabidopsis thaliana*. Cell Diff Dev 28:27–38.

Schultz EA, Haughn GW (1991): *LEAFY*, a homeotic gene that regulates inflorescence development in Arabidopsis. Plant Cell 3:771–781.

Schultz EA, Pickett FB, Haughn GW (1991): The *FLO10* gene product regulates the expression domain of homeotic genes *AP3* and *PI* in Arabidopsis flowers. Plant Cell 3:1221–1237.

Schwarz-Sommer Z, Hujser P, Nacken W, Saedler H, Sommer H (1990): Genetic control of flower development: Homeotic genes in *Antirrhinum majus*. Science 250:931–936.

Schwarz-Sommer Z, Hue I, Huijser P, Flor PJ, Hansen R, Tetens F, Lönnig W-E, Saedler H, Sommer H (1992): Characterization of the *Antirrhinum* floral homeotic MADS-box gene *DEFICIENS*: Evidence for DNA binding and autoregulation of its persistent expression throughout flower development. EMBO J 11:251–263.

Shannon S, Meeks-Wagner DR (1991): A mutation in the Arabidopsis *TFL1* gene affects inflorescence meristem development. Plant Cell 3:877–892.

Sommer H, Beltrán JP, Huijser P, Pape H, Lönnig WE, Saedler H, Schwarz-Sommer Z (1990): *Deficiens*, a homeotic gene involved in the control of flower morphogenesis in *Antirrhinum majus*: The protein shows homology to transcription factors. EMBO J 9:605–613.

Stubbe H (1966): "Genetik und Zytologie von *Antirrhinum* L. sect. *Antirrhinum*." Jena: Gustav Fischer.

Weigel D, Alvarez J, Smyth DR, Yanofsky MF, Meyerowitz EM (1992): *LEAFY* controls floral meristem identity in Arabidopsis. Cell 69:843–859.

Yanofsky MF, Ma H, Bowman JL, Drews GN, Feldmann KA, Meyerowitz EM (1990): The protein encoded by the *Arabidopsis* homeotic gene *agamous* resembles transcription factors. Nature 346:35–39.

Molecular Basis of Morphogenesis, pages 109–133
© 1993 Wiley-Liss, Inc.

8. The Role of Induction in Cell Choice and Cell Cycle in the Developing *Drosophila* Retina

Ross L. Cagan, Barbara J. Thomas, and S. Lawrence Zipursky

Howard Hughes Medical Institute and Department of Biological Chemistry, University of California, Los Angeles, California 90024-1662

INTRODUCTION

General Question of Cell Induction

Induction has been firmly established as one of the key mechanisms which direct cell fate during development of both vertebrates and invertebrates. In particular, remarkable insights have been gained in recent years in our understanding of the role induction plays during cell fate choice in the *Drosophila* retina. The basic morphology of retinal development has now been described for all cell types (Tomlinson, 1985; Cagan and Ready, 1989b; Wolff and Ready, 1991). Combined with detailed genetic studies, this understanding has led to specific hypotheses as to the nature of cell fate choice. By examining the molecules involved, we have begun to test the idea that induction between neighboring cells plays a leading role in the specification of cell fate. This review will focus on one of the best understood cell fate decisions: the directing of an uncommitted precursor cell into the R7 photoreceptor cell pathway due to its contact with an inductive neighbor. The inductive ligand has now been identified, and appears to trigger a developmental pathway with features common to signal transduction pathways found in several other cellular processes. We will also discuss the role of the cell cycle in development, and the possibility that cell cycle regulation also may make use of local inductive signals.

The *Drosophila* eye has several advantages which have made it successful as a system for studying local cell induction. It is a relatively simple system, composed of no more than 20 cell types. Development of the eye proceeds in an ordered manner that has allowed the identification of each cell type as it arises, and since the eye is dispensable, mutations which arrest or perturb this development do not affect the viability of the fly. The combination of a detailed description of eye development with the sophisticated genetic and molecular tools available in *Drosophila* have allowed cell fate induction to be examined from a variety of aspects.

Lineage-directed cell fate requires a separate developmental program for each cell type. But, as has become especially apparent in recent studies involving invertebrate development, induction can produce a variety of cell types using a relatively parsimonious set of rules. These rules determine the competence of cells to be induced as well as which cells will actually develop as a particular cell type. Three aspects appear to be common to most or all inductive processes examined.

1. **Molecular.** A surprisingly small number of molecular classes are involved in most examples of induction studied to date. Indeed, one challenge is to understand how similar molecular pathways direct different inductive responses. Typically, a receptor receives extracellular information from a ligand, and transmits this information by directly (e.g. through steroid receptors) or indirectly (e.g. through receptor kinases) affecting a cell's repertoire of expressed genes. Receptor kinases require activation by an extracellular ligand as well as use of a cell's extensive and complex set of cytoplasmic signal transduction molecules. Cells which contain the appropriate receptors and signal transduction machinery are said to be competent to receive the inductive signal. This review will focus on the interaction of the sevenless receptor kinase with its ligand, and the role of *ras* in transmitting the resulting signal to the nucleus.
2. **Spatial.** Cells are often limited in their access to extracellular signals due to their position within an epithelium or region of the developing animal. Spatial segregation is an important developmental strategy which provides a means of limiting an inductive cue to a few cells. In addition, the ligand itself must be limited in its access, and examples include anchoring the ligand to the extracellular matrix (e.g. fibronectin, TGF-β), limiting diffusion (interleukins, cAMP), and anchoring the ligand to the membrane of neighboring cells. This review will focus on an example of a membrane-bound ligand.
3. **Temporal.** Temporal competence has received growing attention as a further means of restricting a cell's ability to respond to inductive signals. Several lines of evidence suggest cells will only respond to such signals within a limited time frame (e.g. Hart et al., 1989; Watanabe and Raff, 1990; Reh, 1992; Basler and Hafen, 1989; Cagan and Ready, 1989a). This phenomenon is at present poorly understood, but evidence in at least a few examples suggest that cells may contain an internal clock which affects a cell's competence.

BASICS ABOUT EYE DEVELOPMENT

The *Drosophila* eye consists of a well-patterned array of discreet photoreceptive units known as ommatidia. Each ommatidium contains an identical core of eight photoreceptor cells (R1–R8) and four cone cells, and is surrounded

Induction in the Developing *Drosophila* Retina   111

by pigment cells which optically insulate the ommatidium from its neighbors. Like the vertebrate retina, the adult fly retina appears to contain only a small number of photoreceptor cell classes, as defined by their spectral sensitivity and morphology: R1–R6, R7, and R8 (Harris et al., 1976). However, a careful description of cell morphology and expression of histological markers during eye development has revealed a further division of cell types. For example, in the developing eye disc, the two genes *rough* and *sevenup* define three classes: R2 and R5 express *rough,* R1 and R6 express *sevenup*, and R3 and R4 express both (Tomlinson et al., 1988; Mlodzik et al., 1990). In addition, each of these classes begins differentiation at a distinct point during development. Thus, a detailed description of eye development has proven invaluable in better defining the cells involved, as well as in the analysis of mutants.

Ommatidia form within a monolayer epithelium, the eye disc. Cells within the eye disc proliferate during the first two larval periods. Differentiation begins during the final larval stage, proceeding as a wave across the eye epithelium. It is demarcated by the morphogenetic furrow (MF) (Fig. 1), a groove

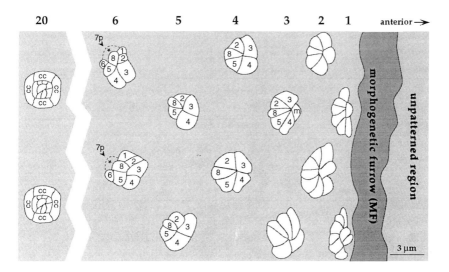

Fig. 1. Tracing of the apical surface of a portion of a developing *Drosophila* eye disc visualized with cobalt sulfide. Cells between ommatidial clusters were omitted for clarity; the two posterior ommatidia (row 20) are schematic additions. In the anterior region of the eye disc (right), cells are proliferating and unpatterned. The basal contraction of cells results in the groove known as the morphogenetic furrow (MF). Posterior to the MF, cells group together into nascent ommatidial clusters. After grouping of the first five photoreceptor cells (R2, R3, R4, R5, and R8), cells derived from the second wave of mitoses form R1 and R6, then R7, to complete each ommatidia's complement of eight photoreceptor cells. Later, four cone cells surround the inner photoreceptor cells. Each row represents approximately 2 hours of development (Campos-Ortega, 1980). Numbers at the top refer to rows of ommatidia. Anterior is to the right; bar is 3 μm. Abbreviations: 1–8, photoreceptor cells R1–R8; R7p, R7 precursor cell; cc, cone cells. (Adapted from Cagan and Zipursky, 1992, with permission of the publisher.)

in the epithelium which is the result of the coordinated contraction of cells along a dorsal-ventral row. The MF sweeps across the eye anteriorly, and represents a point in ommatidial development in which cell division transiently ceases and cells gather together into nascent ommatidial clusters. After the first five cell types have appeared in a row of ommatidia, a second round of cell division provides more than enough cells for the remaining cell types required. After an ommatidium completes development, excess cells are removed through programmed cell death. The result of this orderly development is that a single eye disc contains a gradient of developmental stages: rows of ommatidia near the MF are developmentally less mature than ommatidia further away (Fig. 1). For a more detailed discussion of eye disc development, see reviews by Tomlinson (1988), Ready (1989), Rubin (1991), and Cagan and Zipursky (1992).

Individual cell types in the eye disc are defined by their expression of histological markers and by the order in which they arise in a developing ommatidium (Fig. 1). The first cells to arise are the photoreceptor cells. After cell division ceases at the MF, photoreceptors R2, R3, R4, R5, and R8 group together into the initial five-cell precluster. Expression of most, though not all, neural antigens suggests that R8 is the first cell to differentiate within this group. After a second round of cell division, photoreceptors R1 and R6, and finally R7, are added to the growing ommatidial cluster. After the eight photoreceptor cells have been assembled, the four cone cells are added around the periphery to complete the ommatidial core.

Importance of Induction in the Eye

Using genetic mosaic analysis, Ready et al. (1976) and Lawrence and Green (1979) demonstrated that, despite their stereotyped order of appearance, each photoreceptor cell type in the eye arose without regard to its lineage. This led to the proposal that cells in the eye disc achieve their final fate based on locally provided inductive cues (Tomlinson and Ready, 1987). At least two lines of evidence are consistent with this proposal. First, the fate of precursor cells can be altered by removing the activity of *Notch* (Cagan and Ready, 1989a), a transmembrane protein involved in interactions between cells (Hartenstein and Campos-Ortega, 1984; Fehon et al., 1990; Heitzler and Simpson, 1991). Second, an inductive ligand necessary for R7 development is required exclusively in the neighboring cell, R8. The following section discusses our current understanding of the development of R7, the last photoreceptor cell type added to each ommatidial cluster and currently one of the best understood developmental paradigms.

R7 INDUCTION

Mutations in either of two genes, *sevenless* and *bride of sevenless* (*boss*) result in the R7 precursor cell developing incorrectly as a cone cell (Tomlinson and Ready, 1986) (Fig. 2). Remarkably, no other cell in the fly is known to be affected. This phenotype was especially exciting because it suggested that these two genes play a critical and specific role in the R7 precursor cell's selection of its correct fate. Despite their identical phenotypes, genetic mosaic analysis suggested that the two genes were required in different cells. Sevenless activity is required exclusively in the R7 precursor cell itself (Campos-Ortega et al., 1979; Tomlinson and Ready, 1986), while boss activity is required in the neighboring R8 cell (Reinke and Zipursky, 1988).

sevenless is Required for R7 Specification

The *sevenless* gene encodes a transmembrane protein with a large extracellular domain and an intracellular tyrosine kinase catalytic domain that is required for its function (Hafen et al., 1987; Basler and Hafen, 1988). Thus, the sevenless protein is a member of the receptor tyrosine kinase family. It is composed of two subunits cleaved from a single polypeptide (Simon et al., 1989) (Fig. 3), denoting it as a Class II receptor (Ullrich and Schlessinger, 1990). Although only required in the R7 precursor cell, the sevenless protein is expressed in all cells in the developing disc except R2, R5, and R8 (Fig. 3). Its closest mammalian homologue, *c-ros,* is expressed during development of kidney, gut, and lung (Sonnenberg et al., 1991; Tessarollo et al., 1992). Intriguingly, *c-ros* is expressed during kidney development in the ureter buds, which are involved in direct inductive interactions with the surrounding metanephric mesenchyme.

The discovery that sevenless is a receptor tyrosine kinase suggests a simple model for the mechanism by which it directs the R7 fate: binding of a ligand activates sevenless tyrosine kinase activity; a signal transduction pathway then relays this activity to the nucleus to direct transcription of R7-specific genes. The remainder of this section will examine advances made in exploring this model, including the identification of boss as the ligand to the sevenless receptor, characterization of the signal transduction pathway, and identification of at least one potential transcription factor required to direct R7 development.

boss Encodes a Ligand for Sevenless

The *boss* gene also encodes a transmembrane protein (Hart et al., 1990). Its sequence indicates a large extracellular domain, a smaller cytoplasmic domain, and seven membrane-spanning domains. Antibodies specific for the boss protein indicate that it is expressed exclusively in R8 (Fig. 3), the only cell in which genetic mosaic studies indicated its activity was required. Although

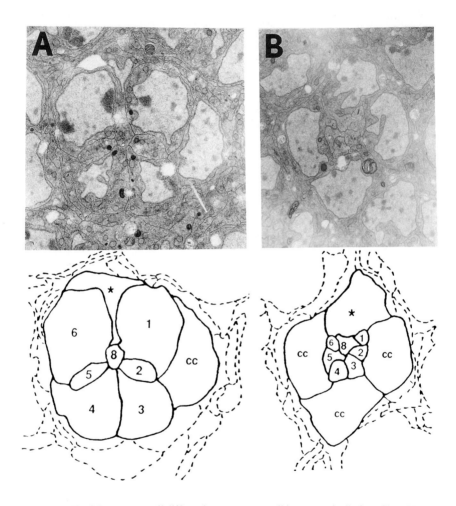

Fig. 2. The R7 precursor cell differentiates as a cone cell in genotypically *boss* flies. Electron micrographs approximately 5 μm below the apical surface. A tracing of each micrograph is presented below. **A:** A *boss* ommatidium approximately 10 rows from the morphogenetic furrow (see Fig. 1). As with the wild type, a single cell (the R7 precursor cell) occupies the niche between R1, R6, and R8 (asterisk). **B:** A *boss* ommatidium approximately 20 rows from the morphogenetic furrow. As the nuclei of the other photoreceptor cells have descended, the nucleus of the R7 precursor cell (asterisk) remains apical with those of the other three cone cells. The apical presence of its nucleus, along with the failure to extend an axon or express neuronal antigens, is the first indication that the R7 precursor cell has begun to differentiate as a cone cell. This phenotype is indistinguishable from that of a *sevenless* ommatidium. Anterior is to the right; bar is 3 μm. Abbreviations: 1–6 and 8, photoreceptors R1–R6, and R8; cc, cone cells.

the boss sequence shows no significant amino acid homology to other identified protein sequences, its seven transmembrane domains would seem to place boss into the G protein receptor family. However, no evidence has been found to date that boss can act as a receptor in R8.

Based on evidence from both tissue culture and in vivo studies, the evidence is now strong that *boss* encodes a ligand of the sevenless receptor (Krämer et al., 1991). Direct evidence that boss and sevenless can bind resulted from aggregation studies in *Drosophila* tissue culture cell lines. Clonal S2 cell lines were produced which expressed high levels of boss or sevenless proteins on their surface (Krämer et al., 1991; Simon et al., 1989). By themselves, each cell line remained as a single cell suspension in the dish. Mixed together, the S2-boss and SL2-sev cell lines formed large aggregates of cells. Aggregation was prevented by the addition of boss-specific or sevenless-specific antibodies or by addition of a soluble form of the extracellular domain of boss, an observation which indicated that aggregation was due to the specific binding of boss to sevenless (Krämer et al., 1991). Boss was internalized into the SL2-sevenless cells in a sevenless-dependent manner (Fig. 4B–D), providing additional evidence for a direct interaction. Addition of S2-boss cells to SL2-sevenless cells has been shown to result in the phosphorylation of the sevenless receptor (H. Krämer and S.L. Zipursky, personal communication), the first step in activation of a receptor tyrosine kinase.

In vivo evidence that boss is a ligand of sevenless comes from a close examination of the distribution of the boss and sevenless proteins within the eye disc during development. Antibodies specific for each of the two proteins indicate that both are localized primarily to apical microvilli (Tomlinson et al., 1987; Banerjee et al., 1987; Krämer et al., 1991). Sevenless is found in most cells in the developing eye disc. By contrast, only R8 contains boss in its apical microvilli (Fig. 3). In addition, since boss immunoreactivity was found exclusively in the perinuclear endoplasmic reticulum of R8, it was concluded that R8 is the only cell to produce the protein. Nevertheless, boss is also found in a multivesicular body (MVB) in the developing R7 precursor cell (Fig. 4A). Multivesicular bodies appear similar to late endosomes, which have been identified in mammalian tissue culture cells as vesicular structures within which ligand/receptor complexes are sorted (e.g., Felder et al., 1990). The localization of boss protein to an MVB in R7 is dependent on the presence of the sevenless protein, as mutations which eliminate *sevenless* expression eliminate internalization of boss as well. Boss is also internalized into R7 precursor cells in flies which express normal levels of an inactivated form of the sevenless receptor. This indicates that boss internalization requires the presence of the sevenless receptor but not its activity.

Internalization of the boss protein exhibits two unusual features: its specificity to the R7 precursor cell (discussed below), and the mechanism of inter-

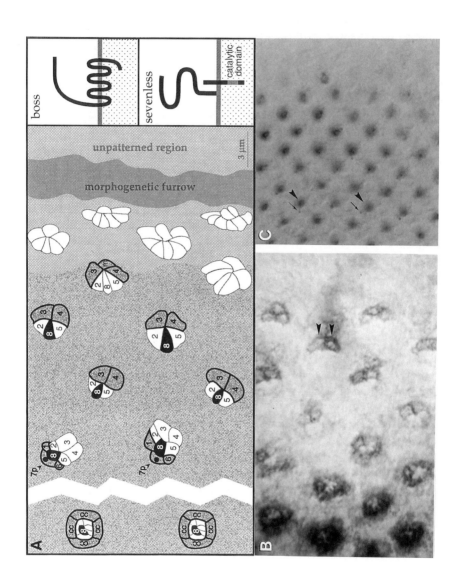

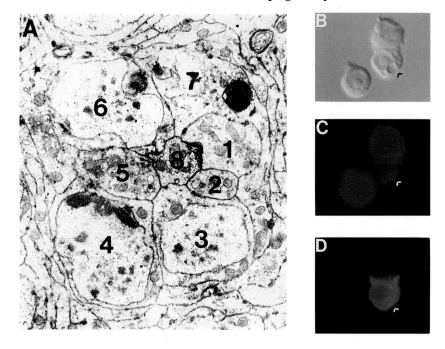

Fig. 4. Boss is internalized into neighboring cells. **A:** Electron micrograph of a single ommatidium, 5 μm below the surface. Boss is visualized with antibody αbossNT1. Most boss protein is localized more apically; a small amount can be seen on R8's membrane (arrowheads). Arrow indicates boss localization in a multivesicular body of the neighboring R7. **B:** Nomarsky view of *boss-* and *sevenless-*expressing *Drosophila* tissue culture cells. Arrowhead indicates large vesicular structure. **C:** Immunofluorescence image of the same cells visualized with Texas Red-conjugated anti-boss antibody. The upper and lower cells are expressing boss. The central cell contains boss in the large vesicle (arrowhead). **D:** Immunofluorescence of the cells visualized with fluoroscein-conjugated anti-sevenless antibody. The central cell is expressing sevenless. High sevenless immunoreactivity is seen in the internal vesicle, indicating this vesicle contains both boss and sevenless. Anterior in (A) is to the right; bar is 3 μm. Abbreviations: 1–8, photoreceptors R1–R8.

Fig. 3. Sevenless and boss immunolocalization in the eye disc. **A:** Sevenless, represented by speckled shading, is expressed in a dynamic pattern in all cells in the developing eye disc except R2, R5, and R8. Boss expression, represented by black shading, is restricted to photoreceptor R8 and a small vesicle in the R7 precursor cell (see Fig. 4). Both proteins are localized primarily to cells' apical microvilli. **Right:** Proposed boss and sevenless structures. Lower shaded region represents intracellular region; extracellular region lies above. **B:** Sevenless localization as visualized with antibody 150C3 (Banerjee et al., 1987). The MF is at the right edge. Arrows indicate R3 and R4, the first photoreceptors to express sevenless. **C:** Boss localization visualized with antibody αbossNT1 (Krämer et al., 1991). The morphogenetic furrow is at the right edge. Arrowheads indicate staining in R8. Arrows indicate vesicular staining in the R7 precursor cell (see Fig. 4). Anterior is to the right; bar is 3 μm. Abbreviations as in Figure 1.

nalization. Most ligands are internalized as soluble factors; for example EGF, TGFα, NDF, TNF, and Steel are synthesized as membrane-bound precursors, which are then cleaved into soluble ligands (e.g., Cohen and Fava; 1985; Wen et al., 1992; Flanagan et al., 1991). However, no evidence was found for cleaved boss product. Instead, antibodies directed to cytoplasmic as well as to extracellular epitopes indicate that the entire boss protein is internalized into the R7 precursor cell (Cagan et al. 1992). The mechanism by which internalization of a protein containing seven membrane-spanning domains might occur is not clear. While not disproven, phagocytosis of R8's microvillar membranes by the R7 precursor cell is not likely, as no large double-membrane structures were detected by electron microscopic analysis. Another possible mechanism is receptor-mediated endocytosis, the mechanism by which soluble ligands are generally internalized (reviewed in Trowbridge, 1991). Interestingly, boss internalization depends on proper activity of the *shibire* gene product, a dynamin homologue which is involved in endocytosis in other cells of the fly (van der Bliek and Meyerowitz, 1991; Chen et al., 1991; Kosaka and Ikeda, 1983).

A second unusual feature of boss internalization is its specificity. Despite expression of the sevenless receptor by most cells in the eye disc, only the R7 precursor cell internalizes boss and activates the *boss/sevenless* pathway. Why do other photoreceptor precursors also in direct contact with R8 fail to internalize boss? When proper development of these precursor cells is disturbed via mutations in the genes *rough* or *Notch*, many respond by expressing R7-specific genes, apparently developing as transformed R7 cells. They will also internalize boss protein early in development (Van Vactor et al., 1991). This indicates that these precursor cells actively block both internalization of the boss protein and activation of the *boss/sevenless* pathway during establishment of their normal cell fate.

Despite the correlation between the internalization of boss and the activation of the *boss/sevenless* pathway, experiments by Basler et al. (1991) suggest that the internalized boss protein itself does not play a role in directly specifying R7 fate in the precursor cell. They showed that an overexpressed truncated form of the sevenless protein can direct R7 fate even in the absence of boss. Although this data rules out an obligate role for internalized boss protein, it remains unclear whether internalization of the activated receptor is required.

Spatial Aspects of R7 Induction

Spatial constraints also play a role in limiting the *boss/sevenless* pathway to a single cell. *Sevenless* itself is expressed in most cells in the developing eye disc. Although this expression is dynamic throughout development (Fig. 2), expressing *sevenless* in all cells via an inducible promotor did not result in the recruitment of additional R7 cells. This indicates that the complex expression pattern is not important for sevenless function. In contrast, the ubiquitous expression of boss produced several additional R7 cells in each ommatidium, as assessed by a variety of genetic and morphological markers (Van Vactor et al., 1991). This transformation required the presence of the sevenless protein, indicating that the *boss/sevenless* pathway was being activated in these cells. The transformations also indicate that part of the pathway's specificity lies in limiting *boss* expression to R8. The R1–R6 cells were unaffected; the additional R7 cells arose from a misrouting of developing cone cells. This suggests that the cone cells, along with the R7 precursor, are competent to respond to the boss inductive signal, but that relegating boss to the surface of R8 keeps it one cell diameter away from the cone cell precursors. Thus, spatial restriction has been achieved by limiting the ligand to a single cell.

Together, these experiments suggest that, with respect to the boss inductive signal, each ommatidium contains four types of cells. The first is R8, which presents the inductive signal to its immediate neighbors. The second is R7, which responds to the signal. The third is the group of cone cells, which are competent to respond to the signal but which are prevented from responding since they are not in contact with the ligand. The final group, composed of R1–R6, is not competent to respond to the inductive signal by virtue of the activation of alternative developmental pathways through genes such as *Notch* and *rough*.

Temporal Aspects of R7 Induction

We have seen the importance of the molecular and spatial aspects of R7 fate induction. A temporal element also appears to be at play. Ubiquitous *boss* expression can redirect cone cell precursors into the *boss/sevenless* pathway. However, these cells are sensitive to the presentation of boss only at the point in developmental time when they would have differentiated as cone cells. Boss expression before or after this point has no effect. This is also true for the R7 precursor cell itself, which differentiates some 20 hours after boss is first expressed in the neighboring R8 cell. Similarly, *sevenless* itself can direct the R7 fate only if expressed in the precursor cell during a limited interval (Basler and Hafen, 1989; Mullins and Rubin, 1991). In support of the idea that the R7 precursor cell has a limited temporal window of competence, experiments with a temperature-sensitive allele of the gene *Notch,* which encodes for a transmembrane protein also required for R7 development, show that loss of *Notch* activity can only block R7 development if activity is removed during the discreet period when the R7 precursor cell would normally select its fate (Cagan and Ready, 1989a). The nature of this apparent internal clock remains a mystery.

Ras and R7 Development

Recent genetic screens have been successful at identifying other components of the *boss/sevenless* pathway (Fig. 5). Most commonly, these screens involved rescue of partially or fully inactive sevenless activity (Simon et al., 1991; Bonfini et al., 1991). Most genes characterized thus far encode *Drosophila* homologues of members of the *ras* regulatory pathway. The *ras* pathway has been previously identified in yeast cells and mammalian tissue culture cells as one of the primary signal transduction pathways employed for both growth and for differentiation (Cantley et al., 1991). Two *Drosophila* homologues of the GTPase $p21^{ras}$ have been identified, but only one, *Ras1*, appears to be involved in the *sevenless* pathway. Loss of *Ras1* activity enhances the effect of reduced *sevenless* activity (Simon et al., 1991). Importantly, when a mutationally activated *Ras1* protein is introduced into the developing eye disc, ectopic R7s arise even in the absence of *sevenless* (Fortini et al., 1992). This provides strong evidence that the sevenless receptor directs R7 fate by activating *ras*, and suggests that *ras* functions downstream of *sevenless* in the signal transduction cascade. Whether sevenless activates the *ras* pathway directly awaits biochemical confirmation.

Ras1 is not active exclusively in the eye, as elimination of *Ras1* activity results in embryonic lethality. Interestingly, *Ras1* also interacts genetically with *Ellipse*, a mutation in the epidermal growth factor receptor homologue *DER1* which affects other aspects of eye development (Simon et al., 1991; Baker and Rubin, 1992). Thus, *Ras1* is active in at least two different receptor tyrosine kinase pathways during eye development, involving *sevenless* and *DER1*. It remains to be determined how activation of the *ras* pathway in different developmental contexts produces different cellular responses.

In other systems studied to date, $p21^{ras}$ is not known to be activated directly by receptor tyrosine kinases, but appears to require an intermediate. Two candidate proteins have been identified which regulate $p21^{ras}$, and homologues of both have been found to be involved in R7 development. Gain-of-function or loss-of-function mutations in the *Son of sevenless* (*Sos*) locus suppress or enhance, respectively, partial loss of sevenless activity (Rogge et al., 1991; Simon et al., 1991; Bonfini et al., 1991) (see Fig. 5). As with *Ras1*, *Sos* activity is also required for fly viability, suggesting it is also active in other developmental processes. *Sos* shows extensive identity with *CDC25*, a guanine nucleotide exchange protein which acts as a direct activator of $p21^{ras}$ in *Saccharomyces cerevisiae* (Broek et al., 1987; Jones et al., 1991).

Another candidate for transmitting sevenless activation to *Ras1* is the product of the *GAP1* gene (Gaul et al., 1992). GTPase activating proteins (GAPs) act as direct negative regulators of $p21^{ras}$ and related GTPase proteins by catalyzing their inactivation (Bourne et al., 1991) (Fig. 5). Interestingly, they can also bind directly to certain receptor tyrosine kinases, making them good can-

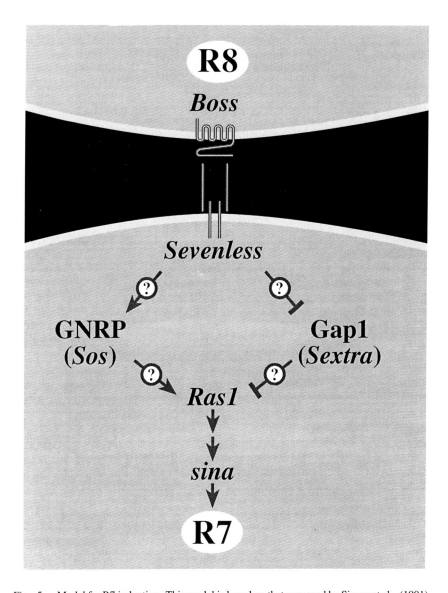

Fig. 5. Model for R7 induction. This model is based on that proposed by Simon et al., (1991) and Gaul et al. (1992). R8 is at the top; R7 below. Binding of the boss inductive ligand to the sevenless receptor activates its tyrosine kinase catalytic activity (sevenless is shown as a dimer). This in turn activates *Ras1*, a *Drosophila* homologue to p21^{ras}, through an intermediary, perhaps *Sos* or *GAP1*. The potential role of these two proteins in *Ras1* regulation is discussed in the text. Finally, activation of *Ras1* eventually results in a change in gene expression (e.g., *sina*) necessary to activate the R7 pathway.

didates to relay signals between a receptor and p21^{ras}. In flies, loss-of-function alleles of *GAP1* result in additional R7 cells, a phenotype strikingly similar to activated *Ras1*, indicating *Gap1* may act as a negative regulator of *Ras1*. However, unlike *Ras1*, *GAP1* activity is not required for embryonic viability.

Is *GAP1* the sole mediator of activity between the sevenless receptor and *Ras1*? Two lines of evidence suggest it probably is not. First, *GAP1* does not appear to contain SH2 domains (Gaul et al., 1992), sites identified in other GAPs as responsible for binding to receptor tyrosine kinases. The second line of evidence comes from examining epistasis in a *sevenless/GAP1* double mutant combination. Mutations in *sevenless* result in loss of R7 cells, whereas loss of *GAP1* activity results in the gain of R7s. However, loss of *GAP1* activity can only occasionally restore the R7 cell in a *sevenless* ommatidium: although some R7s were restored (Gaul et al., 1992), a quantitative analysis using the *GAP1* allele *sextra* found the R7 precursor cell developed as a cone cell in 69 of 70 *sevenless/sextra* ommatidia examined (Rogge et al., 1992). This low penetrance of rescue suggests *GAP1* may be part of a more complex signal transduction machinery. Surprisingly, the *GAP1*-mediated transformation of the cone cells is rarely affected by loss of *sevenless* activity, indicating these cells regulate the *ras* pathway in a manner different from the R7 precursor cell.

Local Induction and Diffusible Factors

Local induction can also make use of diffusible factors. This sort of induction has been best studied in the immune system (e.g., interleukins; Singer, 1992). Less is known of the role of diffusible molecules in eye development, but some evidence suggests that local secretion plays an important role. Three molecules have been identified which are expressed in the developing eye disc and are likely to be secreted. One is the product of the *scabrous* gene, which contains fibrinogen-like repeats. Scabrous is proposed to play a role in the spacing of the initial ommatidial clusters (Baker et al., 1990). A second example is encoded by *decapentaplegic* (*dpp*). Dpp is a member of the TGF-β family of secreted ligands, which have been shown in vertebrates to play an important role in early embryonic decisions (Roberts and Sporn, 1990). In the fly, *dpp* is required throughout development (Spencer et al., 1982), and contains an extensive 3' regulatory region which is required for its complex expression patterns in adult tissues (Blackman et al., 1991). In the eye disc, *dpp* is expressed exclusively in cells in the MF; however, its role in eye development has yet to be carefully examined. Its role in cell proliferation is discussed below. Finally, recent experiments have identified the *argos* gene, an essential gene which appears to act as a secreted negative regulator of cell fate decisions in the eye (Freeman et al., 1992).

Induction in the Developing *Drosophila* Retina 123

Analysis of the *sevenup* mutation has raised the possibility that small secreted molecules could also play a role in specification of cell type. The *sevenup* gene appears to encode two alternatively spliced forms of a homologue of the mammalian COUP steroid receptor. Loss of *sevenup* activity results in ectopic R7 cells, an observation which indicates that these cells are misspecified. The protein is expressed and required genetically in R1, R3, R4, and R6. A reasonable hypothesis for *sevenup* function is that the precursors to these cells receive locally secreted information through the sevenup receptor and that this information prevents them from developing as R7 cells. However, since a ligand for sevenup has not been identified, the possibility of a membrane-bound ligand or ligand-independent *sevenup* activity has not been ruled out.

ASSESSING THE ROLE OF INDUCTION IN THE CELL CYCLE

Can local inductive cues also influence cell cycle control? In the developing vertebrate retina, evidence exists that cell division can be regulated by the complement of cell types present. For example, selective elimination of dopaminergic amacrine cells in the developing frog retina resulted in selective proliferation of these cells in the proliferating marginal zone (Negishi et al., 1987). Furthermore, several examples have been found in which an inductive ligand can act to promote either growth or differentiation, depending on its developmental context (Yarden and Ullrich, 1988; Pardee, 1989).

Cell differentiation and the cell cycle are closely coordinated in the eye disc, leading to the speculation that the cell cycle may also be influenced by local induction. Recall that the first overt signs of differentiation in the eye disc include the cessation of cell division in a band across the disc and the contraction of cells in this region to form the morphogenetic furrow (MF). Emerging from the posterior edge of the MF are morphologically and histochemically distinct groups of cells, which are the precursors of the founding photoreceptor cell neurons in each ommatidial cluster, R2, R3, R4, R5, and R8. Between these regularly spaced clusters, uncommitted cells undergo another single round of cell division to generate the final pool of post mitotic precursor cells.

Questions relating to these early cell cycle events in eye morphogenesis are only beginning to be addressed. However, it has been demonstrated that molecules important for cell cycle regulation in other systems are also expressed in the eye imaginal disc (see below), indicating that regulation of these events may proceed by a common mechanism.

Regulation of Proliferation at the Morphogenetic Furrow

Concomitant with the formation of the MF is the transient termination of cell division beginning at its anterior edge (schematically shown in Fig. 6A). Anterior to this region, cells divide randomly as assayed by incorporation of

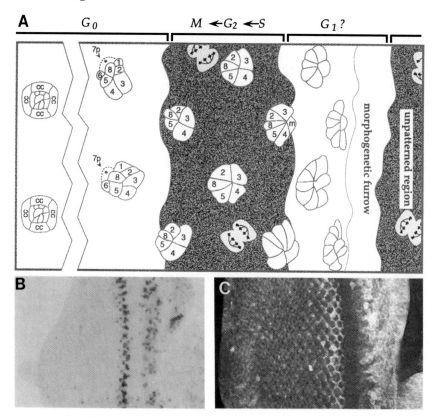

Fig. 6. **A:** Schematic representation of cell cycle events in the eye disc. Anterior is to the right; regions in which cell division is observed are shaded; regions of no cell division are white. Inferred stages of the cell cycle are shown above the figure. Anterior to the morphogenetic furrow (MF), cells cycle randomly. No division is detected within the MF, as assayed by BrDU incorporation and cyclin B expression. At the posterior edge of the MF, clusters begin to form. At row 3, all cells that are not part of a precluster divide again in a sharp wave, indicated by shading. In the most posterior regions of the disc, cells are terminally differentiated. **B:** BrDU labeling in a wild-type eye disc. Position of the MF is indicated by the arrowhead. Eye discs were incubated in vitro for 30 minutes with BrDU: incorporated BrDU was detected immunohistochemically as described (Truman and Bate, 1988). The eye exhibits two regions of labeling. A broad region of unpatterned BrDU incorporation occurs anterior to the MF (some regions of labeled nuclei are not in focus). Within and just posterior to the MF is a gap of BrDU incorporation. Posterior to this gap, a second wave of mitosis is seen as a sharp band of incorporation by cells between clusters. **C:** Cyclin B staining of a wild-type eye antennal disc. Confocal image of a disc stained with antibodies to cyclin B as described (Whitfield et al., 1990). Anterior to the MF (arrowhead), cells express cyclin B fairly ubiquitously. Within and just posterior to the MF, cyclin B expression is not detected. In a region corresponding to the second wave of mitoses described in (B), cyclin B is again seen in a "chicken wire" pattern in the cells surrounding the developing clusters. Cyclin B expression is not detected in more posterior regions of the disc. Anterior is to the right; bar in (C) represents 10 μm for (B) and (C). Abbreviations: 1–8, photoreceptors R1–R8; cc, cone cell.

the nucleotide analogs [^{3}H]-thymidine (Ready et al., 1976) or 5-bromo-2-deoxyuridine (BrDU) as the disc undergoes general proliferative growth (Fig. 6B). Within the MF, no incorporation of BrDU is detected. Emerging from the posterior edge of the MF, DNA synthesis (S phase) is again observed by incorporation of BrDU into groups of cells which undergo a single, further round of cell division. This so-called second wave of mitosis is sufficient to account for the formation of all remaining precursor cells needed to assemble the adult eye; pulse chase analysis has shown that all cells that do not participate in cluster formation undergo another round of cell division in this second wave (Campos-Ortega, 1980; Wolff and Ready, 1991).

Expression of the cell cycle–regulated molecule cyclin B is consistent with the pattern of cell division as assayed by BrDU incorporation (Fig. 6C). *Drosophila* cyclin B has been shown to be highly expressed in cells up to the G_2/M boundary of the cell cycle (G = "gap"; M = mitosis), where it is rapidly degraded, and is not detected in cells in G_1 (Whitfield et al., 1990). In the developing eye disc, cyclin B is expressed fairly ubiquitously in the cells which incorporate BrDU anterior to the MF (Fig. 6C), a finding which supports the notion that these cells are cycling randomly. Immediately anterior to the MF, cyclin B expression is dramatically decreased. This suggests that cells within the MF are in G_1. In the second wave of cell divisions posterior to the MF, the protein is again detected in a small band, in the dividing cells surrounding the developing photoreceptor clusters. Finally, terminally differentiated cells in the most posterior regions of the disc do not express cyclin B.

The sharp anterior boundary of the gap in BrDU incorporation and cyclin B expression at the MF suggests that cells are synchronized at this point to enter G_1. Interestingly, mRNA for the mitotic activating protein *string* is expressed in a line in the eye which parallels the MF (Alphey et al., 1992). The *string* gene is the *Drosophila* homologue of the yeast *Schizosaccharomyces pombe* cell cycle regulator cdc25, a protein phosphatase that activates the cdc2/cyclin complex by dephosphorylation of cdc2 (Edgar and O'Farrell, 1989; Jiminez et al., 1990). In *Drosophila*, *string* activity is required in G_2 during the zygotically regulated embryonic cell divisions to drive entry into mitosis (Edgar and O'Farrell, 1989). It is tempting to speculate that *string* activity is required in the MF to synchronize cells in G_1 by driving randomly cycling cells anterior to the MF into mitosis. Once cells enter G_1 within the MF, they are poised to make a decision to terminally differentiate and become part of the five-cell precluster, or to undergo a further round of cell division in the second mitotic wave. However, it is important to note that the position of *string* expression relative to the G_1 gap in the MF has not been precisely determined.

One interesting possibility is that the inhibition of cell division within the MF is regulated by the TGF-β homologue *decapentaplegic* (*dpp*). Dpp protein is expressed in a line in the MF (Masucci et al., 1990; Blackman et al., 1991). There is a large body of evidence documenting the role of TGF-β in the stimulation or inhibition of proliferation of cells in culture (reviewed by Sporn et al., 1986; Moses et al., 1990). Intriguingly, one aspect of the antiproliferative role of TGF-β is its antagonistic effect on other peptide growth factors, such as EGF. Similarly, in T lymphocytes TGF-β synthesis is induced concomitantly with interleukin-2 (IL-2) upon stimulation, suggesting that TGF-β may oppose the proliferative effects of IL-2 to regulate proliferation. It is intriguing to speculate that *dpp* has a role in inhibiting cell division in the MF in a manner analogous to TGF-β.

Cell Division Posterior to the Morphogenetic Furrow

A locus that appears to play a role in patterning posterior to the MF is the *Drosophila* homologue of the EGF receptor, *DER*. Flies containing the *DER* allele *Ellipse* (*Elp*) have a mispatterned eye, the result of an increase in *DER* activity. The *Elp* phenotype is the result of very few clusters arising behind the MF, although clusters that do form develop normally. Analysis of cell division by BrDU incorporation shows that *Elp/Elp* homozygous discs show a small increase in the number of BrDU incorporating cells posterior to the MF (Baker and Rubin, 1992; Zak and Shilo, 1992). This increase roughly corresponds in number to those cells that would normally participate in precluster formation, and which now proceed down the alternate developmental pathway of cell division (Baker and Rubin, 1992). In vertebrates, cell proliferation is associated with EGF receptor activation (for reviews, see Yarden and Ullrich, 1988; and Pardee, 1989), and one possible explanation for the phenotype observed in *Elp/Elp* mutants is that cells which normally would leave the cell cycle and differentiate respond to the increase in *DER* activity by continuing to divide, and are thus blocked from forming clusters.

Inductive Signals for Cell Division

Immediately posterior to the MF, cells can assume one of two developmental fates: to participate in precluster formation, or to undergo a final, single round of cell division. A model has been proposed to account for this second wave of cell division based on the distribution of BrDU-incorporating cells in developing discs (Wolff and Ready, 1991). This model proposes that DNA synthesis and subsequent mitosis in the second wave of cell division is induced by contacts between precluster cells and their uncommitted neighbors. Since the model is based on examination of S phase cells, it implies that induction must occur prior to DNA synthesis. However, this G_1 regulatory model is not supported by data examining BrDU incorporation in *Elp/Elp* mutant discs, in which S phase occurs apparently normally in the mutant as compared with the wild type. If entry of the uncommitted cells into S phase required a local inductive signal from committed clusters, this signal would not be present in homozygous *Elp* mutant eye discs, which have very few differentiated clusters.

Another possibility is that an inductive signal is required, not for entry into S phase, but for the subsequent entry into mitosis. This model for G_2 regulation is based on observations of cell divisions in *Elp/Elp* discs (Baker and Rubin, 1992). Dividing cells and developing clusters were identified with an antiserum, generated against scabrous protein, which also recognizes an epitope present on chromosomes of metaphase and anaphase cells. Examination of mitoses in *Elp/Elp* mutant discs shows that 29 of 36 of the observed mitoses were associated with developing clusters, as identified by scabrous expression. However, 7 examples were seen of mitoses not associated with a cluster. Therefore, the data do not support an absolute linkage between clusters and mitoses in the mutant. Furthermore, because scabrous is not expressed in all developing clusters, better markers must be utilized to fully assess this model. Finally, it should be noted that the number of mitoses observed associated with ommatidia in the *Elp/Elp* mutant (29/36) is surprisingly high when one considers the total number of mitoses seen at any one time in wild-type discs (Wolff and Ready, 1991); the reason for this difference is unknown.

If an inductive signal required for mitosis is produced by the cluster, it remains to be determined whether this signal is dependent on the identified G_2 mitotic regulator *string*. There are potentially two sites of action for *string* in the eye: to synchronize cells anterior to the MF and to induce the second mitotic wave. However, *string* mRNA expression is described as a single band in the developing eye disc. Therefore, *string* is probably not required in both locations, since its expression in both yeast and *Drosophila* embryos is regulated at the level of transcription (Moreno et al., 1990; Edgar and O'Farrell, 1989). Analysis of protein expression using antibodies to string may shed light on this question.

Is the Second Mitotic Wave Under Inductive Control?

The concept of local induction of mitoses in the eye disc is appealing because of the G_1 arrest at the MF and the resumption of S phase some 10 hours later in the second wave. However, the possibility should be considered that the G_1 gap is merely a result of a synchronization of the cell cycle as cells enter the MF. Indeed, coordinated cells might be expected to leave such a gap, since a simple calculation shows the average length of a cell cycle during the eye disc's proliferative phase is 10 to 12 hours.[1] From this perspective, undergoing another round of cell division would represent a default state, and mechanisms such as local induction could bring cells out of the cell cycle and into a differentiative state. If cell cycle regulation does not depend on cell differentiation, then it should be possible to identify mutations

---

[1]This assumes that the divisions required to produce the 15,000 or so cells in the adult from the 20 embryonic precursors occur uniformly over larval development. A similar estimate was obtained by mitotic recombination analysis (see Postlethwait, 1978).

which affect cell differentiation but not cell cycle. In fact, many mutations which affect early photoreceptor identity—including *rough, split, echinus, Ellipse, Roughened, Star, Rough eye, pebbled, sparking poliert,* and *echinoid*—do not grossly affect the approximate amount or timing of S phase (B.T., unpublished results). However, the role of these genes in local signaling is currently unknown.

Role of Cell Division in Differentiation

The relationship between cell division and regulation of differentiation has received growing interest in recent years with the discovery that molecules which regulate entry into the cell cycle in simple systems such as yeast are also present during development of more complex organisms. Just as intriguing are recent observations that suggest that the position of a cell in the cell cycle may play an important role in the specification of developmental fate. Early work on growing cells in culture (Pardee, 1989), or in the yeast *S. cerevisiae* (reviewed by Nurse, 1985), identified G_1 as a primary control point at which the developmental choice was to divide or differentiate. Recently, G_2 control points have also been identified. For example, entry of cells into mitosis in *S. pombe* is dependent on attaining a critical cell mass in G_2 (Nurse, 1985). In early *Drosophila* embryos, entry into mitosis is regulated by developmental signals that act through *string* (Edgar and O'Farrell, 1989). In addition, it has become clear that cells can respond to signals differently, depending on their position in the cell cycle when they receive the signal. Examples of this include specification of prespore or prestalk fate in *Dictyostelium* (Gomer and Firtel, 1987) and the determination of transplanted neurons in the ferret visual cortex (McConnell and Kaznowsky, 1991).

An intriguing correlation between a cell's position in the cell cycle and its subsequent differentiation also exists early in ommatidial development. Cell determination in the eye probably begins much earlier than previously thought. Differentiation in the eye was originally defined by the expression of cell-type–specific markers such as the neuronal epitope 22C10 (Tomlinson and Ready, 1987). However, histological examination of cobalt or lead sulfide stained discs shows that patterning begins well before the expression of these markers (Tomlinson, 1988; Wolff and Ready, 1991). Similarly, early expression of genes such as scabrous and hairy indicates segregation of cell identity within or before the MF (Baker et al., 1990; Carroll and Whyte, 1989). Indeed, the coordinated nuclear movements within the MF itself are an indication that differentiative events are occurring even at this early stage. It appears likely that the close temporal linkage between expression of early markers, the MF, and the G_1 arrest near the MF represent a common regulatory mechanism.

SUMMARY

The analysis of R7 development reviewed here provides evidence for the importance of local induction during eye development. Identified components include the inductive ligand boss, its receptor sevenless, and components of the *ras* signal transduction pathway. In addition, the state of developmental competence of cells within the developing ommatidium is becoming better defined. This success has been made possible by the accessibility of this system to classical and molecular genetic analyses, the availability of cell-type–specific markers, and the detailed morphological description of eye development. These same features should also prove useful for addressing the relationship between differentiation and regulation of the cell cycle. As our understanding of ommatidial development continues to advance, we can begin to apply this knowledge to developmental systems in vertebrates, which are more complex and less accessible to genetic analyses.

REFERENCES

Alphey L, Jiminez J, White-Cooper H, Dawson I, Nurse P, Glover DM (1992): *twine*, a *cdc25* homolog that functions in the male and female germline of *Drosophila*. Cell 69:977–988.

Baker NE, Mlodzik M, Rubin GM (1990): Spacing differentiation in the developing *Drosophila* eye: A fibrinogen-related lateral inhibitor decoded by *scabrous*. Science 250:1370–1377.

Baker NE, Rubin GM (1992): *Ellipse* mutations in the *Drosophila* homologue of the EGF receptor affect pattern formation, cell division, and cell death in eye imaginal discs. Dev Biol 150:381–396.

Banerjee U, Renfranz PJ, Pollock JA, Benzer S (1987): Molecular characterization and expression of *sevenless*, a gene involved in neuronal pattern formation in the *Drosophila* eye. Cell 49:281–291.

Basler K, Hafen E (1988): Control of photoreceptor cell fate by the *sevenless* protein requires a functional tyrosine kinase domain. Cell 54:299–311.

Basler K, Hafen E (1989): Ubiquitous expression of *sevenless*: Position-dependent specification of cell fate. Science 243:931–934.

Basler K, Christen B, Hafen E (1992): Ligand-independent activation of the sevenless receptor tyrosine kinase changes the fate of cells in the developing *Drosophila* eye. Cell 64:1–20.

Blackman RK, Sanicola M, Raftery LA, Gillevet T, Gelbart WM (1991): An extensive 3' cis-regulatory region directs the imaginal disk expression of decapentaplegic, a member of the TGF-β family in *Drosophila*. Devel 111:657–666.

Bonfini L, Karlovich CA, Dasgupta C, Banerjee U (1991): *The Son of Sevenless* gene product: A putative activator of ras. Science 255:603–606.

Bourne HR, Sanders DA, McCormick F (1991): The GTPase superfamily: Conserved structure and molecular mechanism. Nature 349:117–127.

Broek D, Toda T, Michaeli T, Levin L, Birchmeier C, Zoller M, Powers S, Wigler M (1987): The *S. cerevisiae CDC25* gene product regulates the RAS/adenylate cyclase pathway. Cell 48:789–799.

Cagan R, Ready D (1989a): *Notch* is required for successive cell decisions in the developing *Drosophila* retina. Genes Dev 3:1099–1112.

Cagan R, Ready D (1989b): The emergence of order in the *Drosophila* pupal retina. Dev Biol 136:346–362.

Cagan RL, Krämer H, Hart AC, Zipursky SL (1992): The bride of sevenless and sevenless interaction: Internalization of a transmembrane ligand. Cell 69:393–399.

Cagan RL, Zipursky SL (1992): Cell choice and patterning in the *Drosophila* retina. In Macagno E, Shankland M (eds): "Determinants of Neuronal Identity." San Diego: Academic Press, pp 189–224.

Campos-Ortega JA (1980): On compound eye development in *Drosophila melanogaster*. In Moscona AA, Monroy A (eds): "Current Topics in Developmental Biology," vol 15. New York: Academic Press, pp 347–371.

Campos-Ortega JA, Jurgens G, Hofbauer A (1979): Cell clones and pattern formation: Studies on *sevenless*, a mutant of *Drosophila melanogaster*. Roux's Arch Dev Biol 186:27–50.

Cantley LC, Auger KR, Carpenter C, Duckworth B, Graziani A, Kapeller R, Soltoff S (1991): Oncogenes and signal transduction. Cell 64:281–302.

Carroll SB, Whyte JS (1989): The role of the *hairy* gene during *Drosophila* morphogenesis: Stripes in imaginal discs. Genes Dev 3:905–916.

Carthew R, Rubin G (1990): *seven-in-abstentia*, a gene required for specification of R7 cell fate in the *Drosophila* eye. Cell 63:561–577.

Chen MS, Obar RA, Schroeder CC, Austin TW, Poodry CA, Wadsworth SC, Vallee RB (1991): Multiple forms of dynamin are encoded by *shibire*, a *Drosophila* gene involved in endocytosis. Nature 351:583–586.

Cohen S, Fava RA (1985): Internalization of functional epidermal growth factor: Receptor/kinase complexes in A-431 cells. J Biol Chem 260:12351–12358.

Edgar BA, O'Farrell PH (1989): Genetic control of cell division patterns in the *Drosophila* embryo. Cell 57:177–187.

Fehon RG, Kooh PJ, Rebay I, Regan CL, Xu T, Muskavitch MAT, Artavanis-Tsakonis S (1990): Molecular interactions between the protein products of the neurogenic loci *Notch* and *Delta*, two EGF-homologous genes in *Drosophila*. Cell 61:523–534.

Felder S, Miller K, Moehren G, Ullrich A, Schlessinger J, Hopkins CR (1990): Kinase activity controls the sorting of the epidermal growth factor receptor within the multivesicular body. Cell 61:623–634.

Flanagan JG, Chan D, Leder P (1991): Transmembrane form of the *kit* ligand growth factor can be regulated by alternative spicing and is deleted in the Sl^d mutant. Cell 64: 1025–1035.

Fortini ME, Simon MA, Rubin GM (1992): Signaling by the sevenless protein tyrosine kinase is mimicked by Ras1 activation. Nature 355:559–561.

Freeman M, Klämbt C, Goodman CS, Rubin GM (1992): The *argos* gene encodes a diffusible factor that regulates cell fate decisions in the *Drosophila* eye. Cell 69:963–975.

Gaul U, Mardon G, Rubin GM (1992): A putative ras GTPase activating protein acts as a negative regulator of signaling by the sevenless receptor tyrosine kinase. Cell 68:1007–1019.

Gomer RH, Firtel RA (1987): Cell-autonomous determination of cell-type choice in *Dictyostelium* development by cell-cycle phase. Science 237:758–762.

Hafen E, Basler K, Edstroem JE, Rubin G (1987): *sevenless*, a cell-specific homeotic gene of *Drosophila*, encodes a putative transmembrane receptor with a tyrosine kinase domain. Science 236:55–63.

Harris WA, Stark WS, Walker JA (1976): Genetic dissection of the photoreceptor system in the compound eye of *Drosophila melanogaster*. J Physiol 256:415–439.

Hart A, Krämer H, Van Vactor D, Paidhungat M, Zipursky SL (1990): Induction of cell fate in the *Drosophila* retina: *bride of sevenless* is predicted to contain a large extracellular domain and seven transmembrane segments. Genes Dev 4:1835–1847.

Hart IK, Richardson WD, Heldin CH, Westermark B, Raff MC (1989): PDGF receptors on cells of the oligodendrocyte-type-2 astrocyte (O2-A) cell lineage. Development 105:595–603.

Hartenstein V, Campos-Ortega JA (1984): Early neurogenesis in wild type *Drosophila melanogaster*. Roux's Arch 194:213–216.
Heitzler P, Simpson P (1991): The choice of cell fate in the epidermis of *Drosophila*. Cell 64:1083–1092.
Jimenez J, Alphey L, Nurse P, Glover DM (1990): Complementation of fission yeast cdc2[ts] and cdc25[ts] mutants identifies two cell cycle genes from *Drosophila*: A cdc2 homologue and *string*. EMBO J 9:3565–3571.
Jones S, Vignais M, Broach J (1991): The *CDC25* protein of *Saccharomyces cerevisae* promotes exchange of guanine nucleotides bound to ras. Mol Cell Biol 11:2641–2646.
Kosaka T, Ikeda K (1983): Reversible blockage of membrane retrieval and endocytosis in the garland cell of the temperature-sensitive mutant of *Drosophila melanogaster, shibire*. J Cell Biol 97:499–507.
Krämer H, Cagan RL, Zipursky SL (1991): Interaction of bride of sevenless membrane-bound ligand and the sevenless tyrosine-kinase receptor. Nature 352:207–212.
Lawrence PA, Green SM (1979): Cell lineage in the developing retina of *Drosophila*. Dev Biol 71:142–152.
Masucci JD, Miltenberger RJ, Hoffmann FM (1990): Pattern-specific expression of the *Drosophila decapentaplegic* gene in imaginal disks is regulated by 3′ cis-regulatory elements. Genes Dev 4:2011–2023.
McConnell SK, Kaznowsky CE (1991): Cell cycle dependence of laminar determination in developing cerebral cortex. Science 254:282–285.
Mlodzik M, Hiromi Y, Weber U, Goodman C, Rubin G (1990): The *Drosophila seven-up* gene, a member of the steroid receptor gene superfamily, controls photoreceptor cell fates. Cell 60:211–224.
Moreno S, Nurse P, Russell P (1990): Regulation of mitosis by cyclic accumulation of p80[cdc25] mitotic inducer in fission yeast. Nature 344:549–552.
Moses HL, Yang EY, Pietenpol JA (1990): TGF-β stimulation and inhibition of cell proliferation: New mechanistic insights. Cell 63:245–247.
Mullins MC, Rubin GM (1991): Isolation of temperature-sensitive mutations of the tyrosine kinase receptor sevenless (sev) in *Drosophila* and their use in determining its time of action. Proc Natl Acad Sci USA 88:9387–9391.
Negishi K, Teranishi T, Kato S, Nakamura Y (1987): Paradoxical induction of dopaminergic cells following intravitreal injection of high doses of 6-hydroxydopamine in juvenile carp retina. Dev Brain Res 20:291–295.
Nurse P (1985): Cell cycle control genes in yeast. Trends Genet 1:51–55.
Pardee AB (1989): G_1 events and regulation of cell proliferation. Science 246:603–608.
Postlethwait JH (1978): Clonal analysis of *Drosophila* cuticular patterns. In Ahburner M, Wright TRF (eds): "The Genetics and Biology of Drosophila." New York: Academic Press, pp 359–441.
Ready DF (1989): A mutifaceted approach to neural development. Trends Neurosci 12:101–110.
Ready DF, Hanson TE, Benzer S (1976): Development of the *Drosophila* retina, a neurocrystalline lattice. Dev Biol 53:217–240.
Reh T (1992): Generation of neuronal diversity in the vertebrate retina. In Macagno E, Shankland M (eds): "Determinants of Neuronal Identity" pp 433–467.
Reinke R, Zipursky SL (1988): Cell-cell interaction in the *Drosophila* retina: The *bride of sevenless* gene is required in photoreceptor cell R8 for R7 cell development. Cell 55:321–330.
Roberts AB, Sporn MB (1990): The transforming growth factor-betas. In Sporn, Roberts (eds): "Peptide Growth Factors and Their Receptors." New York: Springer-Verlag, pp 419–472.
Rogge RD, Karlovich CA, Banerjee U (1991): Genetic dissection of a neurodevelopmental pathway: *Son of sevenless* functions downstream of the *sevenless* and EGF receptor tyrosine kinases. Cell 64:39–48.

Rogge R, Cagan RL, Majumdar A, Delaney T, Banerjee U (1992): Neuronal development in the *Drosophila* retina: The *sextra* gene defines an inhibitory component in the development pathway of the R7 photoreceptor cell. Proc Natl Acad Sci USA 89:5271–5275.

Rubin GM (1991): Signal transduction and the fate of the R7 photoreceptor in *Drosophila*. Trends Gen 7:372–377.

Singer SJ (1992): Intercellular communication and cell-cell adhesion. Science 255:1671–1677.

Simon MA, Bowtell D, Rubin G (1989): Structure and activity of the sevenless protein: A protein tyrosine kinase receptor required for photoreceptor development in *Drosophila*. Proc Natl Acad Sci USA 86:8333–8337.

Simon MA, Bowtell D, Dodson GS, Laverty TR, Rubin GM (1991): Ras1 and a putative guanine nucleotide exchange factor perform crucial steps in signaling by the sevenless protein tyrosine kinase. Cell 67:701–716.

Sonnenberg E, Gödecke A, Walter B, Bladt F, Birchmeier C (1991): Transient and locally restricted expression of the *ros1* protooncogene during mouse development. EMBO 10:3693–3702.

Spencer FA, Hoffman FM, Gelbart WM (1982): *Decapentaplegic*: A gene complex affecting morphogenesis in *Drosophila melanogaster*. Cell 28:451–461.

Sporn MB, Roberts AB, Wakefield LM, Assoian RK (1986): Transforming growth factor-β: Biological function and chemical structure. Science 233:532–534.

Tessarollo L, Nagarajan L, Parada L (1992): *c-ros*: The vertebrate homolog of the *sevenless* tyrosine kinase receptor is tightly regulated during organogenesis in mouse embryonic development. Development 115:11–20.

Tomlinson A (1985): The cellular dynamics of pattern formation in the eye of *Drosophila*. J Embryol Exp Morph 89:313–331.

Tomlinson A (1988): Cellular interactions in the developing *Drosophila* eye. Development 104:183–193.

Tomlinson A, Ready DF (1986): *sevenless*: A cell-specific homeotic mutation of the *Drosophila* eye. Science 231:400–402.

Tomlinson A, Ready DF (1987): Neuronal differentiation in the *Drosophila* ommatidium. Dev Biol 120:336–376.

Tomlinson A, Bowtell E, Hafen E, Rubin GM (1987): Localization of the *sevenless* protein, a putative receptor for positional information, in the eye imaginal disc of *Drosophila*. Cell 51:143–150.

Tomlinson A, Kimmel B, Rubin G (1988): *rough*, a *Drosophila* homeobox gene required in photoreceptors R2 and R5 for inductive interactions in the developing eye. Cell 55:771–784.

Trowbridge IS (1991): Endocytosis and signals for internalization. Curr Opin Cell Biol 3:634–641.

Truman JW, Bate M (1988): Spatial and temporal patterns of neurogenesis in the central nervous system of *Drosophila melanogaster*. Dev Biol 125:145–157.

Ullrich A, Schlessinger J (1990): Signal transduction by receptors with tyrosine kinase activity. Cell 61:203–212.

van der Bliek AM, Meyerowitz EM (1991): Dynamin-like protein encoded by the *Drosophila shibire* gene associated with vesicular traffic. Nature 351:411–414.

Van Vactor Jr DL, Cagan RL, Krämer H, Zipursky SL (1991): Induction in the developing compound eye of *Drosophila*: Multiple mechanisms restrict R7 induction to a single retinal precursor cell. Cell 67:1145–1155.

Watanabe T, Raff MC (1990): Rod photoreceptor development in vitro: Intrinsic properties of proliferating neuroepithelial cells change as development proceeds in the rat retina. Neuron 4:461–467.

Wen D, Peles E, Cupples R, Suggs SV, Bacus SS, Luo Y, Trail G, Hu S, Silbiger S, Levy RB, Koski RA, Lu HL, Yarden Y (1992): Neu differentiation factor: A transmembrane glycoprotein containing an EGF domain and an immunoglobulin homology unit. Cell 69:559–572.

Whitfield WGF, Gonzalez C, Maldonado-Codina G, Glover DM (1990): The A- and B-type cyclins of *Drosophila* are accumulated and destroyed in temporally distinct events that define separable phases of the G_2-M transition. EMBO J 9:2563–2572.

Wolff T, Ready DF (1991): The beginning of pattern formation in the *Drosophila* compound eye: The morphogenetic furrow and the second mitotic wave. Development 113:841–850.

Yarden Y, Ullrich A (1988): Growth factor receptor tyrosine kinases. Ann Rev Biochem 57:443–478.

Zak NB, Shilo BZ (1992): Localization of DER and the pattern of cell divisions in wild-type and *Ellipse* eye imaginal discs. Dev Biol 149:448–456.

9. Neurogenesis, Determination, and Migration During Cerebral Cortical Development

Susan K. McConnell, Christine E. Kaznowski,
Nancy A. O'Rourke, Michael E. Dailey, and Jennifer S. Roberts

Department of Biological Sciences, Stanford University, Stanford, California 94305

INTRODUCTION

The patterning of the mammalian central nervous system (CNS) is accomplished through a series of events that include the regionalization of the neural axis, neurogenesis and the production of specific neuronal phenotypes, the migration of young neurons into appropriate positions, and the elaboration of complex dendritic morphologies and formation of specific axonal connections. Attempts to understand the events underlying the assembly of the nervous system during development present significant challenges to the developmental biologist, both because of the enormous range of neuronal phenotypes generated in the CNS and because of the complexity of the cellular processes that guide migrating neurons and growing axons. The early development of the mammalian cerebral cortex presents a fascinating window through which our lab has viewed and explored neuronal determination and migration. The cerebral cortex is formed of layers of neurons, each of which contains neurons with characteristic morphologies, projection patterns, and common cell "birthdays." Here we describe studies in which the developmental potential of cortical progenitor cells have been examined through transplantation. These experiments indicate that early cortical precursors are multipotent: cells removed from host brains early in the cell cycle then transplanted into older hosts yield daughters that change their normal laminar fates. This suggests that factors present in the host environment can influence the development of multipotent progenitors. This multipotency is transient, however. By the time a young neuron undergoes its final mitosis, it has made a commitment to migrate to the layer typical of its birthday. We are also interested in the cellular mechanisms by which young cortical neurons are delivered to their final positions within the cortical layers, and here describe experiments in which we have imaged the movement of migrating cells in living slices of the developing cortex. These experiments have revealed diverse pathways by which young neurons find their way from the site at which they were generated out into the cortical plate.

CORTICAL ORGANIZATION

The adult cortex is essentially a sheet of neurons that covers the surfaces of the two cerebral hemispheres. This sheet is composed of six layers, which can be identified by the characteristic density and morphology of their constituent neurons. Within each of the layers, neurons generally have similar dendritic morphologies, physiological properties and axonal projection patterns (Gilbert, 1977; Gilbert and Kelly, 1975; Gilbert and Wiesel, 1979, 1985; LeVay et al., 1987; Lund, 1973; Lund and Boothe, 1975). As a simple rule of thumb, for primary sensory areas (such as the visual cortex), upper-layer cortical neurons extend long-distance axons to cortical targets, whereas neurons in the deep layers send their axons to subcortical targets (e.g., the thalamus or tectum). Neurons of the different layers sit one on top of another in a stack or cortical column, each of which contains the cellular machinery needed to process information derived from a small area of the sensory world (Hubel and Wiesel, 1962). Because cortical columns are reiterated over and over along the tangential extent of cortex, one can view the adult cortex as composed of a relatively small number of cell types that are each found in hundreds of thousands of copies.

Not only is the cortex organized into layers in the radial domain, but it is also separated in the tangential domain into dozens of functionally distinct areas. Each area processes information from a distinct modality (such as vision or audition), or subserves a particular function (e.g., generating movement). The particular axonal targets of neurons in different areas must reflect this specificity of function, even as they maintain characteristic laminar patterns. For example, neurons in cortical layer 6 of many areas project axons to thalamic targets, but layer 6 cells in the visual cortex specifically innervate the lateral geniculate nucleus, a visual region of the thalamus, whereas similar cells in auditory cortex innervate an auditory thalamic nucleus, the mediate geniculate (Fig. 1). Thus, one can view the laminar structure of the cortex as setting up a common motif for cortical organization, with each area formulating its own variation on this basic theme.

CORTICAL DEVELOPMENT

Cortical neurons are generated in a region called the ventricular zone, which lines the lateral ventricles of the developing neural tube. Proliferating precursor cells generate young neurons that permanently leave the cell cycle; the neurons then migrate through a cell-sparse intermediate zone into the cortical plate, which will form the adult cortical layers (Fig. 2). During their outward migration, neurons are found in close contact with radial glial cells which extend slender radial processes connecting the ventricular and pial surfaces of the brain. Evidence acquired from studies of neuron-glial relationships in vivo

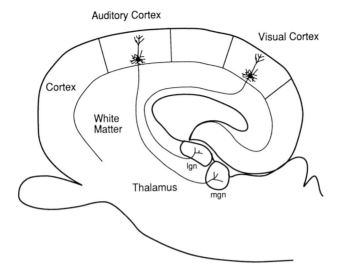

Fig. 1. A sagittal view of the mammalian cerebral cortex, which is composed of functionally distinct areas (such as visual cortex and auditory cortex). Layer-6 neurons throughout neocortex extend axons to the thalamus, but the specific target of these axons varies depending on the area of origin of the projection. Layer-6 neurons in the visual cortex project to the lateral geniculate nucleus (lgn); layer-6 neurons in the auditory cortex extend axons to the medial geniculate nucleus (mgn). (Reprinted from McConnell, 1992a, with permission of the publisher.)

and in vitro suggests that neurons employ the radial glial fibers as guidewires to direct their migration from the ventricle zone into the cortical plate (Rakic, 1971, 1972, 1990; Edmonson and Hatten, 1987; Hatten, 1990; Hatten and Mason, 1990). Cortical neurons are generated in a very orderly fashion: [^{3}H]-thymidine ''birthdating'' studies have shown that the first postmitotic neurons migrate to form the deepest cortical layers; neurons of the more superficial layers are born at subsequently later times and must migrate past older cells to form the upper layers (Angevine and Sidman, 1961; Rakic, 1974; Luskin and Shatz, 1985a). The earliest-generated neurons of the cortex constitute an exception to this ''inside-first, outside-last'' sequence: these earliest-generated cortical neurons migrate and condense into a single layer or ''preplate'' (Fig. 2). The formation of the cortical plate is signalled by the arrival of layer 6 neurons, which splits the preplate in two: above the cortical plate is the marginal zone (layer 1) and below is the subplate (future white matter), composed of a population of neurons that undergo a wave of cell death in late embryonic or early postnatal development (Luskin and Shatz, 1985b; Marin-Padilla, 1971; Shatz et al., 1991).

Over the course of cortical development, proliferating cells generate an enormous variety of neuronal phenotypes, producing cells with different axonal

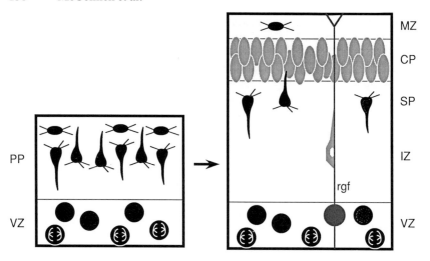

Fig. 2. **Left:** Cortical neurons are generated in the ventricular zone (VZ) which lines the lateral ventricle. The first neurons to leave the cell cycle migrate away from this region and form the preplate (PP), which sits directly under the pial surface of the brain. **Right:** Later in development, neurons destined for the cortical plate (CP) migrate out and divide the preplate into a marginal zone (MZ; future layer 1) and the transient subplate (SP). Radial glial fibers (rgf) are thought to provide a cellular substrate which underlies the movement of migrating neurons through the intermediate zone (IZ) and into the cortical plate (Hatten and Mason, 1990; Rakic, 1990). (Reprinted from McConnell, 1992b, with permission of the publisher.)

projection patterns, neurotransmitters, synaptic inputs, and dendritic morphologies. In general, two types of determinants contribute to the formulation of distinct cellular identities during development. Intrinsic cues can restrict or determine cell fates through the inheritance of specific cytoplasmic factors or patterns of gene expression. Alternatively, extrinsic environmental factors can influence cells whose fates are malleable; such factors include cell–cell interactions, diffusible morphogens, and local positional cues. To delineate the contributions of intrinsic and environmental determinants in the specification of distinct neuronal phenotypes, it is useful to have a way of predicting the normal fate of an undifferentiated cell. For example, in the nematode *Caenorhabditis elegans,* cell lineage reliably predicts the normal destiny of essentially every cell in the animal (Sulston and Horvitz, 1977; Sulston and White, 1980), and in the developing eye of *Drosophila,* the position of a cell developing in wild-type ommatidia accurately predicts its eventual fate (Banerjee and Zipursky, 1990; Zipursky, 1989). In both systems, this predictability of normal cell fates has enabled investigators to directly explore the contribution of cell–cell interactions to normal development. The developing cortex is too large and complex to contain individually identifiable neurons, but the strong correlation between a cell's birthday and its ultimate laminar fate has provided a powerful tool for predicting normal neuronal fates in the radial domain. In the tangential domain, the rules for predicting normal cell fates are less clear: it has been proposed that radial glial cells which provide a conduit for the outward migration of cortical neurons might restrict the tangential movement of cells, thereby creating a ''map'' of the ventricular zone onto the cortical plate (Rakic, 1989). Our work is focused on how the neural epithelium generates a diversity of neuronal phenotypes in the different layers and areas of the cortex, and on how the process of cell migration delivers young neurons into their final positions within the cortical plate.

DETERMINATION OF LAMINAR IDENTITY

We have been intrigued by the correlation between the birthday of a neuron and its ultimate laminar destination—could a cell's birthday in some way determine its fate? One possibility is that cell lineage mechanisms have preprogrammed cortical progenitor cells to first produce neurons of the deep layers, then middle-layer cells, and finally upper-layer neurons, as if some sort of developmental clock were ticking away. Alternatively, environmental signals that change progressively over the course of development could interact with multipotent cortical cells to specify their phenotypes. To distinguish between these possibilities, we have applied a classical test of cell commitment to developing cortical neurons: cells committed to their normal fates will develop in a cell-autonomous manner following transplantation into a novel environment (Stent, 1985). Since the birthday of a cortical neuron predicts its normal laminar destination, it is possible to ask whether cells generated on embryonic day (E) 29 in the ferret brain, cells that ought normally to migrate to layer 6, are committed to this fate by transplanting these cells into older hosts, in which neurons of cortical layers 2 and 3 are being generated (McConnell, 1985, 1988; McConnell and Kaznowski, 1991) (Fig. 3). To this end, embryonic cortical progenitor cells were labeled in vivo with [^{3}H]-thymidine (which is incorporated into DNA during S phase of the cell cycle) on E29. At a variety of time intervals after the thymidine labeling, labeled cells were removed, dissociated, and transplanted back into or near the ventricular zone of the host brain. Figure 4 summarizes the two possible outcomes of this experiment. If young neurons (or their progenitors) have made a commitment to produce layer 6 neurons at the time of transplantation, the labeled cells should migrate into layer 6, the laminar position typical of their birthday, and extend long-distance axons to the subcortical targets. If, on the other hand, multipotent cortical neurons or progenitors respond to local environmental cues that determine their fates, then labeled cells should migrate to the upper cortical layers along with host neurons, and form the cortical projections that are typical of that layer.

The results of these experiments have been intriguing, as they have revealed that the behavior of presumptive layer 6 neurons or their progenitors following

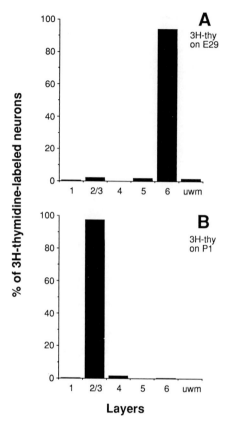

Fig. 3. Histograms showing the normal laminar destinations of ferret visual cortical neurons generated on (A) E29 or (B) P1. In both cases, animals were injected with [^{3}H]-thymidine on the day indicated and allowed to grow to adulthood, when the laminar positions of labeled cells were assessed by autoradiography. **A:** Neurons labeled with [^{3}H]-thymidine on E29 migrate into the deep cortical layers, primarily layer 6. At younger ages (not shown), labeled cells can also be found in the white matter underlying layer 6 (uwm), known during development as the subplate. These cells disappear during postnatal life in a wave of programmed cell death (Shatz et al., 1991). **B:** Neurons labeled with [^{3}H]-thymidine on P1 migrate into the upper cortical layers 2 and 3. (Figure 3B is reprinted from McConnell, 1988, with permission of the publisher.)

transplantation into older host brains depends on their position in the cell cycle at the time of transplantation (McConnell and Kaznowski, 1991) (see Fig. 5). In several developing neural systems, including the eye of *Drosophila* (Banerjee and Zipursky, 1990; Zipursky, 1989) and the vertebrate neural crest (Anderson, 1989; Bronner-Fraser and Fraser, 1989), the final position of a young neuron is a strong determinant of its phenotype—in other words, in these cases local cell–cell interactions play a major role in specifying neuronal phenotypes once cells have moved into place. Thus, we were interested to

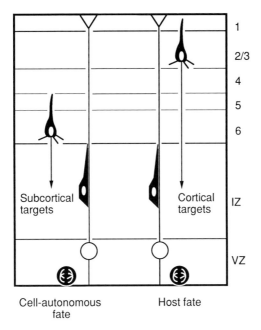

Fig. 4. Two possible outcomes of transplanting presumptive layer 6 neurons into an older host brain, in which upper-layer neurons are being generated. **Left:** Transplanted cells that are committed to their normal laminar fates should develop autonomously within the novel environment, migrating to layer 6 (the destination appropriate for their birthday) and forming subcortical projections. **Right:** If environmental factors determine laminar fate, transplanted neurons should adopt the host fate by migrating to the upper layers 2 and 3, and extending their axons to cortical targets. (Reprinted from McConnell and Kaznowski, 1991, with permission. © 1991 by the AAAS.)

investigate whether newly generated postmitotic neurons are committed to their normal laminar fates before they initiate their migration toward the cortical plate. To specifically explore the state of commitment of newly postmitotic neurons, we labeled cortical progenitors with [^{3}H]-thymidine in the S phase of the cell cycle. The donor cells were left in situ for 24 hours prior to transplantation, which gives the labeled precursors a chance to progress through mitosis after the thymidine injection. Neurons that become forever postmitotic after this division are heavily labeled with [^{3}H]-thymidine, and their behavior can thus be followed selectively after transplantation. Neurons transplanted after their final mitosis (but before initiating migration into the cortical plate) showed a commitment to their normal laminar fates: the vast majority of these cells that migrated into the host cortex were found in the deep cortical layers, the position appropriate for their birthday (McConnell and Kaznowski, 1991) (Fig. 6D). We furthermore found transplanted neurons in the deep layers that formed axonal projections to the thalamus, a normal target of layer 6 neurons

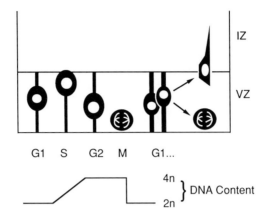

Fig. 5. Cell cycle progression in the embryonic ventricular zone. Cells in G_1 (diploid DNA contents) translocate their nuclei to the top of the ventricular zone upon entry into S phase and the initiation DNA replication. In G_2, after cells have doubled their DNA content, nuclei move to the ventricular surface, where the cells progress through mitosis (M) and DNA contents are halved back to diploid levels. The results of lineage tracing experiments suggest that the predominant mode of cell division in the ventricular zone is asymmetric: one daughter commonly reenters the cell cycle, while the other becomes forever postmitotic and migrates outward toward the cortical plate. Abbreviations: VZ, ventricular zone; IZ, intermediate zone. (Reprinted from McConnell and Kaznowski, 1991, with permission. © 1991 by the AAAS.)

(McConnell, 1988). Thus, in contrast to examples from the fly eye and neural crest, young cortical neurons are committed to their normal laminar fates before they begin to migrate away from the ventricular zone.

This finding raises the possibility that cortical cell fates might be determined through a cell-autonomous, lineage-based mechanism; alternatively, it

---

Fig. 6. Histograms showing the laminar positions of heavily labeled neurons that were labeled in S phase with [^{3}H]-thymidine on E29, allowed to develop for 0 to 24 hours in situ, then removed and transplanted into newborn host brains at four different times after thymidine labeling: (A) 0 hours; (B) 4 hours; (C) 8 hours; (D) 24 hours. Only transplanted neurons that migrated into the host cortex were included in these histograms. Abbreviations: uwm, white matter underlying layer 6 (the adult remnant of the embryonic subplate zone). A: Cells transplanted immediately after thymidine labeling are multipotent, since they have migrated to the upper cortical layers 2 and 3. B–D: Cells transplanted 4 to 24 hours after labeling have undergone a commitment to a deep-layer fate: the majority are in layer 6 or in the subplate (uwm). The fraction of neurons found in layer 6 versus the subplate is variable (B–D), but both of these layers constitute normal destinations of neurons generated on E29. The variability between experiments may be attributable to small variations in determining the gestational age of the donors (± 1 day), which could affect the fraction of subplate neurons generated. In addition, since most subplate neurons die during the first few months of postnatal life, differences in the time at which host animals were sacrificed may influence the fraction of subplate neurons recovered. (Reprinted from McConnell and Kaznowski, 1991, with permission. © 1991 by the AAAS.)

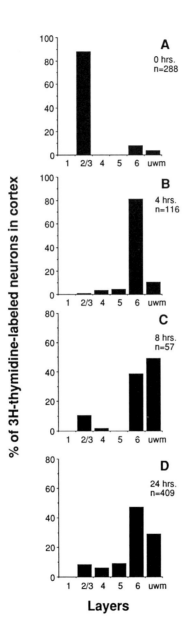

Figure 6.

could be that the environmental cues specify laminar fates, but that they act on cortical progenitor cells before mitosis. We distinguished between these two possibilities by transplanting progenitor cells into older host brains at different stages of the cell cycle, simply by varying the interval between [^{3}H]-thymidine-labeling and transplantation (McConnell and Kaznowski, 1991) (Fig. 7). When deep-layer progenitor cells are transplanted during S phase, they are multipotent: their daughters born in an older environment alter their normal fates and migrate to the upper cortical layers, as do newly generated host neurons (Fig. 6A). Thus environmental cues can exert a determinative effect on the specification of laminar phenotypes, since the fates of early progenitors can be altered if the cells are transplanted early in the cell cycle. The multipotency of progenitors, however, is lost as the cell progresses through the cell cycle: precursors transplanted late in the cell cycle produced daughters that migrated specifically to layer 6, the layer appropriate for their birthday (Fig. 6B,C). Thus laminar commitment occurs prior to the mitotic division that generates the postmitotic cortical neuron.

These experiments suggest that the impact of environmental cues on progenitor cells depends on the position of progenitors in the cell cycle. Precursor cells in mid-S phase are multipotent: exposing these cells to novel environmental signals can trigger a change in the laminar fate of daughters produced in the subsequent cell division. Cell fates become restricted, however, late in the cell cycle (roughly at the time of the transition from S phase into G2). At this point precursors apparently undergo a commitment is to generate a neuron destined for a specific cortical layer. Our results indicate that the layered patterning of cerebral cortical neurons arises through a progressive specification of cell types, presumably accomplished through changes in the nature of environmental cues over time in development. It should be noted, however, that this does not rule out intrinsic changes in cortical progenitors over time, since we have thus far only examined directly the developmental potential of early precursor cells, which are normally fated to produce cells of multiple cortical layers (Walsh and Cepko, 1988; Luskin et al., 1988; Price and Thurlow, 1988; Austin and Cepko, 1990). It remains possible that late progenitors, which normally produce only upper-layer neurons, may have a more restricted developmental potential. This can be tested by transplanting late progenitors into younger brains (G.D. Frantz and S.K. McConnell, work in progress).

What are the environmental determinants that can act to alter normal laminar fates? At this point, nearly anything is possible. One particularly interesting hypothesis (formulated in studies of retinal development) is that the differentiation of neurons generated early in development regulates the production of environmental cues determining laminar phenotype (Reh and Tully, 1986; Anderson, 1989). According to this model, differentiating layer 6 neurons might somehow signal to progenitor cells that they should cease further

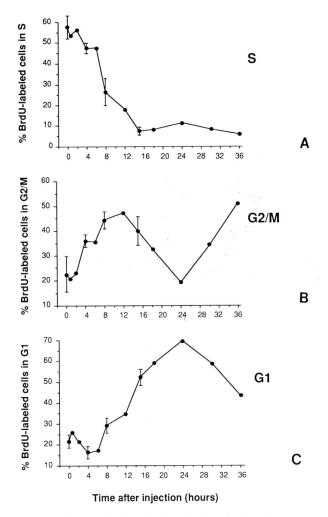

Fig. 7. Progression of E29 cortical ventricular cells through the cell cycle. Shown here are the fraction of 5-bromo-2-deoxyuridine-labeled (BrDU-labeled) cells in three distinct stages of the cell cycle (S phase, G_2/M, and G_1) as determined by their DNA contents, which were measured at a variety of times after the cells were labeled with BrDU. DNA contents were determined by flow cytometry of BrDU-labeled cells, as described in McConnell and Kaznowski (1991). **A:** The fraction of labeled cells in S phase is maximal just after animals were injected with BrDU, which is incorporated into DNA only during replication. This fraction falls with time, as labeled cells enter phase G_2 of the cell cycle. **B:** The fraction of BrDU-labeled cells with G_2/M DNA contents rises steadily over the first 8 hours after labeling, then levels off and falls as labeled cells progress through mitosis and regain diploid DNA levels. **C:** The fraction of labeled cells with G_1 DNA contents begins to rise at aobut 8 hours after injection. Note that essentially no BrDU-labeled cells have completed mitosis and entered the G_1 phase at the 4-hour time point, a time at which cells have undergone a commitment to their normal deep-layer fates in the transplantation assay (Fig. 6). (Reprinted from McConnell and Kaznowski, 1991, with permission. © 1991 by the AAAS.)

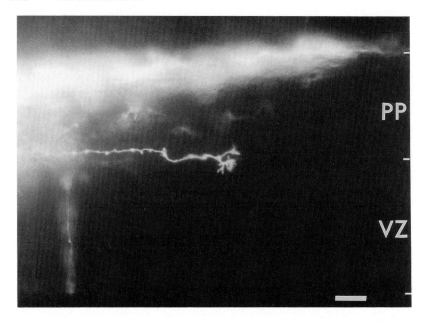

Fig. 8. A DiI-labeled growth cone from the developing cortex of an E24 ferret. The growth cone is located within the nascent intermediate zone, just above the surface of the ventricular zone (VZ), as are most growth cones at this age. Abbreviations: PP, preplate. Scale bar: 25 μm. (Reprinted from Kim et al., 1991, with permission of the publisher.)

production of layer 6 and begin production of the next cortical layer. Currently we have no direct evidence to either support or refute this model in the cortex; however, differentiating cortical neurons may well have the opportunity to communicate directly with progenitor cells. Figure 8 shows that growth cones of neurons in the preplate (and later, in the cortical plate) extend along the top of the ventricular zone, directly above S-phase precursor cells that are in the process of deciding the fates of their daughters (Kim et al., 1991). At this point we do not know whether growth cones provide instructive signals to neuronal progenitors, nor indeed do we know whether there might be a default pathway for differentiation in the absence of active environmental directives.

NEURONAL MIGRATION AND THE DEVELOPMENT OF NEOCORTICAL AREAS

While the determination of neuronal fates in the laminar domain occurs quite early, patterns of cell morphologies and projections vary also across the tangential extent of the cortex. This raises the question of how area-specific phenotypes of neurons are determined during development. Are neurons committed to forming the projections typical of different cortical areas while still

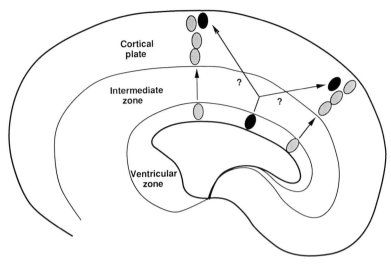

Fig. 9. Two possible patterns of neuronal migration in the developing neocortex. If the clonally related progeny of a single precursor cell in the ventricular zone migrate strictly radially into the developing cortical plate (*stippled cells*), they would maintain a point-to-point map between the ventricular and pial surfaces of the brain. Alternatively, nonradial or tangential modes of migration in any region of the cerebral wall could result in the scattering of clonally related neurons in different cortical areas (*solid cells*). (Reprinted from McConnell, 1992a, with permission of the publisher.)

in the ventricular zone, or are area-specific neuronal identities specified later in cortical development? To address this question, one would like to be able to accurately predict the normal area-specific fate of a premigratory cortical neuron. One hypothesis posits that the cerebral ventricular zone contains a "protomap" of the future cortical areas, and that young neurons migrate radially out to the cortex in a manner which preserves a precise point-to-point mapping between the ventricular and cortical surfaces (Rakic, 1989) (Fig. 9). Such a model would allow (and indeed suggest) that ventricular cells are precommitted to area-specific identities prior to neuronal migration. An alternative hypothesis is that the differences among cortical areas emerge late in development through epigenetic interactions between cortical neurons and (for example) their thalamic afferents. This hypothesis need not depend on a strictly radial migration of cortical neurons (Fig. 9), since a point-to-point mapping of the ventricular surface onto the cortex would not be essential for the specification of differences among cortical areas. Three types of experiments have examined directly the migratory pathways and developmental potential of cortical cells in different areas. First, lineage studies have employed recombinant retroviruses to mark the positions of clonally-related neurons derived from common precursors. Second, we have performed time-lapse imaging studies to visualize the migration of cells in living slices of cerebral cortex. Finally, small pieces of cortex have been transplanted to explore their ability to develop connections and cytoarchitectonic features characteristic of other areas. Evidence from each line of work suggests that young neurons may acquire their area-specific fates only after they have migrated into the cortical plate.

Cell lineage experiments are now feasible in mammalian embryos through the use of replication-incompetent retrovirus vectors that infect precursor cells, integrate stably into the host genome, and are faithfully inherited by the progeny of each infected cell (Sanes et al., 1986; Turner and Cepko, 1987). In the retina and tectum, both of which are layered structures like the cortex, clonally related cells are bunched together in tight radial alignment (Turner and Cepko, 1987; Holt et al., 1988; Wetts and Fraser, 1988; Gray et al., 1990). Surprisingly, clonal boundaries in the cortex have proven much more difficult to ascertain. Although some radially grouped arrays of labeled cells are seen, there are also large numbers of scattered labeled cells (Luskin et al., 1988; Walsh and Cepko, 1988; Austin and Cepko, 1990). These single cells could either be one-cell clones or widely dispersed members of single clones. Walsh and Cepko (1992) have distinguished between these two possibilities by making a library of a hundred different retroviruses, each containing a marker gene encoding β-galactosidase as well as a differently sized insert flanked by common nucleotide sequences; they then infected the developing rat telencephalon with the entire library and used the polymerase chain reaction to genotype individual labeled cells. If the number of clones in a brain is fairly small, statistically it is extremely unlikely that two or more cells containing an identical insert are derived from separate, independent viral infections. The investigators found that roughly half of cortical clones are spread tangentially over wide expanses of cortex, with clonally related cells ending up in completely different cortical areas (Walsh and Cepko, 1992). This evidence suggests either that there is significant tangential migration of progenitor cells within the ventricular zone, or that postmitotic cortical neurons can migrate tangentially during their journey toward the cortical plate, or both. Furthermore, these findings might imply that cortical neurons retain the developmental potential to acquire fates appropriate for the area to which they migrate.

To directly visualize the migratory pathways of young cortical neurons, we have recently begun to study modes of migration in organotypic cultured cortical slices, which support substantial amounts of cell migration (Roberts et al., 1992) and are accessible for time-lapse imaging studies. We have placed slices of neonatal ferret cortex in slice culture (Gähwiler, 1981, 1988) and labeled ventricular cells with the lipophilic fluorescent tracer DiI. As the labeled cells migrated into the intermediate zone, they were imaged over the course of up to 45 hours using time-lapse laser scanning confocal microscopy (O'Rourke et al., 1991, 1992a). Migrating cells exhibited morphological features characteristic of migrating neurons, as described both in fixed tissue through the electron microscope (Rakic, 1971, 1972, 1990), and in living neuron-glial cocultures (Edmonson and Hatten, 1987; Gregory et al., 1988; Hatten and Mason, 1990). Consistent with the prevalent notion that neurons migrate along radial glial fibers out to the cortical plate, the majority (82%) of imaged cells moved in a direction that was at or near radial. When some of the same slices were subsequently immunostained with antibodies to vimentin to reveal radial glial fibers, cells that had been observed to migrate could be seen in close apposition to radial glia, again consistent with glial-guided migration. About 13% of the cells that were imaged, however, migrated with a twist: These cells were found moving along the tangential plane, perpendicular to the radial fibers. A few cells were even observed making abrupt right-angle turns from a radial pathway onto a tangential pathway, or vice versa. We have not yet identified the substrate that supports this tangential cell migration. One possibility is that some radial glial fibers are bent or twisted in the intermediate zone and provide a tangentially directed glial substrate (Gadisseux et al., 1989; Takahashi et al., 1990). It is worth noting, however, that we have imaged cells in the dorsal cortex, where the vast majority of radial glial fibers extend in a relatively straight trajectory from the ventricular to the pial surface (O'Rourke et al., 1992a,b). Another possibility that remains to be addressed is that cells may migrate along axons populating the intermediate zone. This mechanism seems plausible in light of the "neurophilic" modes of cell migration found in several other brain regions (Gray et al., 1990; Rakic, 1990). The orthogonal migration of cells with neuronal morphologies could underlie some or all of the tangential dispersion seen in retroviral lineage tracing studies. It now seems plausible that neuronal migration is not strictly radial, and that once a young postmitotic neuron leaves the ventricular zone on its way out toward the cortical plate, it can follow a fairly complex migratory pathway that departs from a simple straight trajectory. These results are consistent with the interpretation that area-specific fates are determined relatively late in development, after a young neurons has migrated into its final position.

Several studies have assessed directly the developmental potential of neocortical neurons by challenging these cells to change their area-specific fates. To see whether the development of axonal connections typical of different cortical areas is regulated by cell-autonomous cues or through epigenetic interactions, O'Leary and Stanfield began by transplanting small pieces of occipital cortex from embryonic rats into more rostral cortical areas of newborn hosts (Stanfield and O'Leary, 1985; O'Leary, 1989; O'Leary and Stanfield, 1989). They found that the transplanted cortical neurons extended and maintained long-distance axonal projections that were appropriate for the cells' new tangential position rather than of their origin. More recent experiments have revealed that the visual cortex can also develop cytoarchitectonic features typical of other cortical locales: specifically, visual cortex grafted into the somatosensory cortical region develops barrel fields typical of the whisker projections in the rat (Schlagger and O'Leary, 1991). These experiments directly demonstrate the developmental plasticity of the fetal neocortex. Even after the majority of neurons destined for the lower cortical layers have been generated, cortical neurons are capable of altering their area-specific phenotypes following transplantation into novel locations. Interestingly, neocortical regions are not equipotential with phylogenetically older regions of the cortex, such as the limbic cortex. If neocortex is transplanted into a mesocortical region during the middle of neurogenesis, or vice versa, the transplanted region continues to express molecular markers characteristic of its region of origin, not of its new position (Barbe and Levitt, 1991). These results suggest that distinctions between major subdivisions of the forebrain (neocortex, mesocortex, archicortex) may be specified at relatively early times in development.

The nature of the factors that sculpt the cytoarchitectonic features and patterns of connectivity typical of different neocortical regions remains unclear. It is tempting to speculate that afferent inputs from the thalamus somehow stamp a characteristic signature onto different cortical regions by inducing, for example, the formation of barrel fields. It is not clear, however, how ingrowing afferents "know" which region of the cortical plate they ought to invade before overt regional differences have emerged. Clearly, afferents are able to make fine distinctions between potential cortical targets. In the developing primate brain, axons from the lateral geniculate nucleus recognize and innervate visual cortical area V1, an appropriate target, whereas they do not invade the immediately adjacent area V2 (Rakic, 1989). It seems likely that afferents and cortical neurons are capable of exchanging information and influencing one another in a progressive manner over the course of development. One set of players that might mediate these interactions are the subplate neurons that lie below the cortical plate (Shatz et al., 1990). Ingrowing thalamocortical axons accumulate in the region of subplate underlying their ultimate cortical targets and here form transient synaptic connections with subplate neurons (Friauf et al., 1990). When subplate neurons were selectively ablated with a neurotoxin during this time, thalamocortical axons failed to recognize and innervate their appropriate cortical target; instead the axons grew past the normal target and into the white matter beyond (Ghosh et al., 1990). Subplate neurons thus appear well situated to play an important (though as yet poorly understood) role in regulating the earliest interactions between ingrowing thalamic axons and their ultimate cortical targets.

CONCLUSIONS

The results discussed here suggest that the laminar and area-specific fates of neocortical neurons are specified through a series of interactions between cells and their environment. Transplantation experiments reveal that early progenitors of cortical neurons are multipotent and are capable, given appropriate environmental cues, of producing neurons destined for a variety of cortical layers. These experiments also show that the commitment of a cell to its normal laminar fate is made just prior to the mitotic division that generates a young neuron. Somehow this commitment endows the new postmitotic cell with the information it needs to home in on the appropriate cortical layer, and ultimately to form axonal projections to the targets of that layer. In the tangential domain, it seems that there are few early restrictions on area-specific fates. Both lineage experiments and direct observations of migrating cortical cells suggest that the migration of cortical neurons employs a diversity of pathways, and that nonradial modes of cell movement may distribute clonally related neurons into quite disparate cortical areas. Through these studies one can infer the broad developmental potential of neurons to acquire a variety of area-specific identities. Transplantation experiments have revealed directly the malleability of both the axonal connections and cytoarchitectonic features typical of neurons in different neocortical areas. These studies collectively suggest that interactions between neurons and local environmental cues play crucial instructive roles in the determination of cell fates in the developing cerebral cortex.

ACKNOWLEDGMENTS

Much of the work described here was supported by grants from NIH (EY08411), Pew Scholars, Searle Scholars/The Chicago Community Trust, an NSF Presidential Young Investigator Award, a Sloan Research Fellowship, and a Clare Boothe Luce Professorship.

REFERENCES

Anderson DJ (1989): The neural crest cell lineage problem: neuropoiesis? Neuron 3:1–12.
Angevine JB Jr, Sidman RL (1961): Autoradiographic study of cell migration during histogenesis of cerebral cortex in the mouse. Nature 192:766–768.
Austin CP, Cepko CL (1990): Cellular migration patterns in the developing mouse cerebral cortex. Development 110:713–732.
Banerjee U, Zipursky SL (1990): The role of cell-cell interaction in the development of the *Drosophila* visual system. Neuron 4:177–187.
Barbe MF, Levitt P (1991): The early commitment of fetal neurons to limbic cortex. J Neurosci 11:519–533.
Bronner-Fraser ME, Fraser S (1989): Developmental potential of avian trunk neural crest cell in situ. Neuron 3:755–766.
Edmonson JC, Hatten ME (1987): Glial-guided granule neuron migration in vitro: A high resolution time-lapse video microscopic study. J Neurosci 7:1928–1934.
Friauf E, McConnell SK, Shatz CJ (1990): Functional circuits in the subplate during fetal and early postnatal development of cat visual cortex. J Neurosci 10:2601–2613.

Gadisseux JF, Evrard P, Misson JP, Caviness VS (1989): Dynamic structure of the radial glial fiber system of the developing murine cerebral wall. An immunocytochemical analysis. Dev Brain Res. 50:55–67.

Gähwiler BH (1981): Organotypic monolayer cultures of nervous tissue. J Neurosci Meth 4:329–342.

Gähwiler BH (1988): Organotypic cultures of neural tissue. Trends Neurosci 11:484–489.

Ghosh A, Antonini A, McConnell SK, Shatz CJ (1990): Requirement for subplate neurons in the formation of thalamocortical connections. Nature 347:179–181.

Gilbert CD (1977): Laminar differences in receptive field properties of cells in cat primary visual cortex. J Physiol (Lond) 268:391–421.

Gilbert CD, Kelly JP (1975): The projections of cells in different layers of the cat's visual cortex. J Comp Neurol 163:81–106.

Gilbert CD, Wisel TN (1979): Morphology and intracortical projections of functionally characterized neurones in the cat visual cortex. Nature 280:120–125.

Gilbert CD, Wiesel TN (1985): Intrinsic connectivity and receptive field properties in visual cortex. Vision Res 25:365–374.

Gray GE, Leber SM, Sanes JR (1990): Migratory patterns of clonally related cells in the developing nervous system. Experientia 46:929–940.

Gregory WA, Edmondson JC, Hatten ME, Mason CA (1988): Cytology and neuron-glia apposition of migrating cerebellar granules cells in vitro. J Neurosci 8:1728–1738.

Hatten ME (1990): Riding the glial monorail: A common mechanism for glial-guided neuronal migration in different regions of the developing mammalian brain. Trends Neurosci 13:179–184.

Hatten ME, Mason CA (1990): Mechanisms of glial-guided neuronal migration in vivo and in vitro. Experientia 46:907–916.

Holt CE, Bertsch TW, Ellis HM, Harris WA (1988): Cellular determination in the *Xenopus* retina is independent of lineage and birthdate. Neuron 1:15–26.

Hubel DH, Wiesel TN (1962): Receptive fields, binocular interaction and functional architecture in the cat's visual cortex. J Physiol (Lond) 160:106–154.

Kim GJ, Shatz CJ, McConnell SK (1991): Morphology of pioneer and follower growth cones in the developing cerebral cortex. J Neurobiol 22:629–642.

LeVay S, McConnell SK, Luskin MB (1987): Functional organization of primary visual cortex in the mink (*Mustela vison*), and a comparison with the cat. J Comp Neurol 257:422–441.

Lund JS (1973): Organization of neurons in the visual cortex, area 17, of the monkey (*Macaca mulatta*). J Comp Neurol 147:455–496.

Lund JS, Boothe RG (1975): Interlaminar connections and pyramidal neuron organisation in the visual cortex, area 17, of the macaque monkey. J Comp Neurol 159:305–334.

Luskin MB, Pearlman AL, Sanes JR (1988): Cell lineage in the cerebral cortex of the mouse studied in vivo and in vitro with a recombinant retrovirus. Neuron 1:635–647.

Luskin MB, Shatz CJ (1985a): Neurogenesis of the cat's primary visual cortex. J Comp Neurol 242:611–631.

Luskin MB, Shatz CJ (1985b): Studies of the earliest generated cells of the cat's visual cortex: Cogeneration of subplate and marginal zones. J Neurosci 5:1062–1075.

Marin-Padilla M (1971): Early prenatal ontogenesis of the cerebral cortex (neocortex) of the cat (*Felis domestica*). A Golgi study, I. The primordial neocortical organization. Z Anat Entwicklungsgesch 134:117–145.

McConnell SK (1985): Migration and differentiation of cerebral cortical neurons after transplantation into the brains of ferrets. Science 229:1268–1271.

McConnell SK (1988): Fates of visual cortical neurons in the ferret after isochronic and heterochronic transplantation. J Neurosci 8:945–974.

McConnell SK (1992a): The control of neuronal identity in the developing cerebral cortex. Curr Opin Neurobiol 2:23–27.

McConnell SK (1992b): The genesis of neuronal diversity during development of cerebral cortex. Sem Neurosci 4:347–356.
McConnell SK, Kaznowski CE (1991): Cell cycle dependence of laminar determination in developing cerebral cortex. Science 254:282–285.
O'Leary DDM (1989): Do cortical areas emerge from a protocortex? Trends Neurosci 12:400–406.
O'Leary DDM, Stanfield BB (1989): Selective elimination of axons extended by developing cortical neurons is dependent on regional locale: Experiments utilizing fetal cortical transplants. J Neurosci 9:2230–2246.
O'Rourke NA, Daily ME, Smith SJ, McConnell SK (1991): Time-lapse imaging of migrating cells in cortical slices. Soc Neurosci Abstr 17:523.
O'Rourke NA, Daily ME, Smith SJ, McConnell SK (1992a): Diverse migratory pathways in the developing cerebral cortex. Science 258:299–302.
O'Rourke NA, Kaznowski CE, McConnell SK (1992b): Substrates for diverse migratory pathways in the developing cortex. Soc Neurosci Abstr 18:1283.
Price J, Thurlow L (1988): Cell lineage in the rat cerebral cortex. A study using retroviral-mediated gene transfer. Development 104:473–482.
Rakic P (1971): Guidance of neurons migrating to the fetal monkey neocortex. Brain Res 33:471–476.
Rakic P (1972): Mode of cell migration to the superficial layers of fetal monkey neocortex. J Comp Neurol 145:61–84.
Rakic P (1974): Neurons in the rhesus monkey visual cortex: Systemic relationship between time of origin and eventual disposition. Science 183:425–427.
Rakic P (1989): Specification of cerebral cortical areas. Science 241:170–176.
Rakic P (1990): Principles of neural cell migration. Experientia 46:882–891.
Reh TA, Tully T (1986): Regulation of tyrosine-hydroxylase-containing amacrine cell number in the larval frog retina. Dev Biol 114:463–469.
Roberts JS, O'Rourke NA, McConnell SK (1992): Cell migration in cultured cerebral cortical slices Dev Biol 155:396–408.
Sanes JR, Rubenstein JLR, Nicolas J-F (1986): Use of a recombinant retrovirus to study postimplantation cell lineage in mouse embryos. EMBO J 5:3133–3142.
Schlagger BL, O'Leary DDM (1991): Potential of visual cortex to develop an array of functional units unique to somatosensory cortex. Science 252:1556–1560.
Shatz CJ, Ghosh A, McConnell SK, Allendoerfer KL, Friauf E, Antonini A (1990): Pioneer neurons and target selection in cerebral cortical development. Cold Spring Harb Symp Quant Biol 55:469–480.
Shatz CJ, Ghosh A, McConnell SK, Allendoerfer KL, Friauf E, Antonini A (1991): Subplate neurons and the development of neocortical connections. In Lam DM-L, Shatz CJ (eds): "Development of the Visual System." Cambridge, MA: MIT Press, pp 175–196.
Stanfield BB, O'Leary DDM (1985): Fetal occipital cortical neurones transplanted to the rostral cortex can extend and maintain a pyramidal tract axon. Nature 313:135–137.
Sulston JE, Horvitz HR (1977): Postembryonic cell lineages of the nematode *Caenorhabditis elegans*. Dev Biol 56:110–156.
Sulston JE, White JG (1980): Regulation and cell autonomy during postembryonic development of *Caenorhabditis elegans*. Dev Biol 78:577–597.
Takahashi T, Misson JP, Caviness VSJ (1990): Glial process elongation and branching in the developing murine neocortex: A qualitative and quantitative immunohistochemical analysis. J Comp Neurol 302:15–28.
Turner DL, Cepko CL (1987): Cell lineage in the rat retina: A common progenitor for neurons and glia persists late in development. Nature 328:131–136.
Walsh C, Cepko CL (1988): Clonally related cortical cells show several migration patterns. Science 241:1342–1345.

Walsh C, Cepko CL (1992): Widespread dispersion of neuronal clones across functional regions of the cerebral cortex. Science 255:434–440.

Wetts R, Fraser SE (1988): Multipotent precursors can give rise to all major cell types of the frog retina. Science 239:1142–1145.

Zipursky SL (1989): Molecular and genetic analysis of *Drosophila* eye development: *sevenless, bride of sevenless,* and *rough*. Trends Neurosci 12:183–189.

10. Myogenic Factor Gene Expression in Mouse Somites and Limb Buds

Gary E. Lyons and Margaret E. Buckingham

Department of Anatomy, University of Wisconsin Medical School, Madison, Wisconsin 53706; and Department of Molecular Biology, Pasteur Institute, 75724 Paris, France

INTRODUCTION

The question of what regulates determination and differentiation of skeletal muscle cells in vertebrate embryos from a molecular standpoint has received a great deal of attention recently as a result of the identification of a family of genes that are highly conserved during evolution. The MyoD family of myogenic factors, named after the first member of the family to be described (Davis et al., 1987), is expressed in skeletal myocytes from the onset of somite formation from the segmental plate. In this review, we will focus on the temporal and spatial patterns of expression of the four known myogenic factors in developing mouse somites and limb buds, as a first step in understanding what role(s) these factors may play in skeletal muscle formation.

THE MYOD FAMILY OF MYOGENIC REGULATORS

The first two myogenic regulatory factors to be identified, MyoD (Davis et al., 1987) and myogenin (Wright et al., 1989), were cloned by subtractive hybridization of cDNA libraries made from mammalian muscle cell lines. Two other members of this multigene family have been identified in mammals by screening muscle cDNA librariers at low stringency, using MyoD as a probe. These are myf-5 (Braun et al., 1989) and MRF4 (Rhodes and Konieczny, 1989), also known as herculin (Miner and Wold, 1990) and myf-6 (Braun et al., 1990). These genes encode transcription factors which share certain structural and functional properties. Upon transfection of its cDNA into mesodermal stem cells in vitro, each factor has the ability to convert these cells into the myogenic lineage at a high frequency. This was first demonstrated for the C3H10T1/2 cell line, but it is also true for most mesodermal derivatives, and at least to some extent, for other cell lineages (Weintraub et al., 1989; Choi et al., 1990). On this basis, MyoD was originally described as a myogenic determination factor (Davis et al., 1987).

These proteins contain an evolutionarily conserved core consisting of a basic region, which is involved in binding to DNA, and a helix-loop-helix (HLH) domain, which is important for dimerization with other HLH proteins in the nucleus (reviewed in Olson, 1990; Weintraub et al., 1991; Wright, 1992). In vitro assays have shown that formation of heterodimers with proteins such as E12 (Murre et al., 1989a,b), which is found in all cells, is necessary for high-affinity binding to DNA in a site-specific fashion. A consensus sequence for the binding of the myogenic factor heterodimers called the E box, CANNTG, usually occurs in tandem in muscle-specific enhancers which regulate the high-level expression of muscle structural protein genes. This motif is also present in the promoters of many muscle-specific genes.

When a myogenic factor is transfected into tissue culture cells, it appears to induce the myogenic program by activation of its endogenous gene, as well as by upregulation of other endogenous factor genes (see Olson, 1990; Weintraub et al., 1991; Wright, 1992 for review). In cell lines such as CV1, in which transfection of a myogenic factor cDNA does not activate the endogenous program of myogenesis, cotransfection of a factor cDNA with a plasmid construct containing a muscle-specific promoter linked to a reporter gene upregulates expression of the reporter gene. These structural and functional properties shared by MyoD and other members of this group of muscle regulatory factors, suggest that these proteins regulate myogenesis, at least in part, by binding to muscle-specific promoters and enhancers.

These results from experiments with cells grown in vitro suggested that the myogenic factors could be involved with the commitment of stem cells to the myogenic lineage in the developing embryo. We review here a series of in situ hybridization studies which examined the pattern of expression of each of the four members of this multigene family in the developing mouse embryo.

EMBRYONIC ORIGINS OF SKELETAL MUSCLE

In order to provide a context for our discussion or the spatial and temporal patterns of expression of the four different myogenic factor genes during embryonic muscle development, we will summarize the morphological events which occur during skeletal myogenesis. The paraxial mesoderm on both sides of the developing neural tube segments to form epithelial-like balls of cells, called somites, from about 8 days post coitum (p.c.) in a rostrocaudal gradient over a period of about a week (Rugh, 1990; Theiler, 1989). The morphology of the developing somite is grossly similar to that of the chick (Ostrovsky et al., 1988), although the cells of the mouse are much larger and their numbers correspondingly smaller.

The somite is the only known source of skeletal muscle cells in the mouse embryo. Based on chick–quail somite transplantation experiments (Ordahl and LeDouarin, 1992), two apparently distinct phases of skeletal muscle formation occur from separate populations of cells within the somite. The myotome, the first morphologically distinguishable skeletal muscle to form, consists of cells which derive exclusively from the medial half of the somite at the craniomedial corner of the dermatome adjacent to the neural tube (Ede and El-Gadi, 1986; Kaehn et al., 1988). Myotomal cells do not migrate, but differentiate in situ to form the vertebral and the intercostal muscles. Other muscle masses are formed, as in the chick (reviewed in Wachtler and Christ, 1992), from cells which migrate from the lateroventral edge of the dermamyotome (Milare, 1976). Muscles of the head, limbs, and body wall are derived from premyoblast cells which migrate from the lateral half of the somite.

DEVELOPMENTAL EXPRESSION OF MYOGENIC FACTORS IN SOMITES

The first myogenic factor gene transcript detectable by in situ hybridization on sections of mouse embryos is myf-5 (Ott et al., 1991). These mRNAs are located over the dorsomedial corners of newly formed somites (see Fig. 1) which begin to segment around 8 days p.c. (Rugh, 1990; Theiler, 1989). Myf-5 expression is not detected in the segmental plate prior to somite formation. As somites differentiate into dermamyotome and sclerotome, myf-5 transcripts continue to be expressed at a high level in the dorsomedial lip region of the dermamyotome, in cells which are not yet morphologically distinguishable as myocytes. Myotomes begin to form in rostral somites of 8.5-day p.c. mouse embryos. Myogenin mRNAs are first detected in these myocytes (Sassoon et al., 1989) as they develop on the medial surface of the dermatome. As myotomes enlarge and mature myf-5 transcripts decrease in the dorsomedial lip of the dermamyotome and become restricted to myotomal cells (Fig. 1).

Beginning at 9 days p.c., a third member of this family, MRF4, is expressed transiently in differentiating somites. MRF4 transcripts are detected in myo-

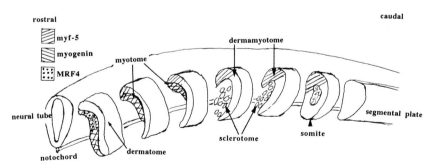

Fig. 1. Schematic representation of the spatial patterns of myogenic factor gene expression in differentiating somites. The developmental sequence of the first three factors detected in somites is presented. MyoD would first appear in a myotome rostral to the one at the far left.

tomes between 9 and 10.5 days p.c. (Bober et al., 1991; Hinterberger, 1991). These mRNAs appear and are downregulated below the level of detection of in situ hybridization in a rostrocaudal gradient. As we will discuss below, this transient expression of MRF4 is unique to muscle cells in the myotomal compartment. It is not detected in muscle cells differentiating in limb buds and other peripheral locations.

The appearance of MRF4 in myotomes may be due to the requirement for two myogenic factors to be present in myocytes for the process of skeletal muscle differentiation. Cusella-De Angelis et al. (1992) have described posttranscriptional regulation of myogenin expression in myotomes. These authors found that although myogenin mRNAs are detected early in myotome formation, the protein is not detected by antibody staining until 2 days later, 10.5 days p.c. If two myogenic regulatory factors are necessary to upregulate the battery of muscle structural genes first detected between 8.5 and 10.5 days p.c., MRF4 would play that role with myf-5 in the absence of myogenin. The presence of cells which are positive for myosin heavy chain but negative for myogenin and MyoD proteins lends support to this hypothesis. The presence of myf-5 and MRF4 proteins has not yet been confirmed in myotomes, since antibodies which recognize these factors in the mouse have not yet been described.

MyoD is the last of the four myogenic factors to appear in mouse embryos. This late developmental appearance of MyoD has only been described in the mouse. In other organisms, such as chicken (Lyons et al. 1991a), quail (Pownall and Emerson, 1992), *Xenopus* (Harvey, 1990) and *Drosophila* (Michelson et al., 1990), the MyoD homologue always appears first. MyoD transcripts (Sassoon et al., 1989) and protein (Cusella-De Angelis et al., 1992) are detected in developing myotomes in the mouse embryo beginning at 10.5 days p.c. As mentioned above, this is also the first stage at which myogenin protein becomes detectable in nuclei of differentiating myocytes.

As MyoD is upregulated, MRF4 transcripts are downregulated in myotomes. Myf-5 mRNAs continue to be detected in myotomes where MyoD transcripts are expressed, but at lower levels compared to those between 8.5 to 10.5 days p.c. These observations suggest that MyoD and myogenin replace myf-5 and MRF4 as the two myogenic factors necessary for myotome maturation and axial muscle formation in embryonic and fetal muscle development. Currently, a number of laboratories are attempting to inactivate different myogenic factor genes by homologous recombination. If our hypothesis that two myogenic factors are required to be expressed for normal differentiation of skeletal muscle in utero is correct, we would predict that the knock-out of one of these genes, for example MyoD, would result in the continued expression of another member of the family, such as MRF4 or myf-5, which is normally downregulated. Knock-out of two myogenic factor genes will probably lead to a lethal phenotype.

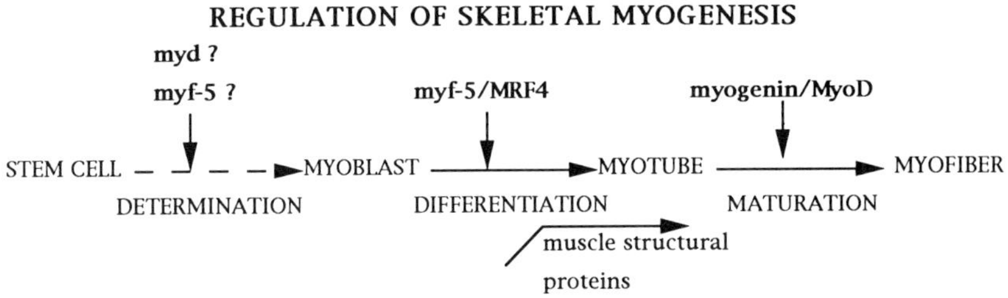

REGULATION OF SKELETAL MYOGENESIS

```
        myd ?
       myf-5 ?              myf-5/MRF4              myogenin/MyoD
          |                     |                        |
          ↓                     ↓                        ↓
STEM CELL ─ ─ ─ ▶ MYOBLAST ─────────▶ MYOTUBE ───────────▶ MYOFIBER
         DETERMINATION       DIFFERENTIATION         MATURATION
                              ╱ muscle structural
                                proteins
```

Fig. 2. A hypothetical scheme for the roles of the different myogenic regulatory factors in the development of somites, myotomes, and axial muscles. Two myogenic factors may be required at each stage to regulate the complex patterns of structural gene activation which have been observed.

Myf-5 is the only myogenic factor whose pattern of expression during embryogenesis is consistent with a potential role in commitment of stem cells to the myogenic lineage (Fig. 2). Another factor, called myd (Pinney et al., 1988), which has not yet been characterized in detail, may also play a role in muscle cell determination. It is unrelated to the MyoD family of genes since there is a lack of hybridization between the two DNA sequences. Myd does have the ability to activate MyoD and myogenin upon transfection into nonmuscle cells, and this suggests that it may act upstream of these genes. Based on their patterns of expression in vivo, MRF4, myogenin, and MyoD appear to play roles in differentiation and possibly maturation of myotubes (Fig. 2) rather than in muscle determination as originally suggested by studies of cells grown in vitro.

It appears that the myogenic factors alone are not sufficient to activate the expression of all of the muscle structural genes. As mentioned above, these factors bind to a consensus sequence in muscle gene promoters and enhancers (reviewed in Olson, 1990; Weintraub et al., 1991; Wright, 1992). Structural genes such as myosin, actin, and muscle creatine kinase (MCK) are activated asynchronously during embryogenesis (see Lyons and Buckingham, 1992). An in situ hybridization study has shown that MCK mRNAs are first detected more than 2 days after the MyoD gene begins to be expressed (Lyons et al., 1991a). MyoD is known to bind to the MCK enhancer and to be required for the high-level expression of this gene (Weintraub et al., 1990). These data suggest that other transcription factors in addition to members of the MyoD family are necessary for the activation of muscle-specific genes during embryonic development. One such factor which has been identified is MEF-2 (Cserjesi and Olson, 1991).

DEVELOPMENTAL EXPRESSION OF MYOGENIC FACTORS IN LIMB BUDS

Cells which form skeletal muscle in limb buds and other peripheral locations such as the branchial arches, head, and body wall are derived from the lateral half of the somite (Jacob et al., 1979; Ordahl and LeDouarin, 1992). In the mouse embryo, elongated cells which appear to be migrating from the ventrolateral edge of dermamyotomes into developing limb buds have been detected by light microscopy (Milaire, 1976). These cells which appear to be determined to become muscle do not express any of the myogenic factor transcripts at a level detectable by in situ hybridization. Sassoon et al. (1989) excised forelimb buds at 9.5 days p.c., a stage at which no myogenic factors are detectable, and grew them in vitro. After several days in culture, cells in these limb buds began to express MyoD and myogenin, suggesting that myogenic cells are present in the limb bud as it forms, but the myogenic program in these cells is not initiated as rapidly as it is in myotomal cells.

The first myogenic factor gene transcript to be detected by in situ hybridization in the forelimb bud is myf-5 at 10.5 days p.c. (Ott et al., 1991). This suggests that premyogenic cells are present in the limb bud for at least 24 hours prior to initiating their myogenic program. In contrast to the developmental sequence of myogenic factor appearance in the myotome, myogenin and MyoD transcripts appear at the same time, approximately 12 hours after myf-5 mRNAs are first detected (Lyons et al., 1991b) (Fig. 3). Both myogenin and MyoD proteins are present in nuclei of dorsal and ventral premuscle masses at this time (Cusella-De Angelis et al., 1992). MRF4 is not expressed transiently in limb buds at a level detectable by in situ hybridization, as it is in myotomes (Bober et al., 1991; Hinterberger et al., 1991).

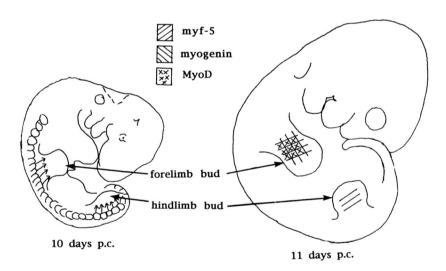

Fig. 3. Schematic representation of myogenic factor gene expression in developing limb buds. Arrows at left represent the path of migrating cells from the somite. Due to the rostrocaudal gradient of muscle formation, expression of myogenic factors in hind limb buds lags behind that of forelimb buds by approximately 12 hours.

TABLE I.
Developmental Expression of Myogenic Factor mRNAs in Myotomes and Body Wall Muscles

Days p.c.	8.5	9.5	10.5	11.5	12.5	13.5	14.5
myf-5	+	+ +	+ + +	+ + +	+ +	+	±
myogenin	+ +	+ +	+ + +	+ + +	+ + +	+ + +	+ + +
MRF4	−	+	+ +	−	−	+	+ +
MyoD	−	−	+ +	+ + +	+ + +	+ + +	+ + +

These differences in the temporal appearance of the myogenic factors between myotomes and limb buds, which are summarized in Tables I and II, suggest that myocytes which originate from the medial and from the lateral halves of somites may have different endogenous programs of muscle-specific gene expression, and thus may represent distinct myogenic lineages. Alternatively, exogenous influences, such as growth factors and extracellular matrix molecules, may play a role in modulating the pattern of gene expression seen in these two populations of myocytes. The switch-graft experiments of Ordahl and LeDouarin (1992) showed that the two halves of newly formed somites are largely interchangeable, so that their ultimate developmental fate is position-dependent. Proximity to the developing neural tube may have a positive regulatory effect on myogenesis in myotomes (Vivarelli and Cossu, 1986). Absence of neural influences and the presence of high levels of growth factors such as basic fibroblast growth factor (bFGF) and other known negative regulatory elements such as the helix-loop-helix protein, Id (Benezra et al., 1990; Wang et al., 1992), may delay the onset of the myogenic program in premyoblasts as they enter the limb bud.

CONCLUSIONS

The patterns of expression of the four members of the MyoD family of myogenic regulators strongly suggest they play important roles in skeletal muscle development in embryos. Differences between the developmental sequence of factors seen in mice and those detected in other organisms suggest that these

TABLE II.
Developmental Expression of Myogenic Factor MRNAs in Forelimb Buds

Days p.c.	8.5	9.5	10.5	11.5	12.5	13.5	14.5
myf-5	−	−	+	+ +	+	±	−
myogenin	−	−	−	+ + +	+ + +	+ + +	+ + +
MRF4	−	−	−	−	−	±	+
MyoD	−	−	−	+ + +	+ + +	+ + +	+ + +

factors may serve overlapping or redundant functions. That they have distinct temporal patterns of gene activation supports the idea that they may also have unique functions at different stages of myogenesis. These factors appear to be necessary, but may not be sufficient, for muscle-specific gene activation during embryogenesis. Their different patterns of expression in myotomes and in peripheral premuscle masses may define distinct lineages of myogenic cells. Their activities are likely to be subject to positive and negative regulation by their interaction with other proteins within muscle nuclei, and by exogenous factors such as extracellular matrix interactions and local levels of mitogens. Experiments designed to express these factors ectopically in vivo or to knock out gene activity by homologous recombination will yield further insight into the function of these factors in embryonic skeletal muscle development.

ACKNOWLEDGMENTS

As a postdoctoral fellow at the Pasteur Institute, G.L. was supported by an NIH/CNRS Fellowship from the Fogarty International Center. During the preparation of this manuscript G.L. was supported by grant IRG-35-33-5 from the American Cancer Society, and by grants from the University of Wisconsin Medical and Graduate Schools. This work was supported by grants from the Pasteur Institute, C.N.R.S., I.N.S.E.R.M., A.F.M., and E.E.C. to M.B.

REFERENCES

Benezra R, Davis RL, Lockshon D, Turner DL, Weintraub H (1990): The protein *Id*: A negative regulator of helix-loop-helix DNA binding proteins. Cell 61:49–59.

Bober E, Lyons G, Braun T, Cossu G, Buckingham M, Arnold H (1991): The myogenic regulatory factor, *myf-6*, shows a biphasic pattern of expression during muscle development. J Cell Biol 112:1255–1265.

Braun T, Buschhausen-Denker G, Bober E, Arnold H (1989): A novel human muscle factor related to but distinct from MyoD1 induces myogenic conversion in 10T1/2 fibroblasts. EMBO J 8:701–709.

Braun T, Bober E, Winter B, Rosenthal N, Arnold H (1990): Myf-6, a new member of the human gene family of myogenic determination factors: Evidence for a gene cluster on chromosome 12. EMBO J 9:821–831.

Choi J, Costa M, Mermelstein C, Chagas C, Holtzer S, Holtzer H (1990): MyoD converts primary dermal fibroblasts, chondroblasts, smooth muscle, and retinal pigmented epithelial cells into striated mononucleated myoblasts and multinucleated myotubes. Proc Natl Acad Sci USA 87:7988–7992.

Cserjesi P, Olson E (1991): Myogenin induces the myocyte-specific enhancer binding factor MEF-2 independently of other muscle-specific gene products. Mol Cell Biol 11:4854–4862.

Cusella-De Angelis M, Lyons G, Sonnino C, De Angelis L, Vivarelli E, Farmer K, Wright W, Molinaro M, Bouchè M, Buckingham M, Cossu G (1992): MyoD1, myogenin independent differentiation of primordial myoblasts in mouse somites. J Cell Biol 116:1243–1255.

Davis R, Weintraub H, Lassar A (1987): Expression of a single transfected cDNA converts fibroblasts to myoblasts. Cell 51:987–1000.

Ede D, El-Gadi A (1986): Genetic modifications of developmental acts in chick and mouse somite development. In Bellairs R, Ede D, Lash J (eds): "Somites in Developing Embryos" New York: Plenum Press, pp 209–224.

Goldhammer D, Faerman A, Shani M, Emerson C (1992): Identification of regulatory elements that control the lineage-specific expression of the myoD gene. Science 256:538–542.

Harvey RP (1990): The Xenopus Myod gene: An unlocalised maternal mRNA predates lineage-restricted expression in the early embryo. Development 108:669–680.

Hinterberger T, Sassoon D, Rhodes S, Konieczny S (1991): Expression of the muscle regulatory factor MRF4 during somite and skeletal myofiber development. Dev Biol 147:144–156.

Jacob M, Christ B, Jacob H (1979): The migration of myogenic cells from somites into the leg region of avian embryos. Anat Embryol (Berlin) 157:291–309.

Kaehn K, Jacob H, Christ B, Hinrichsen K, Poelmann R (1988): The onset of myotome formation in the chick. Anat Embryol (Berlin) 177:191–201.

Lyons GE, Muhlebach S, Moser A, Masood R, Paterson B, Buckingham M, Perriard JC (1991a): Developmental regulation of creatine kinase gene expression by myogenic factors in mouse and chick embryos. Development 113:1017–1029.

Lyons G, Ott MO, Ontell M, Sassoon D, Schiaffino S, Buckingham ME (1991b): Muscle gene expression during embryogenesis in the mouse. In Ozawa E, Masaki T, Nabeshima Y (eds): "Frontiers of Muscle Research." New York: Elsevier-North Holland, pp 99–113.

Lyons GE, Buckingham ME (1992): Developmental regulation of myogenesis in the mouse. Sem Dev Biol 3:243–254.

Michelson A, Abmayr S, Bate M, Arias AM, Maniatis T (1990): Expression of a MyoD family member prefigures muscle pattern in Drosophila embryos. Genes Dev 4:2086–2097.

Milaire J (1976): Contribution cellulaire des somites à la genèse des bourgeons de membres postérieurs chez la souris. Arch Biol (Bruxelles) 87:315–343.

Miner JH, Wold B (1990): Herculin, a fourth member of the *MyoD* family of myogenic regulatory genes. Proc Natl Acad Sci USA 87:1089–1093.

Murre C, Schonleber A, McCaw P, Baltimore D (1989a): A new DNA binding and dimerization motif in immunoglobulin enhancer binding, *daughterless, MyoD* and *myc* proteins. Cell 56:777–783.

Murre C, McCaw P, Vaessin H, Caudy M, Jan L, Jan Y, Cabrera C, Buskin J, Hauschka S, Lassar A, Weintraub H, Baltimore D (1989b): Interactions between heterologous helix-loop-helix proteins generates complexes that bind specifically to a common DNA sequence. Cell 58:537–544.

Olson E (1990): MyoD family: A paradigm for development? Genes Dev 4:1454–1461.

Olson E, Brennan T, Chakraborty T, Cheng TC, Cserjesi P, Edmondson D, James G, Li L (1991): Molecular control of myogenesis: Antagonism between growth and differentiation. Mol Cell Biochem 104:7–13.

Ordahl CP, Le Douarin NM (1992): Two myogenic lineages within the developing somite. Development 114:339–353.

Ostrovsky D, Sanger JW, Lash J (1988): Somitogenesis in the mouse embryo. Cell Diff 23:17–26.

Ott MO, Bober E, Lyons G, Arnold H, Buckingham M (1991): Early expression of the myogenic regulatory gene, *myf-5*, in precursor cells of skeletal muscle in the mouse embryo. Development 111:1097–1107.

Pinney DF, Pearson-White SH, Konieczny SF, Latham KE, Emerson CP (1988): Myogenic lineage determination and differentiation: evidence for a regulatory gene pathway. Cell 53:781–793.

Pownall M, Emerson C (1992): Sequential activation of three myogenic regulatory genes during somite morphogenesis in quail embryos. Dev Biol 151:67–79.

Rhodes SJ, Konieczny SF (1989): Identification of MRF4: A new member of the muscle regulatory factor gene family. Genes Dev 3:2050–2061.

Rugh R (1990): "The Mouse, Its Reproduction and Development." Oxford, UK: Oxford University Publications.

Sassoon D, Lyons G, Wright W, Lin V, Lassar A, Weintraub H, Buckingham M (1989): Expression of two myogenic regulatory factors: Myogenin and MyoD1 during mouse embryogenesis. Nature 341:303–307.

Theiler K (1989): "The House Mouse: Atlas of Embryonic Development." New York: Springer-Verlag.

Vivarelli E, Cossu G (1986): Neural control of early myogenic differentiation in cultures of mouse somites. Dev Biol 117:319–325.

Wachtler F and Christ B (1992): The basic embryology of skeletal muscle formation in vertebrates: the avian model. Sem Dev Biol 3:217–228.

Wang Y, Benezra R, Sassoon D (1992): Id expression during mouse development: A role in morphogenesis. Dev Dynam 194:222–230.

Weintraub H, Tapscott SJ, Davis RL, Lassar A (1989): Activation of muscle-specific genes in pigment, nerve, fat, liver and fibroblast cell lines by forced expression of MyoD. Proc Natl Acad Sci USA 86:5434–5438.

Weintraub H, Davis R, Lockshon D, Lassar A (1990): MyoD binds cooperatively to two sites in a target enhancer sequence: Occupancy of two sites is required for activation. Proc Natl Acad Sci USA 87:5623–5627.

Weintraub H, Davis R, Tapscott S, Thayer M, Krause M, Benezra R, Blackwell T, Turner D, Rupp R, Hollenberg S, Zhuang Y, Lassar A (1991): The myoD gene family: Nodal point during specification of the muscle cell lineage. Science 251:761–766.

Wright W, Sassoon D, Lin V (1989): Myogenin, a factor regulating myogenesis, has a domain homologous to MyoD1. Cell 56:607–617.

Wright W (1992): Muscle bHLH proteins and the regulation of myogenesis. Curr Opin Genet Dev 12:100–110.

11. Myogenic Lineages Within the Developing Somite

Charles P. Ordahl

Department of Anatomy, Cardiovascular Research Institute,
University of California, San Francisco, California 94143-0452

INTRODUCTION: THE SOMITE AS A FUNDAMENTAL UNIT OF CHORDATE STRUCTURE

Segmentation is characteristic of many animal phyla. During vertebrate ontogeny, the first sign of segmentation is the formation of somites: blocks of mesoderm which form immediately lateral to the developing neural tube and notochord. Although somites are transient embryonic structures, the cells of the somite give rise to a variety of adult tissues and structures whose segmental arrangement reflects their somitic origins. The somites, moreover, impose and establish segmental organization upon other embryonic structures, such as the developing central nervous system, where, for example, the segmental arrangement of the spinal nerves is determined by the somite (Bronner-Fraser and Stern, 1991; Keynes and Stern, 1984; Kalcheim and Teillet, 1989; Teillet et al., 1987).

The somite contains at least three distinctive precursor cell populations (Stern et al. 1988; Keynes and Stern, 1988). The sclerotome cells give rise to the axial skeleton (vertebrae) and the ribs. The dermatome cells give rise to cells which migrate away from the somite to form the dermis of the skin over the dorsal body wall. Finally, all of the skeletal muscle of the vertebrate body (excluding most of the muscles of the head) is derived from the somite (Christ et al. 1974; Christ et al., 1977; Chevallier et al., 1977). Two different types of myogenic precursors are derived from the somite: Myotome muscle cells attach the primitive, embryonic vertebrae to one another and are responsible for movement of the intervertebral joints. Migratory myogenic progenitor cells leave the somite to form the muscles of the limbs and limb girdles. Below, I briefly review current ideas about cell lineage and patterning during somitogenesis, with a particular emphasis on the development of these two muscle groups.

DEVELOPMENT OF THE SOMITE

During gastrulation an elongated block of presegmental mesoderm cells collects along both sides of the neural tube in a structure known as the segmental

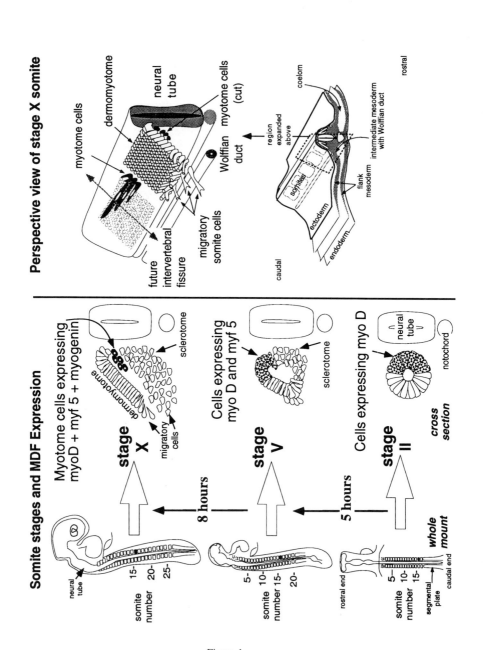

Figure 1.

plate (Fig. 1A). Somites bud off from the rostral end of the segmental plate as new cells are added to the plate caudally from Hensen's node and the primitive streak. The first few somites form soon after neurulation has begun, and somites subsequently form with a periodicity of approximately one pair every 100 minutes (Stern et al., 1988; Keynes and Stern, 1988).

The regularity in the timing of somite formation provides a useful embryonic clock. For example, the total number of somite pairs (typically written with arabic numerals) is an essential component of the systematic staging of early chick embryos (Hamburger and Hamilton, 1951). Moreover, the "age" of each individual somite can also be calibrated according to how recently it has pinched off of the segmental plate. I use roman numerals to designate the age of each somite. In this system, the most recently formed somite is always designated as being at stage "I", the next most recent at stage II, and so on. Thus, as shown in Figure 1, in an embryo with 17 somite pairs, the seventeenth somite is at stage I, the sixteenth at stage II, the fifteenth at stage V, and the eighth at stage X. As the embryo forms an additional pair of new somites, the ages of each of the previously formed somites increases by one roman numeral.

This somite staging provides a framework for systematically describing the process of somitogenesis since similar morphogenetic events occur in succes-

---

Fig. 1. Outline of somite development. **Left:** *Somite stages and MDF expression*. The development of the 16th somite of a chick embryo is followed through development to illustrate stages of somitogenesis. In a 17 somite embryo (bottom), somite 16 is the second most newly-formed somite that is defined as somite stage II. In situ hybridization shows that only myo D mRNA can be detected at this stage and is restricted to the cells in the medial half of the somite. After approximately 5 hours, somite 16 has progressed to stage V (middle) because 3 additional somites have formed behind it. By stage V, both myo D and myf 5 mRNA can be detected but this expression is now limited to the cells in the dorso-medial quadrant of the somite. By somite stage X (top), terminally differentiated myotome cells have begun to appear that express myo D, myf 5 and myogenin at high levels. It is not yet clear if dermomyotome cells continue to express any or all of these mRNAs at low levels. Myogenic determination factor expression patterns are based upon experiments in avian embryos (Pownall and Emerson, 1992). The overall pattern and timing is similar in the mouse but the sequence of myogenic determination gene activation is somewhat different (see Ott et al., 1991). For simplicity I have used the most widely used names for myogenic determination factors. The proper names for the quail factors are: myoD-qmf 1; myogenin-qmf 2; myf 5-qmf 3 (see Pownall and Emerson, 1992). **Right:** *Myotome Formation*. The upper portion shows a perspective view of a stage X somite, illustrating the formation of the early myotome. Postmitotic, terminally differentiated, mononucleate myotome cells remained attached to the dermomyotome epithelium at the rostral end of the somite and grow long processes underneath the epithelium to eventually attach to the caudal end of the somite epithelium. Myotome formation begins with cells most proximal to the neural tube and progresses in a lateral direction, eventually filling the entire area beneath the dermomyotome with myotomal muscle fibers. The lower part of the illustration shows the region expanded in the upper part. This model for myotome formation is based upon the work of B. Christ and colleagues (Kaehn et al., 1988). nt, neural tube; nc, notochord.

sive somites at approximately the same rate. The roman numeral used to identify a somite, therefore, provides information regarding the activity of that somite at that time. The anatomical basis for the following description (see Fig. 2B) is based largely upon the work of B. Christ and his collaborators (Kaehn et al., 1988). This description focuses on the wing-level somites of the chick. While it is generally applicable to all somites, the development of somites at other levels or in other species would require some modifications.

Cells of the newly formed somite (somite I) are not highly organized. During stages II to III, the somite cells organize themselves into a columnar epithelial ball surrounding a central cavity, the somitocoel, typically filled with loosely organized cells. This epithelial somite then begins to reorganize itself in the process of generating the somitic derivatives. The next detectable change in somite structure (about stage IV or V) is an epitheliomesenchymal transition of cells in the ventromedial portion of the somite. At early stages of their migration, these sclerotome cells can be seen to form a bulge on the ventromedial aspect of the somite. Later these cells migrate both dorsally and laterally to surround the axial structures (neural tube and notochord), where they form the cartilage of the developing vertebrae.

The remaining epithelium of the somite will give rise to the myogenic and dermatogenic precursors and has been named the dermamyotome (also written *dermomyotome*). By stage V or VI cells at the rostromedial edge (also referred to as the craniomedial or dorsomedial lip) of the epithelium begin to send out processes caudally, beneath the epithelium of the somite. This is the initiation of myotome formation. These early myotomal cells are anchored at the rostral edge of the epithelium; they simultaneously send a process towards the caudal edge of the epithelium, where they then form a second anchor point. In this way each individual mononucleate myotome cell spans the entire length of the somite. It is important to point out that only postmitotic, mononucleate muscle cells participate in early myotome formation. At much later stages, nonmuscle connective tissue cells and mitotically competent muscle progenitor cells (progenitors of 2° muscle fibers and satellite cells which give rise to, and repair, the muscles of the adult back) enter the myotome. To form the vertebral bodies, each sclerotome unit is cleaved into rostral and caudal portions by the intervetebral fissure. The rostral and caudal portions of adjacent somites then fuse leaving the intervetebral fissures as the future intervetebral joint space. This leaves the myotome fibers spanning the intervertebral joint (Fig. 1C), which they are responsible for moving. The mechanisms involved in the migratory and anchoring processes of the myotome are unknown.

About the same time that the myotome begins forming, a second population of migratory cells can be seen leaving the ventrolateral edges of the somite epithelium. These small, spindle-shaped cells move over the surface of the intermediate mesoderm and Wolffian duct, the latter being primordia of the developing urogenital system (Christ et al., 1991). Approximately 30 to 100 such cells have been observed to migrate from each somite at the wing level (B. Christ, personal communication). Since it is known that limb muscle arises from cells that migrate to the limb from the somite (see above), it is likely that these cells represent the migratory progenitors of limb muscle.

The dermatome cells also leave the somite epithelium, but at a much later stage, to migrate to the area beneath the ectoderm over the dorsal body wall, where they form the dermis of the skin. Dermis in other regions of the body is not derived from the somite. Previously, it was thought that once myotome formation was initiated the myotome cells were self-propagating and, therefore, that only dermatome progenitors remained in the somite epithelium. However, as outlined above, the cells of the myotome proper are permanently postmitotic. Therefore, the "dermatome" epithelium continues to be the progenitor of cells entering the myotome. For this reason it might be more accurate to refer to this structure simply as the *epithelial plate* of the somite, in keeping with earlier nomenclature (Williams, 1910), and with its role as a generative epithelium giving rise to multiple cell types.

MYOGENESIS

Recent advances in molecular analysis of muscle development have led to the discovery of a family of four closely related genes which encode myogenic determination factors (MDFs, recently reviewed in Wright, 1992). These genes (commonly referred to as myoD, myogenin, myf5 and herculin/MRF 4) are homologous at the nucleotide sequence level. All MDFs are equivalent in two regards: First, they are transcriptional activation proteins which bind to conserved DNA sequence motifs within the regulatory regions of a wide variety of muscle-specific promoters. Second, forced expression of any MDF causes many types of nonmuscle cells to respecify their phenotype to myogenesis. Because of these properties it has been proposed that such genes are "master regulatory genes" by which pluripotent mesoderm cells become committed to myogenic differentiation (Weintraub et al., 1991).

Consistent with that proposal, in situ hybridization experiments (Ott et al., 1991; de la Brousse and Emerson, 1990; Pownall and Emerson, 1992) show that MDF expression can first be detected in the medial half of the somite as early as stage II (see Fig. 1B). This is several hours prior to the onset of myotome formation (stage V–VI) and it is generally thought that MDF expression in these epithelial cells is predictive (and possibly determinative) of their ultimate myotomal fate. Interestingly, MDF expression is not observed in the ventrolateral edge of the epithelium, or in migratory cells leaving the somite for the limb muscle beds. The first MDF expression unequivocally attributable to the migratory limb cells occurs 48 hours later, immediately prior to the onset of terminal differentiation of muscle in the limb bud.

This presents an interesting question. Are the migratory myogenic precursors derived from the MDF-positive cells in the medial half of the early somite? If so, then the precursors to the migratory cells would transiently express myogenic determination factors prior to migration. Alternatively, if the migratory myogenic progenitors are derived from a separate somitic cell population perhaps they never express myogenic determination factors until immediately prior to overt differentiation, as occurs in the myotome.

HALF-SOMITE TRANSPLANT EXPERIMENTS

To test the idea that migratory limb muscle precursors might be derived from the MDF-positive cells within the medial half of the newly formed somite, I performed a series of half-somite transplantation experiments in which the lateral or medial halves of newly formed somites were replaced with those of quail (Ordahl and Le Douarin, 1992). Figure 2 outlines the steps involved in a medial half-somite transplantation and Figure 3 A,B shows the basic design of the medial and lateral half-somite experiments. After surgery, embryos were allowed to develop for 4 days until the development of the limb muscle, myotomal muscle, and vertebrae were well established. The distribution of the lateral or medial half-somite cells was then determined by histological analysis using the quail nucleolar marker to identify descendants of the transplanted quail half-somites. The results of those experiments showed that the precursors of the myotome and limb muscle are already segregated in the newly formed somite: Myotomal precursors are derived exclusively from the medial half of the somite. Limb muscle precursors are derived exclusively from the lateral half of the somite. Thus, the migratory precursors to limb muscle are not derived from the MDF-positive cells in the medial half of the newly formed somite.

Are the cells in the lateral half of the somite already determined to form limb muscle at the time the somite is formed? To test this idea I performed switch-graft experiments, in which the lateral half of a chick somite was replaced with the medial half-somite from a quail, and vice versa. Those experiments (Fig. 3C,D) indicated that the cells in either half of the newly formed somite are not determined. Thus, cells within the lateral half-somite, which are normally destined to migrate away from the somite will form vertebral cartilage and myotome if placed in a medial position. Conversely, medial half-somite cells which never migrate to the limb muscle fields will readily do so, if placed in a lateral position.

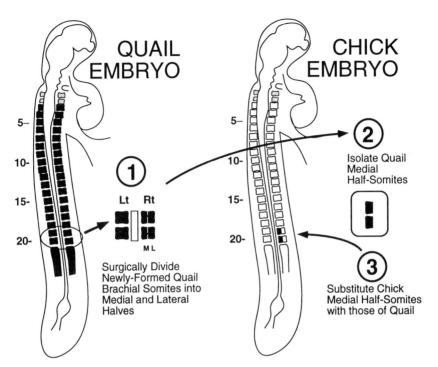

Fig. 2. Replacement of chick medial half-somite with that of a quail. In the three-step operation shown, medial halves of the most recently formed somites from a quail are transplanted homotopically into a chick embryo. Similar approaches were used for transplantation of lateral half-somites and for heterotopic transplantation (see Fig. 3). (Adapted from Ordahl and LeDouarin, 1992.)

How do somite cells become committed to their fate? The timing of the determination process is not yet understood. Preliminary experiments suggest that multipotential cells persist within the somite for considerable lengths of time. One intriguing possibility is that the persistent epithelial plate of the somite (often referred to as the dermatome, see above) actually represents a pool of undetermined cells which, via inductive information from nearby structures, continues to give rise to multiple cell types.

EPAXIAL AND HYPAXIAL DOMAINS

An interesting parallel observation from the half-somite transplant experiments described above is that cells from the lateral and medial halves of the somite appear to distribute themselves according to two fundamental domains of vertebrates: the epaxial and hypaxial domains (see Fig. 4). In fish, the ep-

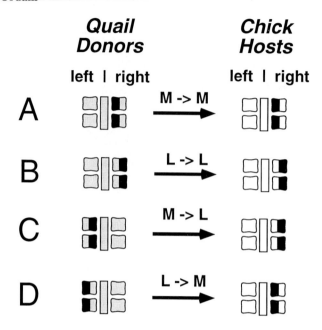

Fig. 3. Homo- and heterotopic half-somite transplantation experiments. Four types of transplantation experiments were performed. **A:** Homotopic medial–medial transplantation experiment. The surgical steps in this experiment are illustrated in Figure 2. **B:** A second series of experiments placed lateral half-somites from a quail into the homotopic position in a chick. **C,D:** Heterotopic transplantation experiments, in which donor somites from the left side of quail embryos were transplanted into the right side of chick embryos. This allowed heterotopic transplantation without rotation of the somite fragments around any axes. (Adapted from Ordahl and Le Douarin, 1992.)

axial and hypaxial domains are roughly equivalent in size and, being composed mostly of muscle, provide for the strong body motions required for locomotion. In higher vertebrates, the hypaxial domain predominates. In man, the epaxial domain contains only the inter vertebral (erector spinae) muscles, the remainder of the body's muscle being hypaxial. The dorsal and ventral rami of the spinal nerves serve the epaxial and hypaxial domains, respectively.

Mesenchymal cells from the lateral half-somite come to reside in the hypaxial domain, while their medial half-somite counterparts remain in the epaxial domain (Fig. 4B). The mesenchymal somite cells from either half-somite appear to be able to wander throughout, and completely fill their respective domains. Near the medial limits of these two domains there is a small zone where the lateral and medial half-somite cells can be seen to mix. However, more laterally, a sharp boundary separates the medial and lateral half-somite cells. This boundary is otherwise invisible but it predicts the location of the boundary between the epaxial and hypaxial do-

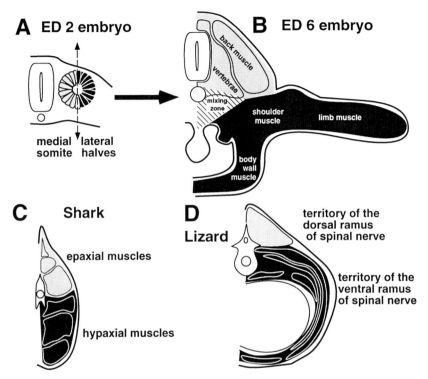

Fig. 4. Medial and lateral half-somite domains correspond to epaxial and hypaxial domains of the vertebrate body. **A:** Diagram of the medial (light shading) and lateral (dark shading) halves of the somite of a 2-day chick embryo. **B:** Shaded areas show the distribution of cells from the medial and lateral somite halves in an embryo of 6 days. A zone of mixing is indicated where medial and lateral half-somite cells intermingle. Lateral to the zone of mixing, a sharp segregation boundary exists between cells derived from medial and lateral half-somites that corresponds to the boundary between the epaxial and hypaxial domains. **C:** Diagrammatic cross section of a mature shark showing the epaxial and hypaxial domains of the body which correspond to the domains of the chick embryo shown in (B). **D:** Similar hypaxial and epaxial domains in the lizard indicate that each is served by a different ramus of the spinal nerve. (Adapted, in part, from Ordahl and Le Douarin, 1992, and from Romer, 1949.)

mains in the adult. Medial half-somite cells could be seen to line up on the dorsal side of this boundary. Lateral half-somite cells line up on the ventral side of this boundary. Such sharp mirror-image distribution of cells derived from the medial and lateral somite halves suggests that this boundary represents a barrier to the movement of somitically derived cells. It would be interesting to know the nature of that barrier.

The half-somite transplant results indicate that the division between epaxial and hypaxial domains is already reflected in medial and lateral halves of the somite at the time of its formation. This separation may arise much earlier because the progenitors of the lateral and medial halves of the somite can be independently labeled by injection of dyes during gastrulation (Selleck and Stern, 1991). Medial- or lateral-half cells can switch their fate if transplanted before, or soon after somite formation (Ordahl and LeDouarin, 1992). As somites age, however, an increasing proportion of cells in either half appear to become "committed" to either the hypaxial or epaxial domain (C.P.O., preliminary unpublished observations). This suggests that the ability to recognize and respect either domain is one which is acquired, much like the acquisition of phenotypic determination as discussed above.

SUMMARY AND SPECULATION

The somite has fundamental importance in vertebrate segmentation and development. Recent application of molecular biology to somite development raised questions which can be addressed using the classical chick-quail chimera technique pioneered by LeDouarin (1973). Using this approach it was revealed that the somite consists of two compartments, medial and lateral. Each compartment arises independently during gastrulation and each has a different developmental fate. Cells from the medial half-somite form structures only within the epaxial domain. This includes the myotome, precursor to the intervertebral muscles of the back. Cells from the lateral half-somite, on the other hand, populate only the hypaxial domain, which includes the muscles of the limb.

These unexpected results raise questions about the relationships between phenotypic determination and the activity of myogenic determination genes. Limb muscle precursors, which must migrate to their destinations, do not appear to express myogenic determination factors either before or during their migratory phase, and this raises questions as to the molecular basis for phenotypic determination in these cells. When do these migratory cells become irreversibly committed to myogenic differentiation? If a cell which has never expressed a myogenic determination factor can be committed to the myogenic phenotype, what factors (genes or otherwise) are responsible for that state of determination?

A number of interesting questions also arise regarding the emergence of the epaxial and hypaxial domains of the embryo. First, do the medial and lateral half-somite cells establish these domains themselves, or are they responding to extrinsic cues from nonsomitic structures? Second, since cells in the newly formed somite are plastic with respect to these domains, how and when do lateral and medial half-somite cells acquire the ability to recognize and/or establish these domains? Third, why does the embryo maintain apparently separate lineages and fates for the lateral and medial halves of the somite, when both halves are interchangeable with one another at early stages? Such questions suggest that the distribution and determination of myogenic precursors of the somite may be governed by global forces within the embryo and that these also establish the dorsoventral axes in the vertebrate body plan.

ACKNOWLEDGMENTS

I would like to thank Professor Nicole LeDouarin, and her associates at the Institut d'Embryologie, for their generous support and encouragement during my sabbatical there. I would also like to thank Professor Bodo Christ for his patient explanations on somitogenesis and his critical comments on this review. Finally, I thank Don Fischman for keeping us all aware of the importance of generative epithelia as it pertains to cardiac and skeletal muscle development. The author's work reported here was supported by the American Cancer Society, the American Heart Association, and the National Institutes of Health.

DEDICATION

This paper is dedicated to the memory of my father, Colonel Stafford N. Ordahl, August 14, 1907–December 26, 1991.

REFERENCES

Bronner-Fraser M, Stern C (1991): Effects of mesodermal tissues on avian neural crest cell migration. Dev Biol 143:213–217.
Charles de la Brousse F, Emerson C (1990): Localized expression of a myogenic regulatory gene, qmf1, in the somite dermatome of avian embryos. Genes Dev 4:567–581.
Chevallier A, Kieny M, Mauger A (1977): Limb-somite relationship: Origin of the limb musculature. J Embryol Exp Morph 41:245–258.
Christ B, Jacob H, Jacob M (1974): Uber den ursprung der flugelmuskulature. Experientia 30:1446–1448.
Christ B, Jacob H, Jacob M (1977): Experimental analysis of the origin of the wing musculature in avian embryos. Anat Embryol (Berlin) 150:171–186.
Christ B, Epperlein H-H, Floel H, Wilting J (1991): The somite-muscle relationship in the avian embryo. In Hinchliffe JR (ed): "Developmental Patterning of the Vertebrate Limb." New York: Plenum Press, pp 265–271.
Hamburger V, Hamilton H (1951): A series of normal stages in the development of the chick embryo. J Morphol 88:49–92.
Kaehn K, Jacob H, Christ B, Hinrichsen K, Poelmann R (1988): The onset of myotome formation in the chick. Anat Embryol (Berlin) 177:191–201.
Kalcheim C, Teillet M (1989): Consequences of somite manipulation on the pattern of dorsal root ganglion development. Development 106:85–93.
Keynes R, Stern C (1984): Segmentation in the vertebrate nervous system. Nature 310:786–789.
Keynes R, Stern C (1988): Mechanisms of vertebrate segmentation. Development 103:413–429.
Le Douarin N (1973): A feulgen-positive nucleolus. Exp Cell Res 77:459–468.
Ordahl C, Le Douarin N (1992): Two myogenic lineages within the developing somite. Development 114:339–353.

Ott M, Bober E, Lyons G, Arnold H, Buckingham M (1991): Early expression of the myogenic regulatory gene, myf-5, in precursor cells of skeletal muscle in the mouse embryo. Development 111:1097–1107.

Pownall ME, Emerson C (1992): Sequential activation of three myogenic regulatory genes during somite morphogenesis in quail embryos. Dev Biol 151:67–79.

Romer AS (1949): "The Vertebrate Body." Philadelphia: WB Saunders, p 254.

Selleck M, Stern C (1991): Fate mapping and cell lineage analysis of Hensen's node in the chick embryo. Development 112:615–626.

Stern C, Fraser S, Keynes R, Primmett D (1988): A cell lineage analysis of segmentation in the chick embryo. Development 104 (Suppl):231–244.

Teillet M, Kalcheim C, Le Douarin N (1987): Formation of the dorsal root ganglion in the avian embryo: Segmental origin and migratory behavior of neural crest progenitor cells. Dev Biol 120:329–347.

Weintraub H, Davis R, Tapscott S, Thayer M, Krause M, Benezra R, Blackwell T, Turner D, Rupp R, Hollenberg S, Zhuang Y, Lassar A (1991): The myoD gene family: Nodal point during specification of the muscle cell lineage. Science 251:761–766.

Williams L (1910): The somites of the chick. Am J Anat 11:55–100.

Wright W (1992): Muscle basic helix-loop-helix proteins and the regulation of myogenesis. Curr Opin Genet Dev 2:243–248.

12. Positional Specification During Muscle Development

Uta Grieshammer, David Sassoon, and
Nadia Rosenthal

Department of Biochemistry, Boston University School of Medicine,
Boston, Massachusetts 02118

INTRODUCTION

The generation of complex arrangements of tissue types during animal development can be initiated by the establishment of general regional differences in embryonic precursor cell populations. In vertebrates, the development of the vertebral column provides a model system to study mechanisms of positional specification. The somites, which represent the first metameric units to appear during embryogenesis, give rise to segmented vertebrae, and intervertebral and intercostal muscles, as well as all other body muscles. Somites initially develop as a series of apparently identical repeated structures along the rostrocaudal axis of the embryo, but later give rise to organs that vary morphologically and functionally with respect to their position along this axis. A current focus of vertebrate embryological studies is to determine when and how regional differences are established during somite development. The homeobox containing (Hox) genes governing the positional specification of the sclerotome, a subset of cells in each somite which participates in the formation of vertebrae, are already well characterized (Holland and Hogan, 1988; Kessel et al., 1990; Kessel and Gruss, 1990, 1991; Le Mouellic et al., 1992), although the downstream gene targets during chondrogenic differentiation are not known. In contrast, although structural gene regulation has been extensively studied during myogenic differentiation, less is known about the regulatory pathways that determine the positional identity of the myogenic precursor cells in somites, the myotomal cells. The discovery of a gradient of expression of a muscle-specific transgene during somite development (Grieshammer et al., 1992) provides a marker of positional identity in the myotome. This transgene system can be used to study the mechanisms that lead to the establishment and maintenance of regional differences of myogenic precursor cells.

THE ORIGIN OF SKELETAL MUSCLES IN THE VERTEBRATE EMBRYO

During vertebrate organogenesis, two rods of paraxial mesoderm, the segmental plates, give rise to the somites, which are formed and mature in a rostral to caudal direction along the anteroposterior (AP) axis of the vertebrate embryo. Initially, somites appear as epithelial spheres which differentiate into the sclerotome, the dermatome, and the myotome (Rugh, 1968; Tam and Beddington, 1987; Keynes and Stern, 1988). Although each somite forms these same three compartments, the tissues arising from them (vertebrae, dermis of the trunk, and skeletal muscles, respectively) vary in their final morphology and function. Since the rostrocaudal position of somite-derived structures in the adult directly reflects the position of the somites from which they originated in the embryo (Bagnall et al., 1989), it is likely that regional specification occurs, enabling somite-derived cells to adopt the fate appropriate for their location along the AP axis. We will review below what is known about positional differences in myotomal cells and the derived skeletal muscles, and will discuss the establishment of these differences during development.

POSITIONAL DIFFERENCES IN MYOGENIC LINEAGES

Adult skeletal muscles display a remarkable diversity, both in their morphology (muscle shape and lamination) and in their function (fiber type composition). These phenotypic markers can be used to assign specific identities to individual muscles, but do not vary systematically with rostrocaudal position. Nevertheless, several lines of evidence suggest that the relative position of each somite in the embryo imparts regional specification upon the resulting muscles. Lineage analysis of myogenic cells, which migrate from the somite to their final position in the skeletal musculature of the chick embryo, reveals a correlation between the rostrocaudal level of somitic origin and the position of the resulting muscle along the AP axis (Chevallier et al., 1977a; Chevallier, 1979; Lance-Jones, 1988a; Bagnall et al., 1989). Furthermore, the rostrocaudal position of motoneurons in the spinal cord corresponds to the position of somitic origin of the target muscles they innervate (Landmesser, 1978; Smith and Hollyday, 1983; reviewed in Fetcho, 1987). In certain cases the information underlying the specificity of this innervation may persist in the muscle itself, since adult rat intercostal muscles from different rostrocaudal levels have been shown to be reinnervated by sympathetic preganglionic axons in a positionally selective manner after transplantation to a common site (Wigston and Sanes, 1982, 1985). This suggests that in the adult, target recognition by the nerve is an intrinsic positional property of certain muscle groups and is not dependent on neighboring tissues.

Studies aimed at pinpointing the nature and timing of positional specification in myogenic precursor populations have been complicated by the fact that body muscle groups are generated by two distinct mechanisms. The majority of trunk and limb muscle arises from myogenic cells which have migrated away from their somite of origin. In contrast, axial muscle groups, including overtly segmented muscles such as intervertebral muscles, are derived directly from differentiation of the myotome in situ (reviewed in Ordahl and Le Douarin, 1992; see Ordahl, Chapter 11, this volume). Embryonic chick-quail chimeric studies have revealed that the migratory myogenic cells respond to extrasomitic tissues for the establishment of positional identity, at a stage of development when the somites are already irreversibly committed to the formation of region-specific vertebrae (Kieny et al., 1972). Specifically, segmental plate reversals and somite exchanges along the AP axis showed that myogenic cells giving rise to craniofacial, thoracic, and limb muscles are not specified to migrate to a predetermined target region, or to give rise to muscles with region-specific morphology (Chevallier et al., 1977b; Chevallier, 1979; Noden, 1986; Lance-Jones, 1988a). Further experiments showed that the fiber type composition of specific wing muscles is independent of the rostrocaudal origin of the myogenic cells in somites (Butler et al., 1988). Finally, the rostrocaudal orientation of motoneuron projection patterns onto wing or hind limb muscles is not affected by the reversal of presomitic mesoderm (Keynes et al., 1987; Lance-Jones, 1988b), so that axon guidance cues which lead to positionally matched innervation patterns in the undisturbed embryo can be respecified in the muscle lineage, or are not intrinsic to the myotomal cells. Taken together, these studies demonstrate that the positional specification of migratory myogenic cells is not irreversibly determined in the somite of origin, but rather relies on signals received from surrounding tissues.

In contrast, the positional identity of segmented axial and body wall muscles which derive directly from the myotomal muscle masses may be already irreversibly determined at an early stage. Segmental plate reversals and somite transplantations in the chick embryo have demonstrated that region-specific phenotypes in axial muscles are established prior to segmentation, as shown by the development of ectopic trunk muscles which assume an identity according to the positional origin of the graft (Murakami and Nakamura, 1991). It therefore appears that distinct mechanisms exist to generate the positional identity of specific myogenic lineages, with respect to the morphology and location of the resulting muscles in the body axis.

The inherent differences between myotomal and lateral muscles are underscored by recent evidence that they are derived from distinct subsets of somitic cells. Chick-quail somite manipulations by Ordahl and LeDouarin (1992; see also Ordahl, Chapter 11, this volume) demonstrated that cells from the medial half of the myotome differentiate in situ to give rise to myotomal muscle masses, whereas cells from the lateral half migrate away from the somite to give rise to the limb musculature. Myotomal muscle cells are dependent on the neural tube or the notochord for their survival, whereas limb and body wall muscles persist in the absence of the axial organs (Rong et al., 1992). Thus the two myogenic lineages giving rise to medial versus lateral muscles are physically distinct in the somite and are regulated independently.

A MOLECULAR MARKER FOR POSITIONAL SPECIFICATION OF MUSCLE PRECURSORS

What markers exist to identify the different precursors of these different muscle groups? Classical definitions of positional identity in myogenic lineages (e.g., muscle morphology and fiber type composition, myogenic cell migration patterns, and motoneuron projection patterns onto skeletal muscles) all rely on scoring a phenotype in the resulting muscle. Ideally, molecular markers would facilitate the dissection of earlier stages in the specification of muscle precursor populations. We recently reported that the rostrocaudal specification of myogenic lineages in developing mouse somites is reflected by the expression pattern of a muscle-specific transgene (Grieshammer et al., 1992), which is comprised of the reporter gene chloramphenicol acetyl transferase (CAT) under the control of the myosin light chain (MLC) 1 promoter and the MLC 1/3 enhancer (Donoghue et al., 1988; Rosenthal et al., 1990). In situ hybridization analysis of MLC-CAT transgenic embryos and quantitative assays of CAT activity in dissected somites revealed a rostrocaudal gradient of MLC-CAT expression in several independently derived transgenic lines (Grieshammer et al., 1992). The gradient is established during the maturation of myotomal muscle masses (Fig. 1), and leads to low expression levels in rostral somites and increasingly higher expression levels in progressively caudal somites as they mature (Figs. 2 and 3). Adult animals maintain a rostrocaudal gradient of MLC-CAT transgene expression, which is most pronounced in the intercostal muscles, and is attributable to the MLC regulatory sequences in the transgenic construct (Donoghue et al., 1991, 1992; Grieshammer et al., 1992). The MLC promoter and enhancer fragment flanking the CAT sequences in the transgene originate from either end of the MLC1/3 locus, and are normally separated by approximately 24 kilobases (kb). Since endogenous MLC1 transcripts do not accumulate in a rostrocaudal gradient in the embryo or in the adult (Donoghue et al., 1991; Grieshammer et al., 1992), it remains to be seen how the gradient observed in transgenic mice relates to the role played by these MLC regulatory sequences in their native context. Nevertheless, the rostrocaudal gradient of MLC-CAT transgene expression in these animals may fortuitously reflect positional cues in myotomal precursors. Notably, the early positional differences in MLC-CAT gene expression arising in the myotomal muscle masses of transgenic mouse embryos are maintained in their segmented intercostal and intervertebral muscle derivatives, whereas CAT activity levels of other muscles, such as those of the limb and the neck, are reassigned in their final position in the body (Donoghue et al., 1991; Grieshammer et al.,

Positional Specification in Muscle Development 181

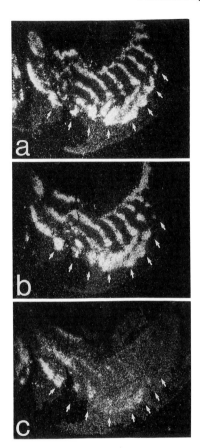
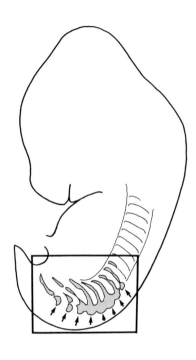

Fig. 1. MLC-CAT transgene transcripts in the myotomes are graded at 12.5 days p.c. Expression of (a) myogenin, (b) MLC1 and (c) CAT transcripts in adjacent sagittal sections from the lumbar region of a 12.5-day p.c. MLC-CAT transgenic embryo. Arrows mark myotomal muscle masses and corresponding intervertebral muscles included in the plane of the section (more rostral myotomes fall outside of the sections due to twisting of the embryo). Hybridizations of the myogenin probe to flanking sections confirmed that material from all somites was present in section (c) (data not shown), indicating that the decrease in CAT transcripts in the more rostral myotomes is not an artifact of sectioning. In situ hybridizations were carried out as described in Grieshammer et al. (1992).

1992). The MLC-CAT expression patterns can therefore be used as early molecular markers, in analyses aimed at unmasking the mechanisms which underlie the regional specification of axial versus lateral muscle precursor cells.

Once established, the graded expression pattern of the MLC-CAT transgene appears to be independent of the context of the intact embryo. By culturing dissociated somites we determined whether myotomal cells maintain their positional identity in vitro, as marked by the graded expression pattern of the

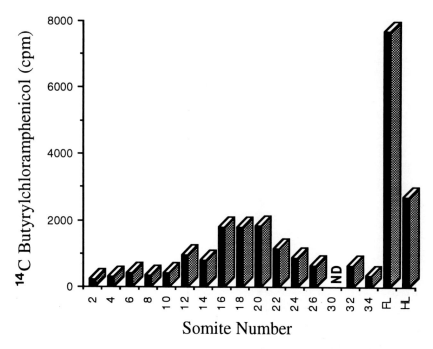

Fig. 2. Quantitative analysis of MLC-CAT activity in the somites of transgenic embryos. Transgenic CAT activity in somites combined from three 11.5-day p.c. embryos. Somites were dissected in subsets of two, CAT assays were performed on tissue homogenates, adjusted to contain equal amounts of total protein in each homogenate. CAT enzyme activity was determined as counts per minute of [^{14}C]chloramphenicol converted to the butyrylated form, as described in Grieshammer et al. (1992). The decreased CAT activity in the most caudal somites is due to their relative immaturity at this embryonic stage (see scheme in Fig. 3).

MLC-CAT transgene. We demonstrated that dissociated somite cells from transgenic mouse embryos establish the graded CAT expression profile when differentiated over 72 hours of culture (Grieshammer et al., 1992). These experiments were performed on somites 10.5 days *post coitum* when they had already matured into sclerotome, dermatome, and myotome (Sassoon et al., 1989). Since the dissections included the neural tube and the notochord, the maintenance of the positional identity in dissociated myogenic cells may not necessarily be a cell-autonomous property, but could be due to influences from those neighboring tissues as well as other somitic cells. Cell mixing experiments, in which dissociated cells from rostral and caudal somites from transgenic and wild-type embryos, respectively, were cocultured for several days, did not result in any perturbation of the initial level of MLC-CAT transgene expression (data not shown), and suggest that the maintenance of the gradient in vitro is probably due, not to the graded production of a diffusible molecule by

Positional Specification in Muscle Development

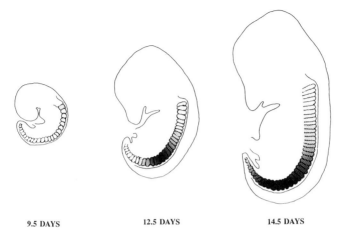

Fig. 3. Generation of the gradient in MLC-CAT transgene expression. Schematic representation of the gradient in MLC-CAT transgene expression as it is generated during embryonic development. 9.5 days p.c: Transcription of the MLC-CAT transgene (shaded area) is first activated in rostral somites, presumably by the action of myogenic factors. 12.5 days p.c: MLC-CAT transcription is increased to higher levels in more caudal somites as they mature and express increasing levels of myogenic factors, whereas the most caudal somites are too immature to express muscle-specific genes. 14.5 days p.c.: Myogenic factors are expressed in the most caudal somites and drive transcription of the MLC-CAT transgene to high levels in these compartments. Scheme is derived from in situ hybridization and quantitative CAT enzyme analysis of transgene expression at different embryonic stages (see Grieshammer et al., 1992).

neural tissues or other somitic cells, but to an autonomous property of the myogenic cells in the cultures.

WHAT ARE THE POSITIONAL CUES GUIDING MUSCLE PRECURSORS?

The experiments described above demonstrate that different myogenic lineages are positionally specified by several mechanisms in the embryo. Prior to somitogenesis the presomitic mesoderm is already committed to give rise to some muscles with region-specific phenotypes arising from the medial myotome, whereas migrating myogenic precursor cells arising from the lateral myotome are still influenced by neighboring tissues to attain positional identities in their ultimate destination. In either case the question arises as to what are the source and the nature of the positional cues. Since subpopulations of somitic cells have not yet been used in transplantation studies along the AP axis, it is not known if all positionally specified cells within a newly formed somite are already autonomous with respect to their identity.

Alternatively, only a small subpopulation of somitic cells is irreversibly specified and influences other cells within the somite. Several lines of evidence suggest that myogenic lineages are dependent on nonmuscle tissues for their positional specification. Connective tissue components might play an important role in the patterning of skeletal muscles (Chevallier and Kieny, 1982; Christ et al., 1983; Noden, 1986, 1988; Lance-Jones, 1988c; Murakami and Nakamura, 1991). The developmental origin of connective tissues is not uniform and has been traced to the neural crest, the lateral plate mesoderm (somatopleure), or the somites for individual muscles (Christ et al., 1977, 1983; Noden, 1986, 1988; Murakami and Nakamura, 1991). Transplantations of lateral plate mesoderm and overlying ectoderm show that these tissues control regional muscle patterns in the limb (Jacob and Christ, 1980; Lance-Jones, 1988c). Furthermore, a positive correlation can be made between the positional specification of muscles and the developmental origin of their associated connective tissues: Muscles arising from transplanted somites maintain their original regional phenotype only in cases where their associated connective tissue was somite-derived (Noden, 1986; Murakami and Nakamura, 1991). The fact that connective tissue components from extrasomitic origins appear to influence the patterning of muscles derived from migratory cells could explain why these myogenic precursor cells are not specified in the somite. A good target tissue for the isolation of molecules involved in mechanisms of regional specification of migrating myogenic cell populations might therefore be the connective tissue components which contribute to different muscle groups in the lateral musculature, such as those in the limb. It is possible, however, that specification of muscles derived from myotomal masses may reflect general mechanisms of axial specification in embryonic mesoderm. If this is the case, the MLC-CAT transgene may constitute a downstream target for the action of the Hox code, in which overlapping expression domains of multiple Hox genes appear to specify regional identities along the embryonic AP axis. A candidate molecule for a positional regulator of mesodermal tissues is retinoic acid, which has been shown to cause homeotic transformations of vertebrae, concomitantly changing hox gene expression in mouse embryos (Kessel and Gruss, 1991; Kessel, 1992). Perturbation of the MLC-CAT transgene gradient in the myotomal compartments, by appropriate ectopic application of retinoic acid to transgenic embryos, may reveal a potential relationship between Hox gene action and positional specification in myotomal precursors.

CONCLUSIONS

The studies reviewed above suggest that regional specification of skeletal muscles in vertebrate embryos involves distinct mechanisms for different myogenic cell lineages in the developing somites. Specifically, embryonic manipulations have shown that those muscle precursors which derive from the medial myotome appear to be positionally determined at a relatively early stage. The fact that myotomal muscle lineages display a graded expression pattern of the muscle-specific MLC-CAT transgene suggests that myotomal cells are labeled early in embryogenesis according to their location along the rostrocaudal axis. The extent to which myotomal cells are dependent on other tissue types in order to maintain the graded MLC-CAT expression profile, and the precise developmental stage at which they become autonomous with respect to this positional phenotype, will provide clues for the identification of the regulatory pathways which confer positional information on developing somites. In this respect, the DNA sequences included in the MLC-CAT transgene represent a target for potential transcriptional regulators in the myotome, and can therefore serve as a probe for the characterization and isolation of molecules involved in the positional specification of myogenic cells.

ACKNOWLEDGMENTS

We thank S. Hauschka and C. Ordahl for helpful discussions. This work was supported by grants from the Council for Tobacco Research to D.S. and N.R., and by grants from the NIH (RO1-HD27585-01A2) to D.S. and (RO1-AG08920) to N.R. U.G. is a recipient of a Boston University Graduate Student Research Award. N.R. is an Established Investigator of the American Heart Association.

REFERENCES

Bagnall KM, Higgins SJ, Sanders EJ (1989): The contribution made by cells from a single somite to tissues within a body segment and assessment of their integration with similar cells from adjacent segments. Development 107:931–943.

Butler J, Cosmos E, Cauwenbergs P (1988): Positional signals: Evidence for a possible role in muscle fiber-type patterning of the embryonic avian limb. Development 102:763–772.

Chevallier A (1979): Role of the somitic mesoderm in the development of the thorax in bird embryos. J Embryol Exp Morph 49:73–88.

Chevallier A, Kieny M, Mauger A (1977a): Limb-somite relationship: Origin of the limb musculature. J Embryol Exp Morph 41:245–258.

Chevallier A, Kieny M, Mauger A, Sengel P (1977b): In Ede DA, Hinchliffe JR, Balls M (eds): "Vertebrate Limb and Somite Morphogenesis." London and New York: Cambridge University Press, pp 421–432.

Chevallier A, Kieny M (1982): On the role of the connective tissue in the patterning of the chick limb musculature. Roux's Arch Dev Biol 191:277–280.

Christ B, Jacob HJ, Jacob M (1977): Experimental analysis of the origin of the wing musculature in avian embryos. Anat Embryol 150:171–186.

Christ B, Jacob M, Jacob HJ (1983): On the origin and development of the ventrolateral abdominal muscles in the avian embryo. Anat Embryol 166:87–101.

Donoghue MJ, Ernst H, Wentworth B, Nadal-Ginard B, and Rosenthal N (1988): A muscle-specific enhancer is located at the 3' end of the myosin light chain 1/3 locus. Genes Dev 2:1779–1790.

Donoghue MJ, Merlie JP, Rosenthal N, Sanes JR (1991): Rostrocaudal gradient of transgene expression in adult skeletal muscle. Proc Natl Acad Sci USA 88:5847–5851.

Donoghue MJ, Morris-Valero R, Johnson YR, Merlie JP, Sanes JR (1992): Mammalian muscle cells bear a cell-autonomous, heritable memory of their rostrocaudal position. Cell 69:67–77.

Fetcho JR (1987): A review of the organization and evolution of motoneurons innervating the axial musculature of vertebrates. Brain Res Rev 12:243–280.

Grieshammer U, Sassoon DA, Rosenthal N (1992): A transgene target for positional regulators marks early rostrocaudal specification of myogenic lineages. Cell 69:79–93.

Holland PWH, Hogan BLM (1988): Expression of homeobox containing genes during mouse development: A review. Genes Dev 2:773–782.

Jacob HJ, Christ B (1980): On the formation of muscular pattern in the chick limb. In Merker HJ, Nau H, Neubert D (eds): "Teratology of the Limbs." Berlin, New York: de Gruyter, pp 89–97.

Kessel M (1992): Respecification of vertebral identities by retinoic acid. Development 115:487–501.

Kessel M, Gruss P (1990): Murine developmental control genes. Science 249:374–379.

Kessel M, Gruss P (1991): Homeotic transformations of murine vertebrae and concomitant alteration of hox codes induced by retinoic acid. Cell 67:89–104.

Kessel M, Balling R, Gruss P (1990): Variations of cervical vertebrae after expression of a hox-1.1 transgene in mice. Cell 61:301–308.

Keynes RJ, Stirling RV, Stern CD, Summerbell D (1987): The specificity of motor innervation of the chick wing does not depend upon the segmental origin of muscles. Development 99:565–575.

Keynes RJ, Stern CD (1988): Mechanisms of vertebrate segmentation. Development 103:413–429.

Kieny M, Mauger A, Sengel P (1972): Early regionalisation of the somitic mesoderm as studied by the development of the axial skeleton of the chick embryo. Dev Biol 28:142–161.

Lance-Jones C (1988a): The somitic level of origin of embryonic chick hindlimb muscles. Dev Biol 126:394–407.

Lance-Jones C (1988b): The effect of somite manipulation on the development of motoneuron projection patterns in the embryonic chick hindlimb. Dev Biol 126:408–419.

Lance-Jones C (1988c): Motoneuron axon guidance: Development of specific projections to two muscles in the embryonic chick limb. Brain Behav Evol 31:209–217.

Landmesser L (1978): The development of motor neuron projection patterns in the chick hind limb. J Physiol 284:391–414.

Le Mouellic H, Lallemand Y, Brulet P (1992): Homeosis in the mouse induced by a null mutation in the hox-3.1 gene. Cell 69:251–264.

Murakami G, Nakamura H (1991): Somites and the pattern formation of trunk muscles: A study in quail-chick chimera. Arch Histol Cytol 54:249–258.

Noden DM (1986): Patterning of avian craniofacial muscles. Dev Biol 116:347–356.

Noden DM (1988): Interactions and fates of avian craniofacial mesenchyme. Development 103 (Suppl):121–140.

Ordahl CP, Le Douarin NM (1992): Two myogenic lineages within the developing somite. Development 114:339–353.

Rong PM, Teillet M-A, Ziller C, Le Douarin NM (1992): The neural tube/notochord complex is necessary for vertebral but not limb and body wall striated muscle differentiation. Development 115:657–672.

Rosenthal N, Berglund E, Wentworth B, Donoghue MJ, Winter B, Braun T, Bober E, Arnold H (1990): A highly conserved enhancer downstream of the human MLC1/3 locus is a target for multiple myogenic factors. Nucleic Acids Res 18:6239–6245.

Rugh R (1968): "The Mouse, Its Reproduction and Development." Oxford: Oxford University Press.

Sassoon DA, Wright W, Lin V, Lassar AB, Weintraub H, Buckingham M (1989): Expression of two myogenic regulatory factors, myogenin and MyoD, during mouse embryogenesis. Nature 341:303–307.

Smith CL, Hollyday M (1983): The development and postnatal organization of motor nuclei in the rat thoracic spinal cord. J Comp Neurol 220:16–28.

Tam PPL, Beddington RSP (1987): The formation of mesodermal tissues in the mouse embryo during gastrulation and early organogenesis. Development 99:109–126.

Wigston DJ, Sanes JR (1982): Selective reinnervation of adult mammalian muscle by axons from different segmental levels. Nature 299:464–467.

Wigston DJ, Sanes JR (1985): Selective reinnervation of intercostal muscles transplanted from different segmental levels to a common site. J Neurosci 5:1208–1221.

13. From Cartilage to Bone—The Role of Collagenous Proteins

Phyllis Lu Valle, Olena Jacenko, Masahiro Iwamoto, Maurizio Pacifici, and Bjorn R. Olsen

Department of Anatomy and Cellular Biology, Harvard Medical School, Boston, Massachusetts 02115 (P.L.V., O.J., B.R.O.); Department of Anatomy and Histology, University of Pennsylvania, School of Dental Medicine, Philadelphia, Pennsylvania 19104 (M.I., M.P.)

INTRODUCTION

The morphogenesis of vertebrate long bones is a two-stage process: first, the formation of a cartilage model; second, the replacement of this cartilage by bone and marrow during endochondral ossification. The first process is one of cell differentiation and pattern formation. The second process is one of cartilage maturation (hypertrophy), matrix degradation, vascular invasion, and ossification. Little is at present known about the identity of genes that regulate chondrocyte differentiation and how they are controlled, and the role of patterning genes in cartilage morphogenesis is only beginning to be understood. In contrast, the understanding of the processes of cartilage hypertrophy and endochondral ossification is much more advanced. For example, we know in some detail the molecular structure and regulation of the major constituents of the extracellular matrix in cartilage, how these constituents change during cartilage hypertrophy, and the molecular mechanisms responsible for their degradation in ossification centers and growth plates. The information concerning collagenous proteins secreted by chondrocytes and their roles during cartilage-to-bone transition is particularly detailed. As structural components of high tensile strength, these proteins are important for maintaining the form of cartilaginous skeletal rudiments and for providing a rigid scaffold around chondrocytes in articular and growth plate cartilage.

COLLAGENOUS COMPONENTS OF CARTILAGE
COLLAGEN FIBRILS

The collagen fibrils of cartilage contain molecules of the fibrillar collagens II and XI, and the fibril-associated collagen IX (for review, see Jacenko et al., 1991). Type II molecules represent the major components, while type XI

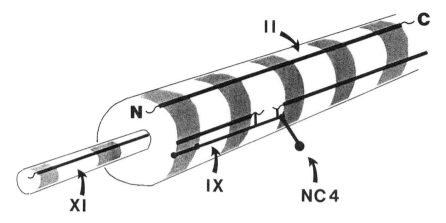

Fig. 1. The molecular composition of cartilage collagen fibrils. The proposed central core of type XI collagen molecules is surrounded by type II collagen molecules. Type IX collagen molecules are located on the surface of type II collagen, with their NC4 domains oriented away from the fibril surface.

molecules may form a central core (Mendler et al., 1989) (Fig. 1). The reason for this heterotypic structure is not entirely clear, but it is possible that type XI molecules somehow play a role in the regulation of fibril diameters. Type IX collagen molecules are located on the surface of the fibrils (Vaughan et al., 1988), and become cross-linked to type II collagen molecules within the fibril (Eyre et al., 1987; van der Rest and Mayne, 1988). Based on the positions of specific lysyl residues within types IX and II polypeptides that are involved in such cross-links, it is possible to define the relative positions and orientations of types II and IX molecules (Eyre et al. 1992) (Fig. 1).

Recently, several laboratories have demonstrated that mutations or deletions in the triple-helical domain of mouse (Garofalo et al., 1991; Metsaranta et al., 1992) or human (Vandenberg et al., 1991) $\alpha 1(II)$ collagen causes a phenotype in transgenic mice which is characterized by skeletal defects, including chondrodysplasia and abnormal limb and craniofacial development. The cause of these skeletal deformities has been attributed to depletion of type II collagen fibrils in cartilage. These data strongly imply that type II collagen is required for normal skeletal development.

Type IX molecules are composed of three genetically distinct polypeptide subunits, $\alpha 1(IX)$, $\alpha 2(IX)$, and $\alpha 3(IX)$. Since each subunit contains three domains with Gly-X-Y repeats, the trimeric molecules contain three triple-helical (COL) domains separated by non-triple-helical (NC) sequences (Jacenko et al., 1991; see references therein). The non-triple-helical sequence (NC3) that separates the amino-terminal and central triple-helical domains is interesting in that it contains an attachment site (a seryl residue) for a glycosaminoglycan side chain in the $\alpha 2(IX)$ chain (McCormick et al., 1987), and it is of different length (12, 17, and 15 residues), respectively, in the $\alpha 1(IX)$, $\alpha 2(IX)$ (McCormick et al., 1987), and α3(IX) chains (Brewton et al., 1992; Har-El et al., 1992). As a result, this region in the molecule represents a kink or hinge so that the amino-terminal triple-helical domain projects as an arm into the perifibrillar matrix (Fig. 1). At the tip of this arm, the α1(IX) chain encodes a globular domain, NC4, in cartilage (Jacenko et al., 1991; see references therein).

Cartilage collagen fibrils are therefore decorated with amino-terminal globular α1(IX) domains along their surface, and it has been proposed that these domains may be important for the interaction of the fibrils with other matrix components such as proteoglycans (Vasios et al., 1988). Consequently, mutations in type IX collagen may cause defects in the assembly of cartilage in embryos and/or alterations in the mechanical stability of cartilage in articular surfaces and growth plates in the adult. This hypothesis has recently been tested in transgenic mice. First, Nakata et al. (1993) have introduced an α1(IX) gene construct expressing a truncated α1(IX) chain into mice. Tissue-specific expression of the transgene in cartilage and eyes was obtained by placing the gene under the control of the type II collagen promoter-enhancer. An in-frame deletion in the central part of the α1(IX) DNA led to the synthesis of shortened α1(IX) chains that were disulfide-bonded with endogenous chains to form trimeric molecules unable to fold properly in their triple-helical domains. Light and electron microscopy of offspring showed pathological changes similar to those of osteoarthritis in articular cartilage of the knee joints, and animals homozygous for the transgene developed a mild chondrodysplasia.

Second, Faessler et al. (1993) have inactivated the α1(IX) collagen gene in mouse embryonic stem cells and generated mice homozygous for the inactivated allele. As in the mice with a dominant negative mutation in α1(IX), the cartilage of the α1(IX) "knock-outs" is apparently assembled normally, but articular knee-joint cartilage appears to develop degenerative changes with age.

Based on these in vivo studies, it appears that type IX collagen is not essential for the developmental assembly of cartilage, but plays a role in maintaining a matrix that can withstand mechanical stress. The precise molecular interactions of type IX molecules that allow this fibril-associated collagen to serve in this stabilizing role are presently not known. The changes observed in the transgenic mice described above are consistent with an interaction between type IX and aggregates of aggrecan, the major proteoglycan component of hyaline cartilage. Direct evidence for binding of type IX domains to proteoglycans, however, is not available.

THE α1(X) COLLAGEN GENE STRUCTURE AND REGULATION

Type X collagen is localized within the hypertrophic zone of growth plates and in ossification centers (Jacenko et al., 1991; see references thercin). Thus it is not synthesized by chondrocytes until they have undergone maturation. This process results in a hypertrophic chondrocyte phenotype which is characterized by an increase in alkaline phosphatase activity (Fell and Robinson,

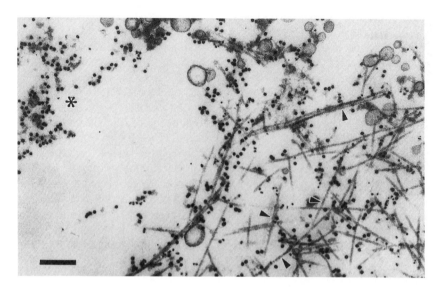

Fig. 2. Type X collagen molecules localize to the surface of type II collagen fibrils. Double labeling of type II collagen (5 nm gold) and type X collagen (15 nm gold) in chick cartilage with monoclonal antibodies adsorbed to colloidal gold. Matlike aggregates are identified with an asterisk. Arrowheads indicate some fibrils containing both types X and II collagen. Bar = 0.2 µm. (Reprinted from Schmid and Linsenmayer, 1990, with permission of the publisher).

1929), the presence of parathyroid hormone receptors (Barling and Bibby, 1985), and expression of type X collagen (Schmid and Conrad, 1982; Gibson et al., 1982; Schmid and Linsenmayer, 1983; Gibson et al., 1984; Capasso et al., 1984). The phenotype can also be distinguished from the proliferative chondrocyte phenotype by an absence of insulin-like growth factor I (IGF-1) (Nilsson et al., 1986) and c-myc (Quarto et al., 1992) expression.

Type X collagen molecules are homotrimeric proteins with triple-helical domains that are about 130 nm in length. The 59-kda α1(X) chick polypeptide chain consists of a 52-amino-acid residue globular amino-terminal domain, a 460-residue triple-helical central domain, and a 162-residue globular carboxy-terminal domain (Lu Valle et al., 1988). Type X collagen molecules do not form fibrils; however, they have been localized by immunoelectron microscopy to the vicinity of type II collagen fibrils in hypertrophic cartilage (Poole and Pidoux, 1989; Schmid and Linsenmayer, 1990) (Fig. 2). In addition, type X collagen has been localized in matlike aggregates in the pericellular area of hypertrophic chondrocytes (Schmid and Linsenmayer, 1990). Both conventional and confocal microscopy have demonstrated the presence of a pericellular capsule consisting of type X collagen surrounding hypertrophic chondrocytes (Gibson et al., 1986; Lu Valle et al., 1992). Based on the striking structural similarity of type X collagen to type VIII collagen, it has been

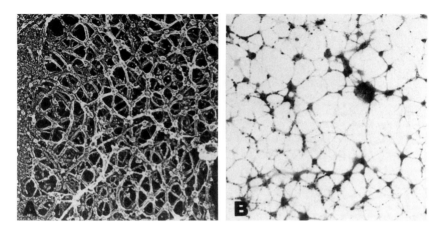

Fig. 3. The hexagonal lattice formed in vitro by type VII collagen and in vivo in the pericellular matrix surrounding hypertrophic chondrocytes. **A:** Bovine corneal epithelial cells incubated in the presence of β-aminopropionitrile form an extracellular matrix consisting of type VIII collagen shown here by freeze-etch replica. Magnification × 57,000. **B:** Hypertrophic cartilage matrix from the growth plate of yucatan swine was treated with 0.1% trypsin for 30 minutes at 37°C, and fixed in 2% paraformaldehyde, 2% glutaraldehyde, 0.7% ruthenium hexamine trichloride. Magnification × 24,000. (Reprinted from Jacenko et al., 1991, with permission of the publisher).

suggested (Gordon and Olsen, 1990) that type X collagen molecules may form an hexagonal lattice work similar to the structure formed by type VIII collagen molecules in Descemet's membrane (Sawada et al., 1984; Jacenko et al., 1991) (Fig. 3A), and reminiscent of the pericellular lattice seen in the hypertrophic zone of the swine growth plate after brief trypsin digestion (C. Farnum, personal communication; Jacenko et al., 1991) (Fig. 3B). In support of this hypothesis, in vitro aggregation studies of type X collagen have shown that it can form an hexagonal latticework; this may represent the molecular organization of the matlike structure observed by immunoelectron microscopy (Kwan et al., 1991) (Fig. 4).

The type X collagen gene has a condensed structure. Unlike fibrillar collagens, it consists of only three exons, with one of the exons encoding the entire triple-helical and carboxy-terminal domains. In the chicken gene, these exons are separated by relatively small introns of 670 base pairs (bp) (between exons 1 and 2) and 2000 bp (between exons 2 and 3) (Lu Valle et al., 1988) (Fig. 5). This condensed structure is conserved among the bovine (Thomas et al., 1991), murine (Apte et al., 1992), and human (Thomas et al., 1991; Apte et al., 1992) α1(X) genes.

The type X collagen gene is transcriptionally activated in hypertrophic chondrocytes (Lu Valle et al., 1989). Type X collagen expression is therefore restricted to hypertrophic cartilage. In chick embryonic sterna, synthesis of type

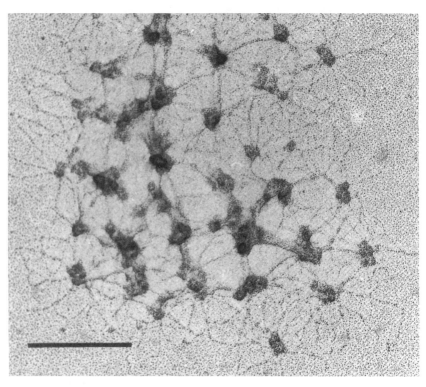

Fig. 4. Type X collagen molecules form an hexagonal lattice in vitro. Chick type X collagen, purified from the media of primary hypertrophic chondrocyte cultures, forms these aggregates after incubation in phosphate buffered saline at 34°C. This is a rotary shadowed image viewed by electron microscopy. (Reprinted from Kwan et al., 1991, with permission of the publisher).

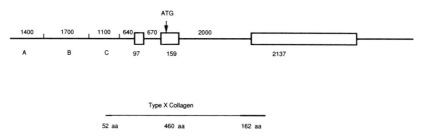

Fig. 5. The chicken type X collagen gene and protein. The type X collagen gene consists of three exons (top, open rectangles), separated by two introns. 4700 bp of upstream regulatory sequence are also included in this gene map and are divided into four fragments called (5'–3') A,B,C, and 640. The translation start (ATG) is identified by the arrow in the second exon. The type X collagen protein contains 460 aa of triple-helical sequence (bottom, thin line) flanked by noncollagenous amino-terminal (52 aa) and carboxy-terminal (162 aa) domains (thick lines).

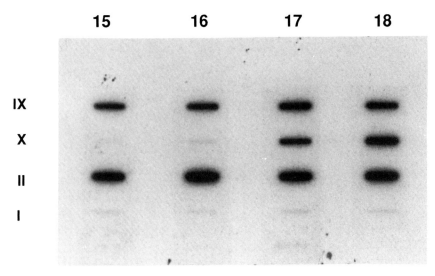

Fig. 6. Run-off transcription assays. Nuclei isolated from chick embryonic cephalic sterna from day 15, 16, 17, and 19 embryos were metabolically labeled with [^{32}P]-UTP. Labeled RNA was then isolated and hybridized to immobilized cDNAs for chick collagens types α2(I) (Ninomiya and Olsen, 1984), α1(II) (Ninomiya et al., 1984), α1(IX) (Ninomiya and Olsen, 1984), and α1(X) (Lu Valle et al., 1988). See text for details.

X collagen mRNA is restricted to the cephalic portion which undergoes endochondral ossification, and is first detected by run-off transcription assay at embryonic day 16 (Lu Valle et al., 1992) (Fig. 6). Between embryonic days 16 and 18, the rate of type X collagen transcription increases by approximately 30-fold. In contrast, the rate of transcription of types II and IX collagen mRNA remain relatively constant. Since hypertrophic chondrocytes are removed as vascular invasion occurs, and cartilage is replaced by bone and marrow during endochondral ossification, the expression of the α1(X) gene is transient, although the type X protein remains in the vicinity of the growth plate for some time (Iyama et al., 1991).

The molecular mechanisms responsible for the hypertrophic chondrocyte-specific transcriptional activation of the α1(X) collagen gene are becoming more clear. Analysis of 4700 bases upstream of the transcription start in the chicken type X gene, and examination of promoter activity in this region and the first intron by reporter gene assays, suggest that the restricted transcription of the gene is controlled by multiple, upstream elements which act together to effectively silence the otherwise strong, nonspecific (with respect to cell type) type X collagen promoter (Lu Valle et al., 1993). In one set of experiments, promoter fragments of 640 bp (including 558 bp of 5' sequence and most of the first exon) and 1500 bp (including the 640 bp above plus the remainder of

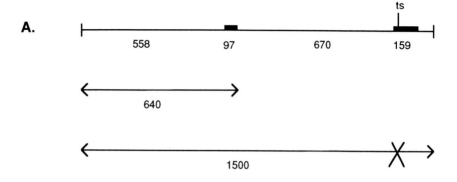

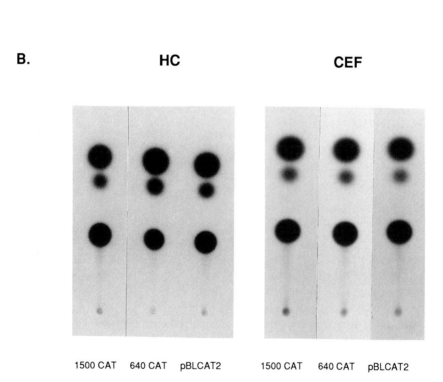

Fig. 7. Type X promoter/CAT assays in hypertrophic chondrocytes (HC) and fibroblasts (CEF). **A:** 640 bp (including 558 bp of 5′ sequence and the first exon) or 1500 bp (including the above-described 640 bp as well as the first intron, the second exon with the translation start-site deleted, and 50 bp of the second intron) were used to drive the expression of CAT in HC and CEF. ts, translation start site; X, deletion of ts. **B:** 1500 CAT and 640 CAT demonstrated high CAT activity in both HC and CEF, comparable to that of the positive control CAT vector, pBLCAT2, in which CAT expression is under the control of the HSV thymidine kinase promoter.

the first exon, the first intron, the second exon in which the translation start-site had been mutated, and 50 bp of the second intron) (Fig. 7A) were used to drive the expression of chloramphenicol acetyl transferase (CAT) in primary cultures of chick embryonic hypertrophic chondrocytes from day 18 cephalic sterna, as well as chick embryonic fibroblasts from day 15 tendons. Chloramphenicol acetyl transferase activity assays demonstrated that the 1500 bp and the 640 bp promoter fragments induced high activity, comparable to the promoter activity of the herpes simplex virus (HSV) thymidine kinase promoter, in both hypertrophic chondrocytes and fibroblasts (Fig. 7B). Three conclusions can be drawn from these experiments. The first is that the $\alpha1(X)$ collagen gene contains an unusually strong promoter within a 550-bp region upstream of the transcription start. The second is that, unlike the promoters of some other collagen genes, notably those of $\alpha1(I)$, $\alpha2(I)$, $\alpha1(II)$, and $\alpha1(IV)$ collagen, the first intron in the type X collagen gene has no enhancing effect on the type X collagen promoter. The third conclusion is that this portion of the type X collagen gene does not show cell specificity, since it is highly active in both hypertrophic chondrocytes which express type X collagen, and in fibroblasts which do not express type X collagen.

In further experiments, the activity of the 4100-bp region upstream of the 640-bp fragment was examined. This 4100-bp region was divided into three fragments conveniently labeled fragments A, B, and C (5' to 3', respectively, see Fig. 5). The effect of these fragments, singly and in combination, on the promoter activity of the 640-bp fragment was examined by CAT activity assays in hypertrophic chondrocytes from embryonic day 18 cephalic sterna, immature chondrocytes from the same stage caudal sterna, and embryonic fibroblasts from skin. While no upstream fragment had any significant effect on the promoter activity of the 640-bp fragment in hypertrophic chondrocytes, all three fragments, either alone or in combinations, had dramatic inhibitory effects on promoter activity in immature chondrocytes and embryonic fibroblasts. Fragment C had the weakest effect, reducing promoter activity 2.5-fold in immature chondrocytes and 2-fold in embryonic fibroblasts. Fragment B reduced activity 5-fold in immature chondrocytes. In fibroblasts, both fragment B and A reduced activity 2.5-fold. Combinations of fragments (i.e., AB, BC, or ABC) reduced promoter activity in both non–type X collagen-expressing cell types 7- to 9-fold (see Table I), to a background level which was close to values obtained with a negative control CAT vector (data not shown). We conclude therefore that the effect of the upstream fragments on type X collagen promoter activity is to effectively silence transcription of the type X collagen gene in non–type X collagen-expressing cells.

The fragment ABC was also placed upstream of the HSV thymidine kinase promoter in the positive control CAT vector, pBLCAT2, and CAT activity was assayed in embryonic fibroblasts. To our surprise, fragment ABC had no inhibitory effect on the activity of this promoter (see Table I). Therefore, se-

TABLE I.
Effect of Type X Collagen 5' Flanking Regions on CAT Activity in Cultured Cells

	640	A	B	C	AB	BC	ABC	CAT2	ABC
HC	100	nd	93	91	79	80	74	nd	nd
IC	100	nd	21	39	14	11	11	nd	nd
CEF	100	40	39	48	14	13	13	100	131

5'-flanking regions upstream of the 640 bp type X collagen promoter fragment repress CAT activity in non–type X expressing cells in a promoter-specific manner. The 4100 bp adjacent and upstream to the 640 bp fragment of the type X collagen gene was divided into three fragments of (5'–3') 1400 pb, 1700 pb, and 1100 pb, called fragments A, B, and C, respectively (see Fig. 5). These fragments, singly or in combination, were placed 5' of the 640 bp promoter fragment in CAT constructs. Chloramphemical acetyl transferase activity was assayed as described (Lu Valle et al., 1993) in HC, IC, and CEF. Any combination of more than one fragment (AB, BC, or ABC) reduced the activity of 640-bp type X promoter fragment by 7- to 9-fold in IC and CEF, which are both non–type X collagen-producing cells. Individual fragments (A, B, or C) had less negative effect than combinations of fragments on the activity of the type X promoter in IC and CEF. Fragment ABC had no reductive effect on the activity of the HSV thymidine kinase promoter (CAT2) when assayed in CEF. HC, hypertrophic chondrocytes; IC, immature chondrocytes; CEF, chick embryonic fibroblasts.

CONSTRUCT FOR DOMINANT INTERFERENCE IN TRANSGENIC MICE

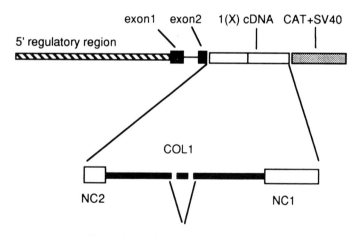

Fig. 8. The chicken α1(X) DNA construct designed to generate a dominant negative phenotype in transgenic mice. Four constructs were made by insertion of the chicken α1(X) cDNA which contained an in-frame deletion of 21 or 293 amino acids within the triple-helical domain, between the 5' portion of the chicken α1(X) gene and CAT. The 5' portion contains repressor elements located within 1600 bp, or 4700 bp, 5' of the transcription start-site.

quences within the 640-bp proximal promoter fragment of the type X collagen gene are apparently required for the silencing activity of fragment ABC to be effective.

MUTATIONS IN THE α1(X) COLLAGEN GENE IN TRANSGENIC MICE

Transient transfection experiments with the chicken type X-5' regulatory region–CAT constructs showed that the upstream silencing elements were also active in rat chondrocytes. We have reasoned, therefore, that chicken α1(X) gene constructs would function in mouse cells as well, and have used such constructs to introduce negative dominant mutations in type X collagen in transgenic mice. The constructs include the chicken α1(X) cDNA, SpLX (Lu Valle et al., 1988), with a central in-frame deletion of 21 or 293 amino acid residues, inserted between the 5' portion of the chicken α1(X) gene and the CAT gene sequences that contain splice signals and a polyadenylation site from SV40. The 5' portion of the chicken α1(X) collagen gene contains 5' regulatory regions comprised of repressor elements located within 1600 bp (corresponding to fragment C and 640) or 4700 bp (corresponding to fragment ABC and 640) of the transcription start-site (see Fig. 5), as well as exon 1, intron 1, and the 5' half of exon 2 (Fig. 8). Injection of construct DNA into fertilized mouse eggs (by DNX, Inc., Bethseda, MD, on NIH-subsidized transgenic service) yielded 20 independent founders that were positive for the chicken transgene. These founders were bred to establish 20 transgenic lines, each with the transgene inserted at different genome positions. To date, 8 of these lines (representing 3 different constructs), showed similar abnormalities in tissues undergoing the cartilage-to-bone transition during endochondral ossification.

About 20% of the transgenic pups became hunchbacked and exhibited paresis of hind limbs at about 17 days after birth, and died within 3 days, probably due to respiratory failure (Fig. 9). X-ray analysis revealed an accentuated neck lordosis and kinking of the vertebral column in the thoracic/cervical region. Histological examination of long bones and vertebral bodies from genotypically positive mice, showed compression of the growth plates and a reduced number and size of bony trabeculae with calcified cartilage cores (Figs. 10 and 11). In 20% of these mice a striking alteration in the bone marrow was also evident, and was characterized by a predominance of mature erythrocytes, and a reduction of white blood cells. These mice also had a dramatically reduced thymus, with the cortical region virtually devoid of lymphocytes. As mentioned above, this phenotype is seen in several independent lines carrying transgenes with short (21 residues) or long (293 residues) deletions. We believe therefore that these abnormalities are caused by the expression of the transgene and not by gene disruption of an endogenous gene at the integration sites.

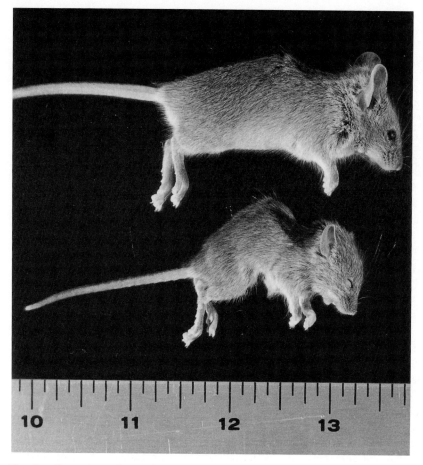

Fig. 9. Comparison of normal (top) and mutant (bottom) littermates at day 19 after birth, from transgenic line 3-2. Note hunched back and inset eyes in the mutant.

The structural abnormalities within the growth plates of the affected mice support the hypothesis that type X collagen primarily plays a structural and/or supportive role within hypertrophic cartilage. The constructs used for microinjection were designed so that expression of the truncated chicken $\alpha1(X)$ polypeptide would compete with endogenous mouse type X chains for assembly and would interfere with the triple-helical folding of trimers. Whether this is the molecular basis for the observed phenotypic changes is not yet known; it is also conceivable that trimeric molecules containing truncated chains are secreted into the matrix and interfere with the formation of type X collagen oligomers and polymers.

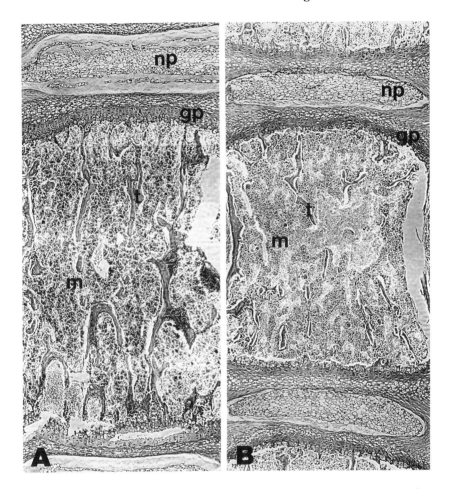

Fig. 10. Hematoxylin and eosin staining of 5-μm frontal sections of lumbar vertebrae from littermates from transgenic line 5-1, at day 21 after birth. **A:** Normal **B:** Mutant. Note compressed growth plate (gp), reduced trabeculae (t), and altered bone marrow (m) in the mutant. np, nucleus pulposus of intervertebral disc.

SUMMARY AND CONCLUSIONS

Skeletal morphogenesis is a complex process dependent on the combined action of genes that regulate cell differentiation patterns, cell proliferation and maturation, and the production and metabolism of extracellular matrices. We have shown here how alterations in two specific matrix genes can effect skeletal morphogenesis. A major challenge for the future will be to define the molecular connection between such matrix effector genes and regulatory genes that control patterns of cell differentiation.

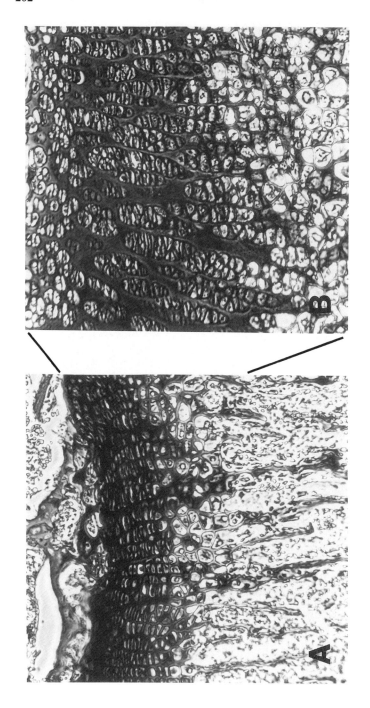

Fig. 11. Hematoxylin and eosin staining of 5-μm longitudinal sections of tibial growth plate from littermates at day 21 after birth, from transgenic line 3-2. **A:** Mutant. **B:** Normal (identical magnification). Note the difference in the thickness of growth plates, as indicated by lines.

ACKNOWLEDGEMENTS

This work was supported in part by NIH grants AR36819 and AR36820, and National Research Service Awards AR08030 (to O.J.) and AR07922 (to P.L.V.), and an Arthritis Investigator Award from the National Arthritis Foundation (to P.L.V.). DNX, Inc., the source of the transgenic mice, is supported by NICHHD contract NO-HD-0-2911.

REFERENCES

Apte SS, Seldin MF, Hayashi M, Olsen BR (1992): Cloning of the human and mouse type X collagen genes and mapping of the mouse type X collagen gene to chromosome 10. Eur J Biochem 206:217–224.
Barling PM, Bibby NJ (1985): Study of the localization of ^{3}[H]bovine parathyroid hormone in bone by light microscopic autoradiography. Calcif Tissue Int 37:441–446.
Brewton RG, Ouspenskaia MV, van der Rest M, Mayne R (1992): Cloning of the chicken α3(IX) collagen chain completes the primary structure of type IX collagen. Eur J Biochem 205:445–449.
Capasso O, Quarto N, Descalzi-Cancedda F, Cancedda R (1984): The low molecular weight collagen synthesized by chick tibial chondrocytes in the extracellular matrix both in culture and in vivo. EMBO J 3:823–827.
Eyre DR, Apon S, Wu J-J, Erikson LH, Walsh KA (1987): Collagen type IX: Evidence for covalent cross-linkages to type II collagen in cartilage. FEBS Lett 220:337–341.
Eyre DR, Wu J-J, Woods P (1992): Cartilage-specific collagens. Structural studies. In Kuettner K et al. (eds): "Articular Cartilage and Osteoarthritis." New York: Raven Press, pp 119–131.
Faessler R, Schnegelsberg P, Dausman J, Muragaki Y, Shinya T, McCarthy MT, Olsen BR, Jaenisch R (1993): Mice lacking α1(IX) collagen develop osteoarthritis. Nature Genetics (submitted).
Fell HB, Robison R (1929): The growth, development, and phosphatase activity of embryonic avian femora and limb buds cultivated in vitro. Biochem J 23:767–784.
Garofalo S, Vuorio E, Metsaranta M, Rosati R, Toman D, Vaughan J, Lozano G, Mayne R, Ellard J, Horton W, de Crombrugghe B (1991): Reduced amounts of cartilage collagen fibrils and growth plate abnormalities in transgenic mice harboring a glycine-to-cysteine mutation in the mouse type II procollagen α1-chain gene. Proc Natl Acad Sci USA 88:9648–9652.
Gibson GJ, Beaumont BW, Flint MH (1984): Synthesis of a low molecular weight collagen by chondrocytes from the presumptive calcification region of the embryonic sterna: The influence of culture with collagenase. J Cell Biol 99:208–246.
Gibson GJ, Bearman CH, Flint MH (1986): The immunoperoxidase localization of type X collagen in chick cartilage and lung. Collagen Res Rel 6:163–184.
Gibson GJ, Schor SL, Grant ME (1982): Effects of matrix macromolecules on chondrocyte gene expression: Synthesis of a low molecular weight collagen species by cells cultured within collagen gels. J Cell Biol 93:767–774.
Gordon MK, Olsen BR (1990): The contribution of collagenous proteins to tissue-specific matrix assemblies. Curr Opin Cell Biol 2:833–838.
Har-El R, Sharma YD, Aguilera A, Ueyana N, Wu J-J, Eyre D, Juricic L, Chandrasekeran S, Li M, Nah H-D, Upholt WB, Tanzer ML (1992): Cloning and development expression of the α3 chain of chicken type IX collagen. J Biol Chem 267:10070–10076.
Iyama K-I, Ninomiya Y, Olsen BR, Linsenmayer TF, Trelstad RL, Hayashi M (1991): Spatiotemporal pattern of type X collagen gene expression and collagen deposition in embryonic chick vertebrae undergoing endochondral ossification. Anat Rec 229:462–472.

Jacenko O, Olsen BR, Lu Valle P (1991): Organization and regulation of collagen genes. Crit Rev Euk Gene Expression 1:327–353.

Kwan AP, Cummings CE, Chapman JA, Grant ME (1991): Macromolecular organization of chicken type X collagen in vitro. J Cell Biol 114:597–604.

Lu Valle P, Ninomiya Y, Rosenblum N, Olsen BR (1988): The type X collagen gene: Intron sequences split the 5' untranslated region and separate the coding regions for the non-collagenous amino-terminal and triple-helical domains. J Biol Chem 263:18378–18385.

Lu Valle P, Hayashi M, Olsen BR (1989): Transcriptional regulation of type X collagen during chondrocyte maturation. Dev Biol 133:613–616.

Lu Valle P, Daniels K, Hay ED, Olsen BR (1992): Type X collagen is transcriptionally activated and specifically localized during sternal cartilage maturation. Matrix 12:404–413.

Lu Valle P, Iwamoto M, Pacifici M, Olsen BR (1993): Multiple, promoter-specific negative elements restrict type X collagen gene expression to hypertrophic chondrocytes. J Cell Biol (in press).

McCormick D, van der Rest M, Goodship J, Lozano G, Ninomiya Y, Olsen BR (1987): Structure of the glycosaminoglycan domain in the type IX collagen-proteoglycan. Proc Natl Acad Sci USA 84:4044–4048.

Mendler M, Eich-Bender SG, Vaughn L, Winterhalter KH, Bruckner P (1989): Cartilage contains mixed fibrils of collagen types II, IX, and XI. J Cell Biol 108:191–197.

Metsaranta M, Garofalo S, Decker G, Rintala M, de Crombrugghe B, Vuorio E (1992): Chondrodysplasia in transgenic mice harboring a 15-amino acid deletion in the triple-helical domain of proα1(II) collagen chain. J Cell Biol 118:203–212.

Nakata K, Ono K, Miyazaki J-I, Olsen BR, Muragaki Y, Adachi E, Yamura K-I, Kimura T (1993): Osteoarthritis associated with mild chondrodysplasia in transgenic mice expressing α1(IX) collagen chains with a central deletion. Proc Natl Acad Sci USA 90:2870–2874.

Nilsson A, Isgaard J, Lindahl A, Dahlstrom A, Skottner A, Isaksson O (1986): Regulation by growth hormones of number of chondrocytes containing IGF-1 in rat growth plate. Science 233:571–574.

Ninomiya Y, Olsen BR (1984): Synthesis and characterization of cDNA encoding a cartilage-specific short collagen. Proc Natl Acad Sci USA 81:3014–3018.

Ninomiya Y, Showalter AM, van der Rest M, Seidah NG, Chretien M, Olsen BR (1984): Structure of the carboxyl peptide of chicken type II procollagen determined by DNA and protein sequence analysis. Biochemistry 23:617–624.

Poole AR, Pidoux I (1989): Immunoelectron microscopic studies of type X collagen in endochondral ossification. J Cell Biol 109:2547–2554.

Quarto R, Dozin B, Tacchetti C, Robino G, Zenke M, Campanile G, Cancedda R (1992): Constitutive myc expression impairs hypertrophy and calcification in cartilage. Dev Biol 149:169–176.

Sawada H, Konomi H, Nagai Y (1984): The basement membrane of bovine corneal endothelial cells in culture with β-aminoproprionitrile: Biosynthesis of hexagonal lattices composed of a 160 nm dumbbell-shaped structure. Eur J Cell Biol 35:226–234.

Schmid TM, Conrad HE (1982): Metabolism of low molecular weight collagen by chondrocytes obtained from histologically distinct zones of the chick embryo tibiotarsus. J Biol Chem 257:12451–12457.

Schmid TM, Linsenmayer TF (1983): A short chain (pro)collagen from aged endochondral chondrocytes: Biochemical characterization. J Biol Chem 258:9504–9509.

Schmid TM, Linsenmayer TF (1990): Immunoelectron microscopy of type X collagen: Supramolecular forms within embryonic chick cartilage. Dev Biol 138:53–62.

Thomas JT, Cresswell CJ, Rash B, Nicolai H, Jones T, Solomon E, Grant ME, Boot-Handford RP (1991): The human collagen X gene. Biochem J 280:617–623.

Thomas JT, Kwan AP, Grant ME, Boot-Handford RP (1991): Isolation of cDNA encoding the complete sequence of bovine type X collagen. Biochem J 273:141–148.

van der Rest M, Mayne R (1988): Type IX collagen proteoglycon from cartilage is covalently cross-linked to type II collagen. J Biol Chem 263:1615–1618.

Vandenberg P, Khillan JS, Prockop DJ, Helminen H, Kontusaari S, Ala-Kokko L (1991): Expression of a partially deleted gene of human type II procollagen (COL2A1) in transgenic mice produces a chondrodysplasia. Proc Natl Acad Sci USA 88:7640–7644.

Vasios G, Nishimura I, Konomi H, van der Rest M, Ninomiya Y, Olsen BR (1988): Cartilage type IX collagen proteoglycan contains a large amino-terminal globular domain encoded by multiple exons. J Biol Chem 263:2324–2329.

Vaughan L, Mendler M, Huber S, Bruckner P, Winterhalter KH, Irwin MH, Mayne R (1988): D-periodic distribution of collagen type IX along cartilage fibrils. J Cell Biol 106:991–997.

14. Pattern Formation and Limb Morphogenesis

Lewis Wolpert and Cheryll Tickle

Department of Anatomy and Developmental Biology, University College and Middlesex School of Medicine, London W1P 6DB, UK

INTRODUCTION

The developing limb is an excellent system to study the mechanisms that lead to skeletal pattern. Extensive classical embryological experiments on chick wing bunds have identified some of the cell interactions involved, and recent work that has now begun to unravel the molecular basis of these cell interactions points to a major rule for retinoids, growth factors, and homeobox genes (reviewed Tabin, 1991). There are good prospects that a focus on the developing limb will lead to an understanding of the molecular basis of pattern formation in this part of the vertebrate embryo.

The central problem is the relationship between genes and fingers, or more generally, how genes control the pattern and the form of the limb. The problem is difficult in itself, but is made more so because of the limited possibilities for genetic analysis in vertebrates. While there are a number of mutations affecting limb development in mice they are not easily amenable to experimental analysis as in the chick. Fortunately, however, genes identified in organisms, as seemingly remote from vertebrates as *Drosophila,* have been used to identify genes in limb development (reviewed Duboule, 1991).

In posing problems about limb development, we shall make three general assumptions: (1) the mechanisms involved in limb development are essentially similar to those used elsewhere in the embryo. No special developmental mechanisms were invented during evolution to bring about limb development. (2) Limb development is controlled by a single positional field. The assumption is that the pattern of cartilage, muscle, tendon, and ectodermal cell differentiation is controlled by the same patterning mechanism; so, whatever mechanisms control the pattern of cartilage also control the pattern of muscle. This is consistent with the observation that operations which result in a mirror-image pattern of cartilage elements also result in a similar mirror-image pattern of muscles and tendons, and even in pigment cell patterns (Richardson et al., 1990). (3) Patterning of the elements of the limb—cartilage, muscle, tendons, ectoderm—occurs long (many hours) before overt differentiation takes place. Thus the pattern of cartilage elements would be specified long before the early signs of cartilage differentiation, such as condensation, take place.

The limb develops from a small bud of undifferentiated mesenchyme cells encased in ectoderm, with a thickening, the apical ectodermal ridge, at its distal end (Fig. 1). The mesenchyme cells have two origins: lateral plate mesenchyme, which gives rise to the connective tissues including cartilage, and cells from the somites, which migrate into the early limb-forming region and give rise to myogenic cells. The skeleton of the chick wing is first laid down as cartilage and later ossifies, so the pattern of cartilage cell differentiation defines the basic limb skeleton (Fig. 1). The digits are morphologically distinct and are arranged in a sequence, digit *2*, digit *3*, and digit *4*, running from anterior to posterior.

Two major sets of interactions are involved in the development of limb pattern (reviewed Tickle 1991). The first is a set of epithelial-mesenchymal interactions between the apical ectodermal ridge and the underlying mesenchyme. The apical ectodermal ridge is required for outgrowth of the bud and maintains a region of undifferentiated mesenchyme cells at the tip of the elongating bud. This region of cells is known as the progress zone and, as cells leave this zone, they form the sequence of structures along the proximal-distal axis of the limb. When the apical ridge is surgically removed from a wing bud, further outgrowth is inhibited, and truncated limbs develop. This is related to the proposed mechanism for specifying position along the proximal-distal axis, namely that the cells positional values are determined by how long the cells spend in the progress zone (Wolpert et al., 1979).

A second major set of interactions in the limb bud involves the polarizing region, a small group of mesenchyme cells at the posterior margin of the bud.

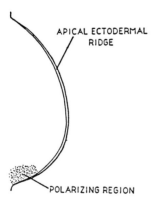
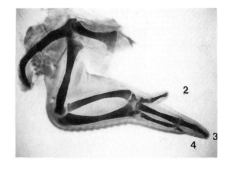

Fig. 1. Diagram (left), showing early chick wing bud and two main signaling regions, the apical ectodermal ridge and the polarizing region. Whole mount photograph (right), showing skeletal pattern of chick wing first laid down as cartilage. Note three digits, digit *2*, digit *3*, digit *4* (anterior to posterior).

The polarizing region interacts with cells in the progress zone to control the pattern of structures that develop across the anteroposterior axis of the wing. The signaling of the polarizing region can be demonstrated by a simple grafting experiment in chick wing buds. When the polarizing region of one wing bud is grafted to the anterior margin of a second bud, there is a dramatic change in wing pattern (Fig. 2). Six digits are formed, instead of three, and these are arranged in a mirror-image symmetrical fashion to give the pattern *4 3 2 2 3 4* (Fig. 2); the muscles and tendons are similarly duplicated. Following grafts to early buds, the more anterior element in the "forearm" is also frequently affected and may be converted into an ulna. The action of the polarizing region graft is therefore to respecify anterior cells to form posterior structures (Tickle, 1991).

The signaling of the polarizing region can be understood in terms of a model in which the polarizing region cells are the source of a diffusible morphogen, which becomes distributed in a concentration gradient across the limb bud (Tickle, 1991). Cells at different distances from the polarizing region would be exposed to different concentrations of morphogen, and this would specify their position and so determine their subsequent development. The signal from the polarizing region does not exert its effect immediately. A polarizing region graft takes about 15 hours to bring about any change in limb pattern and if it is removed earlier the limb develops normally.

This model is consistent with two further observations. First, it accounts for the pattern of digits found when either one or two polarizing regions are placed at varying positions along the anteroposterior axis. For example, if a

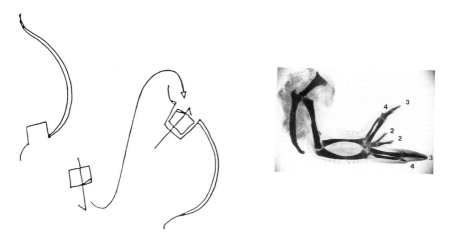

Fig. 2. Polarizing region from one wing bud is grafted to the anterior margin of a second wing bud. The wing that results (left) has 6 digits in mirror-image, symmetrical pattern: *4 3 2 2 3 4*.

polarizing region is placed near the centre of the limb, the resulting pattern is typically *2 3 4 4 3 4*. And if two polarizing regions are grafted, one at the centre and the other anteriorly, then the pattern is *4 3 4 4 4 3 4* (Wolpert and Hornbruch, 1981). Secondly, if the signal from the polarizing region is attenuated by, for example, grafting in reduced numbers of polarizing region cells, then digits which are thought to be specified at a high concentration of the morphogen are lost first. Thus, with increasing attenuation, the pattern changes from *4 3 3 3 3 4*, to *3 2 2 3 4*, to *2 2 3 4*, to *2 3 4* (Tickle, 1981). However, the model has difficulty in accounting for the specification of the humerus (Wolpert and Hornbruch, 1987, see below).

It is of great interest that tissues which signal in other systems are also capable of providing a positional signal when grafted to the anterior margin of the limb bud. The two most striking examples are Hensen's node from both chick (Hornbruch and Wolpert, 1986; 1991) and mouse embryos (Izpisúa-Belmonte, et al., 1992a; Hogan et al., 1992), and the floor plate of the developing chick spinal cord (Wagner et al., 1990). Hensen's node is able to induce an additional axis in early embryos, and the floor plate seems to provide a positional signal in the spinal cord that is formally very similar to the action of the polarizing region in the limb.

It should be noted that grafts of a polarizing region to the posterior margin of another limb have no effect on the pattern. Removal of the polarizing region results, in general, in the failure of posterior structures to develop, though anterior digits may still do so.

RETINOIC ACID AND THE SIGNAL FROM THE POLARIZING REGION

Local application of all-*trans* retinoic acid to chick wing buds mimics the effect produced by polarizing region grafts (Tickle, 1991). When a bead soaked in an appropriate dose of retinoic acid is implanted at the anterior margin of the bud, anterior cells of the wing bud are respecified to form posterior structures. This results in the mirror-image duplicated pattern, with six digits arranged in a mirror-image symmetrical pattern *4 3 2 2 3 4* (Fig. 2). Lower doses result in an attenuated signal, with effects as described above.

Experiments in which either the retinoid source or the polarizing region graft is removed after various lengths of time, show that the additional digits are specified by retinoic acid treatment in a two-step process. In the first phase, lasting about 12 hours, no additional digits are produced. In the second phase, additional digits are sequentially specified. An additional digit 2 requires an exposure of about 14 hours to either retinoic acid or to a polarizing region graft, while an additional *3* and *2* require about 16 hours, and complete reprogramming of the anterior cells to give *4 3 2* pattern requires about 18 to 20 hours. It should be noted that retinoic acid only respecifies the fate of those anterior cells in the progress zone.

A number of lines of evidence point to the operation of an endogenous retinoid-based signaling system in normal limb development. Thaller and Eichele (1987) showed that retinoic acid is present in nanomolar concentrations in chick wing buds and is distributed unevenly across the limb bud, with the highest concentration (50 nM) present posteriorly, in the part of the bud encompassing the polarizing region. Retinol, the precursor of retinoic acid, is also present in the limb bud, but at much higher concentrations, and is uniformly distributed. An antibody that recognizes retinoids, including both retinol and retinoic acid, labels cells in chick limb buds. Furthermore, posterior cells can metabolize retinol to retinoic acid in vivo, and biologically active retinoic acid could, therefore, be generated in a controlled fashion. Limb bud cells express molecules that are involved in detecting and responding to retinoids (reviewed Maden 1991; Ruberte et al., 1991). In the developing chick limb, cellular retinoic acid binding protein (CRABP) is expressed at high levels while cellular retinal binding protein (CRBP) is only present in low amounts. CRABP is present in high amounts in distal mesenchyme cells of the progress zone and appears to be distributed in a gradient, with maximal expression anteriorly and falling-off posteriorly. It has been suggested that this graded distribution of CRABP might amplify the difference in concentration of free retinoic acid across the limb bud. There are also spatial and temporal patterns of expression of transcripts of the nuclear retinoic acid receptors in developing limb buds. Retinoic acid receptor (RAR) alpha and RAR gamma gene transcripts are uniformly distributed in the early mouse limb bud, including the progress zone, whereas RAR beta expression is only expressed in cells that lie in a proximal-anterior position. This expression of RAR beta in the chick is position-dependent (Schofield et al., 1992).

A controversial issue concerns the precise role of retinoic acid in the signaling system that leads to the development of anteroposterior pattern. One possibility is that retinoic acid is the morphogen produced by the polarizing region, and its enrichment in the posterior region of the bud where the polarizing region is located is consistent with this idea. However, there is no evidence that cells directly respond to the local concentration of retinoic acid and that this determines their subsequent development.

It has recently been suggested that retinoic acid is not itself a morphogen but leads to the production of a second signal that provides positional information (reviewed Ragsdale and Brockes, 1991). A graft of a retinoic acid bead to the anterior margin results in the upregulation of expression of RAR beta in the adjacent area, whereas a polarizing region graft does not have this effect (Noji et al., 1991). In addition, the action of retinoic acid may primarily be the specification of another polarizing region (Wanek et al., 1991). This question will not be fully resolved until the train of events that follow retinoic acid application has been completely elucidated.

The situation has also become more complicated because other retinoids, which are morphogenetically active, have been identified in chick wing buds.

For example, the chick wing bud contains 3,4-didehydroretinoic acid (Thaller and Eichele, 1990). When this retinoid is applied to the anterior margin of chick wing buds, it is just as potent as retinoic acid in its ability to respecify the fate of anterior limb bud cells and produce mirror-image duplications. The significance of two forms of biologically active retinoid in the limb is not clear, but presumably any scheme invoking a role for retinoic acid in the limb bud should take account of 3,4-didehydroretinoic acid, although the exact nature and extent of its influence in vivo is not yet known. It looks as though the retinoid-based signaling system in the developing limb may be rather complex.

THE RECORDING OF POSITIONAL INFORMATION

Homeobox genes are good candidates to encode the positional value of cells in developing embryos, and at least 17 different homeobox genes have been detected in the chick limb bud using polymerase chain reaction (PCR) techniques. Experimental analysis suggests that members of the Hox-4 cluster of homeobox genes are involved in pattern formation along the anterior-posterior axis of the wing bud (reviewed Duboule, 1991). During normal limb bud development, members of the Hox-4 complex (Hox-4.4 to Hox-4.8) are progressively activated in a 3' to 5' direction, a process which correlates with their organization in the gene cluster. They are expressed in overlapping domains in the limb bud so that the expression pattern of each 5' gene is contained within its 3' neighbor. Thus nearest to the 5' terminus gene, Hox-4.8, has the most posterior and distal expression domain, and cells at different positions across the anterior-posterior axis of the bud express different combinations of Hox-4 genes. It is not unreasonable to think that this pattern of Hox gene expression defines the position of cells along the anterior-posterior axis and is interpreted by the cells in terms of the development of, for example, specific digits. Thus digit 4 would be encoded by the expression of Hox 4.8 and Hox 4.7, while digit 2 would be interpreted in regions where Hox 4.4 is expressed. Of course other Hox genes may be involved. The Hox 1 series is initially expressed in a pattern similar to that of Hox 4, but then is expressed in a well-defined pattern along the proximal-distal axis and so may specify position values along this axis (Yokouchi et al., 1991).

Evidence that the Hox genes encode positional information comes from the observation that, when anterior limb bud cells are respecified to form posterior structures by a polarizing region or retinoic acid, ectopic domains of Hox-4 expression are induced to give a mirror-image pattern of expression which correlates with the subsequent development of the mirror-image duplication of digits. In limb buds treated with retinoic acid at the anterior margin, Hox-4 genes are switched on between 16 and 24 hours after application. The 3' genes are activated first and 5' ones last, so that after 20 hours Hox-4.6 has been induced, but closest to the 5' terminus gene, Hox-4.8, is not expressed until

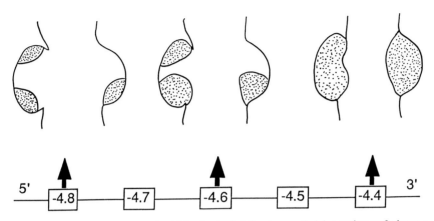

Fig. 3. Hox-4 gene expression in chick wing buds following a polarizing region graft shows a mirror-image pattern. Operated limb bud is on the left; control limb bud is shown on the right.

24 hours after treatment (Izpisúa-Belmonte et al., 1991). This pattern of activation corresponds with the sequential addition of extra digits. The time delay in induction of gene expression suggests that the genes are not directly responding to the retinoid signal.

Polarizing region grafts also result in mirror-image patterns of Hox-4 expression in wing buds (Fig. 3). When mouse/chick heterospecific grafts are carried out, and species-specific probes are used so that expression in graft and host can be distinguished, it was shown, using a mouse polarizing region, that Hox-4 genes are activated in the responding chick host cells (Izpisúa-Belmonte et al., 1992a).

Further evidence for a correlation between Hox-4 expression and anteroposterior patterning has also come from analysis of the chicken talpid[3] mutant. In this mutant, the limbs are polydactylous and a series of up to 10 morphologically similar digits develop from the buds, which are very broad. In talpid[3] buds, the normal pattern of Hox-4 gene expression is changed so that, for example, Hox-4.8, which is normally expressed at the very posterior of the bud, is now expressed all the way across the anterior-posterior axis. The uniform Hox-4 gene expression correlates with the absence of the normal anteroposterior pattern of digits (Izpisúa-Belmonte et al., 1992b).

Very good evidence support for the role of the Hox 4 genes in encoding positional values comes from experiments in which Hox 4.6 is expressed ectopically. By introducing a replication-competent retrovirus containing the mouse Hox 4.6 gene into the early chick limb, all the cells now express this gene. In the hind limb, this results in digit *1* becoming digit *2* and, in the wing, results in the development of an additional digit *2*. These changes in morphology correlate quite well with the changes in positional coding (Morgan et al., 1992).

An important feature of Hox gene expression in the limb (and also along the primary body axis) is that the complex is always transcribed in the 3' to 5' direction. Moreover, there seems to be a "ratchet mechanism," in that once a gene is on, it is not turned off if the tissue is moved to a new site. Thus, using mouse/chick chimeras it was found that when posterior tissue was grafted to an anterior posterior there was no change in its Hox gene expression; by contrast when anterior tissue was grafted posteriorly, new Hox gene expression was activated (Izpisúa-Belmonte et al., 1992a).

Not all homeobox genes respond to the signal from the polarizing region in the same way. For example Xlhbox1 is expressed in the anterior mesenchyme in the region of the bud that will give rise to the shoulder. XlHbox1 expression is extended by retinoic acid and so responds quite differently from the Hox 4 complex. Its pattern of activity is correlated with shoulder girdle abnormalities (Oliver et al., 1990).

INTERACTION BETWEEN THE RIDGE, POLARIZING REGION, AND THE PROGRESS ZONE

It is well established that the apical ectodermal ridge is essential for outgrowth and patterning of the limb bud. A signal from the ridge specifies the progress zone where cells remain labile and continue to proliferate. Patterning along the proximal-distal axis may, as suggested above, be linked to the time the cells spend in the progress zone. In normal limb development, outgrowth of the bud is at first symmetrical, but after stage 20 it becomes enhanced posteriorly in the region where the ridge is best developed.

The interactions between the ridge, progress zone and polarizing gene are quite complex: for instance, a polarizing region is necessary for the maintenance of the ridge but if a polarizing region is placed directly beneath a ridge, the ridge flattens. A plausible set of interactions is as follows: a signal from the polarizing region results in the digital mesoderm maintaining the ridge; the ridge in turn specifies the mesoderm which in turn further maintains it; and the length of the ridge is thus specified by the signal from the polarizing region (Tickle and Brickell, 1991).

The observation that ectopic domains of Hox-4 genes induced by retinoic acid and polarizing region grafts were always found in the distal mesenchyme under the ridge suggested that induction of 5' Hox-4 genes involves cooperation with a signal from the apical ridge. According to this view, the new domain could result from the overlap of the two signals, one from the ridge, and one based on retinoic acid or the grafted polarizing region. Recent experiments in which the apical ridge is removed at various times following retinoic acid treatment support this view (Izpisúa-Belmonte et al., 1992c). No Hox-4 genes are activated when the ridge is removed up to 20 hours after treatment. When the ridge was removed at 24 hours, some 3' genes had been activated but expression of 5' genes was not induced. The presence of the ridge also affects Hox-4 gene expression during normal wing development. When the ridge is removed very early in bud formation, sequential activation of Hox-4 genes is inhibited. Later on in development, removal of the ridge "freezes" the Hox-4 gene expression and further elaboration of the pattern is prevented.

Several molecules have been implicated in the epithelial interactions in the limb bud. They include the homeobox genes Hox-7 and Hox-8. These genes have a distally restricted, position-dependent, pattern of expression in the mesenchyme and this pattern seems to be controlled by a signal from the ridge (Davidson et al., 1991). Similar interactions can be found in micromass cultures (J. Schofield and J. Brown, manuscript in preparation). It is interesting that Hox-8 expression in anterior parts of the mesenchyme is also inhibited by local application of retinoic acid. The nature of the ridge signal is not known, but candidates include proteins, for instance, Wnt5 and TGFb related factors, like, for example, BMP-2 (reviewed Tabin, 1991) and FGF-4 (Niswander and Martin, 1992).

A PREPATTERN MECHANISM?

A prepattern refers to an underlying pattern that reflects the observed pattern. The distinction between generating a pattern by positional information and a prepattern can be illustrated with reference to the digits. To specify the digits by positional information there could be a monotonic concentration gradient (though not necessarily with a diffusible morphogen) and the successive digits would be specified at differing thresholds. There would be no obvious correspondence between the set of positional values and the pattern of digits, the pattern emerging being due to the interpretation of the positional values. By contrast a prepattern mechanism specifies a wavelike concentration of a chemical, and the digits could correspond to the peaks of the wave.

The evidence for a prepattern that would be modified by positional information is indirect and negative. Negative in the sense that it seems most unlikely that a mechanism based on positional information alone is adequate. The most telling observation is the development of cartilaginous structures resembling digits and more proximal elements in reaggregates of limb mesenchyme which lack a localized polarizing region (Pautou, 1973). Our own observations on the development of the humerus when a polarizing region is placed at different positions along the anterior-posterior axis are also not consistent with a positional signal from the polarizing region specifying the humerus (Wolpert and Hornbruch, 1987). Although the humerus is modified by such grafts it is very difficult to either eliminate the humerus or to get two humeri in the region between the host and grafted polarizing region. Yet another difficulty is polydactyly. Polydactyly, which is quite common, could be accounted for by a prepattern mechanism giving an extra "peak" when the bud widens, but is not easily explained in terms of positional information. Even the pattern of muscles might be based on a prepattern mechanism which first divides the initial dorsal and ventral blocks in two, followed by further subdivisions.

A number of models for generating prepattern have been put forward (see Wolpert and Stein, 1984). However, at present, there is no evidence to support such mechanisms, though reaction diffusion mechanisms (Murray, 1989) could in principle provide just the sort of mechanism required. It is also striking that the pattern of stripes of pair rule gene activity in early *Drosophila* development does not result from either a prepattern or reaction diffusion but from interacting gradients, each stripe being specified separately (Pankratz and Jackle, 1990). Even if a prepattern mechanism is present, then positional signals are still required to specify each element.

MORPHOGENESIS

A major problem is how positional specification leads to the development of definitive structures like the overall form of the limb and the cartilaginous elements, as well as their growth. If the basic concepts are correct, the homeobox genes must have downstream effects on all the components of the limb. But at present this connection remains to be established.

The overall form the limb bud presents one problem in form. It is attractive to think that the apical ridge acts in a mechanical manner to impose a dorsoventral flattening. When the ridge disappears a local rounding occurs. But the mechanics of ridge formation need further investigation—what maintains the long contacts between the cells and gives them their very columnar morphology?

The homeobox genes presumably control cartilage differentiation so that it occurs at the appropriate sites. It may be that this control is negative in that all the cells in the progress zone seem to be committed to forming cartilage and do so when placed in micromass culture (Cottrill et al., 1987). But it is not just cartilage differentiation that is involved in the development of the elements. For example, the digits, *2, 3* and *4,* have quite distinct shapes, and we remain ignorant of the cellular basis of these differences. There is also within the long cartilage elements a well-defined pattern of cell differentiation and orientations which is symmetrical with respect to the middle of the element. These lead to a typical element with well-regionalized diaphyseal and epiphyseal regions and the formation of joints is part of this process. Early in cartilage development the cells in the future diaphyseal region become oriented at right angles to the long axis of the element, while cells in the region of the future perichondrium become oriented along the long axis (Rooney et al., 1984). The mechanism of these orientations is not known. The problem of mechanism is accentuated by the failure of carpal and tarsals to form such a structure. Yet in the frog, one of the tarsal elements does develop into a long-bone element.

The mechanism whereby the overall form of the dumbbell shaped cartilage element is generated is not known. The perichondrium has been implicated, the suggesting being that it determines the form of the dilating cartilage within it. It is also not known how very localized changes in shape that characterize different elements are brought about. What, for example, is the mechanism whereby a lateral extension develops at the distal end of the chick tibia which makes contact and fuses with the distal end of the fibula (Archer et al., 1983)? To what extent are the perichondrium, matrix secretion, cell movement, or cell proliferation involved in this process?

A central assumption is that patterning precedes morphogenesis by many hours. For example, it seems that the basic pattern of the proximal limb is laid down by stage 20/21 while the first evidence of cartilage differentiation is not apparent until stage 24/25 some 24 hours later. Yet Oster et al., (1985) have proposed that the pattern of the cartilaginous elements could result from cell mechanics, tractions, or changes in the extracellular matrix at the time of condensation. To show that specification of cartilage does occur prior to condensation, we have constructed double anterior limbs at stages well before condensation of the humerus occurs and we found that in a significant number of cases two humeri develop proximally (Wolpert and Hornbruch, 1990). The overall form of the limb is unaltered and the mechanical model thus cannot account for this result. Condensation is best understood as an early feature of cartilage differentiation.

Growth is an essential feature of limb morphology. It seems that the growth pattern of the individual elements is specified at the same time as the pattern itself is specified (Wolpert, 1981). However, it is not known what this means in cellular or molecular terms. The different patterns of growth of the cartilage elements may require them to be nonequivalent, having different positional values. These positional values may be involved in specifying the different growth programs, which in turn requires an understanding of how growth plates develop and how their intrinsic properties are specified. It has been suggested that the length of the proliferation zone rather than the proliferative rate may be the more important factor in determining growth rate (reviewed Wolpert, 1981).

Another major feature of morphogenesis is the patterning of muscles and tendons. Muscle cells enter the limb from the somites at an early stage of limb development. It seems that the pattern of the muscles is determined by the muscle connective tissue, which determines the pattern of presumptive muscle cell migration. The factors involved are not known but may reflect patterns of differential adhesiveness. Whatever the factors, the pattern is under the same control as the cartilaginous elements, and in the mirror-image limbs that result from polarizing region grafts, the pattern of muscles and tendons is altered in a similar way (Shellswell and Wolpert, 1977). A major problem for the future is to find the links between the specification of positional information in the limb, its recording by Hox genes, and its interpretation in terms of cell differentiation and morphogenesis to give the limb structures.

ACKNOWLEDGMENTS

The authors wish to thank the EEC and MRC for support.

REFERENCES

Archer CW, Hornbruch A, Wolpert L (1983): Growth and morphogenesis of the fibula of the chick embryo. J Embryol Exp Morph 75:101–116.
Cottrill CP, Archer CW, Hornbruch A, Wolpert L (1987): The differentiation of normal and muscle-free distal limb bud mesenchyme in micromass culture. Dev Biol 119:143–151.
Davidson DR, Crawley A, Hill R, Tickle C (1991): Position-dependent expression of two related homeobox genes in developing vertebrate limbs. Nature 352:429–431.
Duboule D (1991): Patterning the vertebrate limb. Curr Opin Genet Dev 1:211–216.
Hogan BLM, Thaller C, Eichele G (1992): Evidence that Hensen's node is a site of retinoic acid synthesis. Nature 359:237–241.
Hornbruch A, Wolpert L (1986): Positional signalling by Hensen's node when grafted to the chick limb bud. J Embryol Exp Morph 94:257–265.
Hornbruch A, Wolpert L (1991): The spatial and temporal distribution of polarizing activity in the flank of the pre-limb-bud stages in the chick embryo. Development 111:725–731.
Izpisúa-Belmonte J-C, Tickle C, Dolle P, Wolpert L, Duboule D (1991): Expression of the homeobox Hox-4 genes and the specification of position in chick wing development. Nature 350:585–589.
Izpisúa-Belmonte J-C, Brown JM, Crawley A, Duboule D, Tickle C (1992a): Hox-4 gene expression in mouse/chicken grafts of signalling regions to limb buds reveals similarities in patterning mechanisms. Development 115:553–560.
Izpisúa-Belmonte J-C, Ede DA, Tickle C, Duboule D (1992b): The mis-expression of posterior Hox-4 genes in *Talpid* (ta^3) mutant wings correlates with the absence of antero-posterior polarity. Development 114:959–963.
Izpisúa-Belmonte JC, Brown JM, Duboule D, Tickle C (1992c): Expression of Hox-4 genes in the chick wing links pattern formation to the epithelial-mesenchymal interactions that mediate growth. EMBO J 11:1451–1457.
Maden M (1991): Retinoid-binding proteins in the embryo. Semin Dev Biol 2:161–170.
Morgan, BA, Izpisúa-Belmonte J-C, Duboule D, Tabin C (1992): Targeted mix-expression of Hox 4.6 in the avian limb bud causes apparenthomeotic transformations. Nature 358:236–240.
Murray, J. (1989) "Mathematical Biology." Berlin: Springer.
Niswander L, Martin GR (1992) Fgf-4 expression during gastrulation, myogenesis, limb and tooth development in the mouse. Development 114:755–768.
Noji S, Nohno T, Koyama E, Muto K, Ohyama K, Aoki Y, Tamura K, Ohsugi K, Ide H, Taniguchi S, Saito T (1991): Retinoic acid induces polarizing activity but is unlikely to be a morphogen in the chick limb bud. Nature 350:83–86.
Oliver G, De Robertis EM, Wolpert L, Tickle C (1990): Expression of a homeobox gene in the chick wing bud following application of retinoic acid and grafts of polarizing region tissue. EMBO J 9:3093–3099.
Oster GF, Murray JD, Maini PK (1985): A model for chondrogenic condensations in the developing limb: The role of extracellular matrix and cell tractions. J Embryol Exp Morph 89:93–112.
Pankratz MJ, Jackle H (1990): Making stripes in the *Drosophila* embryo. Trends Genet 6:287–292.

Pautou MP (1973): Analyse de la morphogenese du pied des oiseaux à l'aide de melange cellulaires interspecifiques, I. Etude morphologique. J Embryol Exp Morph 29:175–196.

Ragsdale CW, Brockes JP (1991): Retinoids and their targets in vertebrate development. Curr Opin Cell Biol 3:928–934.

Richardson MK, Hornbruch A, Wolpert L (1990): Mechanisms of pigment pattern formation in the quail embryo. Development 109:81–89.

Rooney P, Archer C, Wolpert L (1984): Morphogenesis of cartilaginous long bone rudiments. In Trelstad R (ed): "The Role of the Extracellular Matrix in Development." New York: Alan R. Liss, Inc., pp 305–322.

Ruberte E, Kastner P, Dolle P, Krust A, Leroy P, Mendelsohn C, Zelent A, Chambon P (1991): Retinoic acid receptors in the embryo. Semin Dev Biol 2:153–160.

Schofield JN, Rowe A, Brickell PM (1992): Position-dependence of retinoic acid receptor-β gene expression in the chick limb bud. Dev Biol 152:344–353.

Shellswell GB, Wolpert L (1977): The pattern of muscle and tendon development in the chick wing. In Ede DA, Hinchliffe JR, Balls M (eds): "Vertebrate Limb and Somite Morphogenesis." Cambridge: Cambridge University Press, pp 71–86.

Tabin C (1991): Retinoids, homeoboxes and growth factors: Towards molecular models for limb development. Cell 66:199–217.

Thaller C, Eichele G (1987): Identification and spatial distribution of retinoids in the developing chick limb bud. Nature 397:625–628.

Thaller C, Eichele G (1990): Isolation of 3,3-didehydroretinoic acid, a novel morphogenetic signal in the chick wing bud. Nature 345:815–819.

Tickle C (1981): The number of polarizing region cells required to specify additional digits in the developing chick wing. Nature 289:295–298.

Tickle C (1991): Retinoic acid and chick limb development. Development Suppl 1:113–121.

Tickle C, Brickell PM (1991): Retinoic acid and limb development. Semin Dev Biol 2:189–198.

Wanek N, Gardiner DM, Muneoka K, Bryant SV (1991): Conversion by retinoic acid of anterior cells into ZPA cells in the chick wing bud. Nature 350:81–83.

Wagner M, Thaller C, Jessell T, Eichele G (1990): Polarizing activity and retinoic synthesis in the floor plate of the neural tube. Nature 345:819–822.

Wolpert L (1981): Cellular basis of skeletal growth during development. Br Med Bull 37:215–219.

Wolpert L, Hornbruch A (1981): Positional signalling along the anteroposterior axis of the chick wing. The effect of multiple polarizing region grafts. J Embryol Exp Morph 63:143–159.

Wolpert L, Hornbruch A (1987): Positional signaling and the development of the humerus in the chick limb bud. Development 100:333–338.

Wolpert L, Hornbruch A (1990): Double anterior chick limb buds and models for cartilage rudiment specification. Development 109:961–966.

Wolpert L, Stein WD (1984): Positional information and pattern formation. In Malacinski GM, Bryant SV (eds): "Pattern Formation." New York: Macmillan.

Wolpert L, Tickle C, Sampford M (1979): The effect of cell killing by X-irradiation on pattern formation in the chick limb. J Embryol Exp Morph 50:175–185.

Yokouchi Y, Sasaki H, Kuroiwa A (1991): Homeobox gene expression correlated with the bifurcation process of limb cartilage development. Nature 353:443–445.

15. The Bone Morphogenetic Proteins in Cartilage and Bone Development

John M. Wozney, Joanna Capparella, and Vicki Rosen

Genetics Institute, Inc., Cambridge, Massachusetts 02140

INTRODUCTION

The formation of the embryonic skeleton is a highly complex process which is closely associated with establishment of the embryonic body plan. During skeletal formation, multiple cell types progress through a series of sequential interactions that rely both on signals for cell differentiation and on signals involved in conveying positional information, both of which, at present, are poorly understood. For example, bone formation is known to occur through two distinct pathways, which have been designated endochondral bone formation and intramembranous bone formation, whose end products are identical. The first step in each of these bone formation pathways is the condensation of mesenchymal cells into tightly packed groups. If endochondral bone formation takes place, the condensed mesenchyme differentiates into cartilage and the subsequent growth of the cartilage is in the shape of the bone that will replace it. The transition from cartilage to bone occurs after the cartilage has been infiltrated by vascular elements. Once bone formation begins, cartilage remains only at the ends of bones, providing an area of growth and allowing for joint function. In contrast, if intramembranous bone formation takes place, the condensing mesenchyme differentiates directly into bone-forming cells without formation of a cartilage intermediate.

Our interest in the regulation of embryonic skeletal formation stems from our work defining the signals in the adult organism which lead to de novo bone formation, a process regulated by bone morphogenetic proteins (BMPs). The events that occur when BMPs are implanted in adult animals parallel the bone differentiation pathways utilized by the embryo, so that the BMPs may also be important regulators of embryonic skeletal formation.

THE BONE MORPHOGENETIC PROTEINS

Bone morphogenetic protein activity describes the ability of a variety of substances to induce cartilage and bone formation in adult animals (Urist, 1965).

Extracts of bone, as well as other sources such as osteosarcoma and epithelial tissues and extracts, have been shown to have bone-inductive activity when implanted in animals (Neuhof, 1917; Huggins, 1931; Anderson et al., 1964; Urist, 1965; Wlodarski, 1969). Through our efforts to characterize the BMP activity found in bovine bone (Wang et al., 1988), it is now known that a set of related proteins, the BMPs, are responsible for this bone induction.

The amino acid sequences of the individual proteins called BMP-2 through BMP-8 indicate that they are related to the transforming growth factor-β (TGF-β) superfamily of growth factors and, as such, have certain common structural characteristics (Wozney et al., 1988; Wozney, 1989; Celeste et al., 1990; Celeste et al., 1992; Özkaynak et al., 1990). For example, each protein has a hydrophobic leader sequence necessary for secretion of the protein from the cell and a relatively large propeptide region; the mature proportion of the protein residues at the carboxy-terminus of the precursor peptide. Comparing the amino acid sequences of the mature regions of the proteins, the BMPs can be divided into three subgroups, the first consisting of BMP-2 and BMP-4; the second, BMP-5 through BMP-8; and a third group, which contains BMP-3.

Many of the BMPs have now been produced using recombinant systems, allowing for production of substantially pure individual BMP proteins, and also for analysis of the in vivo activities of each BMP. A detailed report of the in vivo activity of a BMP was that of BMP-2 produced using Chinese hamster ovary cells (Wang et al., 1990). In this recombinant system, BMP-2 (or for that matter, any BMP) is synthesized and processed in a complex manner. The BMP-2 precursor polypeptides dimerize, are glycosylated, and upon secretion, the mature region is processed away from the propeptide, allowing for secretion of an active homodimeric BMP-2 molecule. Recombinant human BMP-2 (rhBMP-2) has been shown to have all of the activities associated with the bone-derived BMP material, which is known to be a mixture of BMP proteins. Substantial characterization of rhBMP-2 in vivo activity was first performed using a rat ectopic assay system (Sampath and Reddi, 1981), in which rhBMP-2 is combined with a carrier and then implanted at a subcutaneous site in rats. By removing implants at various times following implantation and processing them for histology, we were able to conclude that rhBMP-2 induces the endochondral bone formation sequence typical for long-bone formation. Implantation of BMP-2 resulted in the differentiation of presumably uncommitted mesenchymal cells into chondroblasts (cartilage cells), maturation of these cells into hypertrophic chondrocytes, calcification of the cartilage tissue, invasion of the implant site with vasculature, and finally bone formation. As the bone forms, the cartilage and the carrier material are resorbed, hematopoietic marrow forms, and the end result of the process is a living piece of bone tissue. In this system, cartilage formation is always observed in conjunction with bone formation, though no stable cartilage is produced. Increasing the amount of rhBMP-2 implanted can dramatically reduce the time needed for bone formation to be observed; however, the volume of bone formed is always limited to the quantity of carrier implanted. Subsequent studies have shown that other individual BMP molecules produce very similar results when tested in this system, though the specific activities of each BMP may vary.

The in vivo sequence of events induced by a BMP is very complex and points in the process at which BMP acts directly are still to be determined. It is possible that a BMP may only initiate this process, with other locally available growth factors or systemic hormones completing the process. One manner in which this issue has been addressed is through the analysis of the activities of BMPs on cell types in vitro. Perhaps as expected given their activities in vivo, BMPs have been shown to induce differentiation of primary cells and established cell lines into cells with chondroblast-like phenotypes. Thus primary chick and murine limb bud cells, immortalized murine limb bud cell lines, and the multipotent embryonic C3H10T1/2 cells increase expression of alkaline phosphatase and the cartilage specific molecules aggrecan and type II collagen when treated with BMPs (Vukicevic et al., 1989; Harrison et al., 1991; Hiraki et al., 1991; Carrington et al., 1991; Chen et al., 1991a; Rosen et al., 1991; Wang et al., 1993; Yamaguchi et al., 1992). Bone morphogenetic proteins have also been shown to induce expression of bone cell phenotypic markers, including osteocalcin (a specific marker of mature osteoblasts), in some of these same systems, and also in multipotent stromal cell lines such as W-20 and calvarial-derived cells (Yamaguchi et al., 1991; Thies et al., 1992; Chen et al., 1991b; Maliakal et al., 1991; Takuwa et al., 1991).

BONE MORPHOGENETIC PROTEINS AND DEVELOPMENT

The BMPs induce, in animals, a bone formation sequence similar to that observed during embryonic long-bone development. In addition, in vitro, BMPs act on precursor cells to differentiate them into both cartilage and bone cells. Therefore, it is possible that BMPs may act as inductive signals in the formation of the embryonic skeletal system. Support for a role for the BMPs in embryonic development comes from a variety of sources and experimental systems. First, the BMPs have homologues in a large number of species. For example, in *Drosophila,* where only two TGF-β family members are found, *decapentapletic (dpp)* is the homologue of the BMP-2 and BMP-4 genes, while *60A* is the homologue of the BMP-5 through BMP-8 genes (Padgett et al., 1987; Wharton et al., 1991; Doctor et al., 1992). The assertion that the BMP-2, BMP-4, and the *dpp* genes are derived from a common ancestor comes not only from their sequence similarity, but also from their strikingly similar gene structures. *dpp* is known from genetic experiments to be required at multiple stages during *Drosophila* development: both early in development, for dorsal-ventral specification, and later, in the formation of the imaginal disks which lead to appendage formation. Additional support for a role for the BMPs in embryonic development has come from mammalian systems (see below).

In Situ Analysis of BMP-2, BMP-4, and BMP-6 mRNAs

One manner in which to gain insight into the possible role of BMPs in embryonic skeletal development is to examine whether their mRNAs are present in a spatial and temporal pattern consistent with the formation of specific structures, by in situ hybridization analysis. The localization of BMP-2, BMP-4, and BMP-6 mRNAs during embryonic development of the mouse has been reported (Rosen et al., 1989; Lyons et al., 1989; Lyons et al., 1990; Pelton et al., 1990). Both BMP-2 and BMP-4 can be found early in the development of the limb bud in the thickened epithelium surrounding the bud. At 10.5 days *post coitum* (p.c.), both mRNAs are localized to the apical ectodermal ridge (AER), an area required for continued distal outgrowth of the limb. Additionally, BMP-4 mRNA is found distributed in a graded fashion throughout the rest of the limb mesenchyme. Later, by 12.5 days p.c. these BMPs are no longer found in the epithelium but rather are present in the interdigital mesenchyme, surrounding the areas where the digits will form. On the other hand, BMP-6 mRNA is not found in these early stages of limb development, but is found after the skeletal system has begun to form, in areas where hypertrophic cartilage is seen. Later in bone development BMP-2 and BMP-4 are also found in the periosteal layer surrounding the newly formed bones. In addition to localization in the limb, BMP-2 and BMP-4 mRNAs are found in the developing facial processes. These specific localizations of BMP mRNAs in the developing limb bud at specific times and positions suggests that the molecules may play an important role in its development.

In Situ Analysis of BMP-3, BMP-5, and BMP-7 mRNA Expression

We have begun analysis of the expression of several of the other BMP molecules during formation of the embryonic mouse skeleton. In this study, we looked at embryos at 10.5, 13.0 and 15.5 days p.c. only. The localizations we report here pertain only to formation of the skeleton.

Using a BMP-3 antisense probe, we observed no BMP-3 mRNA in areas destined to become skeletal elements in 10.5 days p.c. embryos. For example, in the limb bud, only background hybridization levels were observed. This is in sharp contrast to the localizations reported for BMP-2 and BMP-4 at this stage, and may suggest that BMP-3 (like BMP-6) is not involved in the apical ectodermal ridge–mesenchymal interactions ongoing at this time. By 13 day p.c., we are able to find BMP-3 mRNA in individual cartilage cells populating the developing limb skeleton, although the relative level of BMP-3 mRNA is low, and there does not appear to be a gradient or pattern to the observed localization. Several days later, at 15.5 days p.c., the BMP-3 mRNA localizes intensely in areas of cartilage undergoing hypertrophy and also in regions where we observed a transition from cartilage to bone (Fig. 1). Once a periosteum forms around the newly laid down bones, we find a uniform and fairly significant amount of BMP-3 mRNA in the periosteal layer and an absence of BMP-3 mRNA in other areas of the skeleton.

Although our in situ technique is not quantitative, BMP-5 mRNA, in general, appears to be present in greater abundance that BMP-3 mRNA at the same developmental stages. In sections through 10.5 days p.c. embryos, there is diffuse distribution of BMP-5 mRNA throughout limb bud mesenchymal tissues (Fig. 2A). This is, again, in contrast to the patterns of expression of BMP-2 and BMP-4, which are found in the AER as well as the underlying mesenchyme at this stage. At 13.5 days *p.c.*, BMP-5 mRNA is present in low

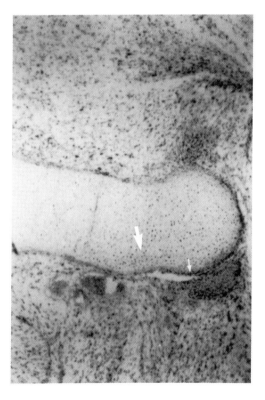

Fig. 1. Localization of BMP-3 mRNA in 15.5 d.p.c. mouse embryos using in situ hybridization with [^{35}S]cRNA probes. Shown is a 5-micron longitudinal section through the hind limb, photographed in bright-field at 10× magnification. Silver grains are present over areas of cartilage undergoing hypertrophy (large arrow) and also over the periosteum (small arrow).

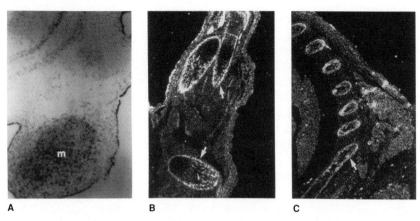

Fig. 2. Localization of BMP-5 mRNA in the skeletal tissues of mouse embryos using in situ hybridization with [^{35}S]cRNA probes. **A:** A 5 micron section through the hind limb of a 10.5 d.p.c. embryo, photographed in a bright-field at 10× magnification. Diffuse labeling is seen over the mesenchymal cells (m) of the limb bud. A higher concentration of grains is found in areas that will condense and form cartilage. **B:** Photomicrograph of a 5-micron longitudinal section through a 15.5 d.p.c. mouse embryo, using dark-field imaging at 4× magnification. Intense labeling is seen in the periosteum of the newly formed shoulder and arm bones (large arrows) and in the areas undergoing the transition from cartilage to bone (small arrows). **C:** BMP-5 mRNA localization in the periosteum (large arrow) and cartilage/bone interface (small arrow) of the ribs of a 15.5 d.p.c. mouse embryo. Dark-field imaging at 4× magnification.

but discernable amounts in all skeletal areas, including the cartilaginous condensations of the appendicular skeleton, the mesenchyme of the jaw, and the vertebral bodies. By 15.5 days *p.c.*, as with BMP-3, we observed a highly specific localization of BMP-5 mRNA in the periosteum of newly formed bones (Fig. 2B,C). This localization may suggest a role for BMP-5 in the maintenance of the periosteal progenitor cell populations.

As with BMP-5, we observe BMP-7 mRNA throughout the mesenchyme of the 10.5 days *p.c.* limb bud but absent in the AER. By 13.0 days *p.c.*, BMP-7 mRNA is present in many of the areas undergoing skeletal morphogenesis. A typical localization is shown in Figure 3A, a section through the developing hind limb. In this figure, the BMP-7 mRNA is contained in the cartilage cells undergoing differentiation. At 15.5 days *p.c.*, as with other BMPs studied to date, BMP-7 is very specifically found in the periosteum of the newly formed bones (Fig. 3B,C,D).

Localization of BMP mRNAs in Other Tissues

Various BMP mRNAs are also found to be specifically localized in many other tissues in the developing embryo, especially in areas that require mesenchymal–epithelial interactions for morphogenesis. These include the heart,

BMPs in Cartilage and Bone Development 227

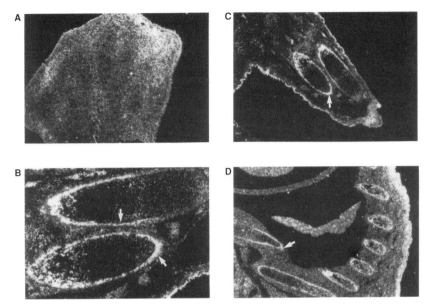

Fig. 3. Localization of BMP-7 mRNA in the skeletal tissues of mouse embryo using in situ hybridization with [^{35}S]cRNA probes. **A:** A 5-micron section through the hind limb of a 13.0 d.p.c. mouse embryo, photographed in dark-field at 10× magnification. Hybridization is seen throughout the cartilaginous cell population. **B, C:** Photomicrographs of a longitudinal section through the hind limb of a 15.5 d.p.c. mouse embryo. Dark-field imaging at 4× (C) and 10× (B) magnification. Note the labeling in the periosteum (large arrows). **D:** Photomicrograph of the rib area of a 15.5 d.p.c. embryo, dark-field imaging at 4× magnification. Intense labeling is seen over the periosteum of each rib (large arrow).

kidney, whisker follicles, and skin. In the formation of teeth, another highly mineralized tissue, both BMP-2 and BMP-4 have been found at specific sites in the tooth bud and tooth germ at different times.

CONCLUSIONS

Although the BMPs were originally discovered as a result of their ability to induce the formation of cartilage and bone in the adult animal, several arguments point to their involvement in the formation of the embryonic skeletal system, as well as other tissue systems in the embryo. In this paper, we have presented additional evidence regarding the expression of individual BMP mRNAs in the developing skeleton. While our in situ hybridization analysis was not quantitative in nature, we believe the levels of BMP-3, BMP-5, and BMP-7 mRNA expression are significantly lower than those of the TGF-βs at the same stages of embryogenesis. In general, in all of the systems studied to date, each BMP mRNA appears to have a distinct yet overlapping area of expression, similar to results observed for TGF-β isoforms. It is unclear at present whether these kinds of results are indicative of functional redundancy for BMP molecules; genetic experiments using techniques that eliminate expression of specific BMPs will be needed to address this question. The overlapping expression patterns of the BMPs also provide for the possibility that BMP heterodimer formation occurs; different BMP heterodimers could have unique functions, thus providing additional specificities for individual molecules. Again, in situ techniques do not provide direct evidence to address this question.

For each BMP studied, we observed a transition in its localization, from the cells actually undergoing differentiation, to the cells thought to remain as progenitor cells for these populations. For example, in the limb, the BMP mRNAs appear first in the mesenchymal cells, then in chondrogenic populations, and finally in periosteal cells, after morphogenesis is completed. This may suggest a shift in function for the specific BMP at certain developmental stages or may imply both autocrine and paracrine roles for these factors. It will be interesting to determine if the same BMP mRNAs remain in their periosteal locations in older embryos, in neonatal animals, and in adult mice, as well as to observe any changes in localizations during normal skeletal growth and remodeling, and during active skeletal repair. These types of studies may provide us with a better understanding of the roles of individual BMPs during the life cycle of the skeleton.

Multiple BMPs have now been localized both within the developing skeleton and in other morphogenetically active areas of the embryo. In addition, embryonic cells have been shown to respond to BMPs in vitro. These data, coupled with the genetic analysis of BMP genes in other species, present a strong case for the involvement of BMPs in the formation of the skeleton and indeed, the entire embryonic body plan.

REFERENCES

Anderson HC, Merker PC, Fogh J (1964): Formation of tumors containing bone after intramuscular injection of transformed human amnion cells (FL) into cortisone-treated mice. Am J Pathol 44:507–519.

Carrington JL, Chen P, Yanagishita M, Reddi AH (1991): Osteogenin (bone morphogenetic protein-3) stimulates cartilage formation by chick limb bud cells in vitro. Dev Biol 146:406–415.

Celeste AJ, Iannazzi JA, Taylor RC, Hewick RM, Rosen V, Wang EA, Wozney JM (1990): Identification of transforming growth factor β family members present in bone-inductive protein purified from bovine bone. Proc Natl Acad Sci USA 87:9843–9847.

Celeste AJ, Taylor R, Yamaji N, Wang J, Ross J, Wozney J (1992): Molecular cloning of BMP-8: A protein present in bovine bone which is highly related to the BMP-5/6/7 subfamily of osteoinductive molecules. J Cell Biochem Suppl 16F:100.

Chen P, Carrington JL, Hammonds RG, Reddi AH (1991a): Stimulation of chondrogenesis in limb bud mesoderm cells by recombinant human bone morphogenetic protein 2B (BMP-2B) and modulation by transforming growth factor β1 and β2. Exp Cell Res 195:509–515.

Chen TL, Bates RL, Dudley A, Hammonds RG Jr, Amento EP (1991b): Bone morphogenetic protein-2b stimulation of growth and osteogenic phenotypes in rat osteoblast-like cells: Comparison with TGF-β1. J Bone Min Res 6:1387–1393.

Doctor JS, Jackson PD, Rashka KE, Visalli M, Hoffmann FM (1992): Sequence, biochemical characterization, and developmental expression of a new member of the TGF-β superfamily in *Drosophila melanogaster*. Dev Biol 151:491–505.

Harrison ET Jr, Luyten FP, Reddi AH (1991): Osteogenin promotes reexpression of cartilage phenotype by dedifferentiated articular chondrocytes in serum-free medium. Exp Cell Res 192:340–345.

Hiraki Y, Inoue H, Shigeno C, Sanma Y, Bentz H, Rosen DM, Asada A, Suzuki F (1991): Bone morphogenetic proteins (BMP-2 and BMP-3) promote growth and expression of the differentiated phenotype of rabbit chondrocytes and osteoblastic MC3T3-E1 cells in vitro. J Bone Min Res 6:1373–1385.

Huggins CB (1931): The formation of bone under the influence of epithelium of the urinary tract. Arch Surg 22:377–408.

Lyons KM, Pelton RW, Hogan BLM (1989): Patterns of expression of murine Vgr-1 and BMP-2a RNA suggest that transforming growth factor-β-like genes coordinately regulate aspects of embryonic development. Genes Dev 3:1657–1668.

Lyons KM, Pelton RW, Hogan BLM (1990): Organogenesis and pattern formation in the mouse: RNA distribution patterns suggest a role for bone morphogenetic protein-2A (BMP-2A). Development 109:833–844.

Maliakal JC, Hauschka PV, Sampath TK (1991): Recombinant human osteogenic protein (hOP-1) promotes the growth of osteoblasts and stimulates expression of the osteoblast phenotype in culture. J Bone Min Res 6:S251.

Neuhof H (1917): Fascia transplantation into visceral defects. Sur Gynecol Obstet 24:383–427.

Özkaynak E, Rueger DC, Drier EA, Corbett C, Ridge RJ, Sampath TK, Oppermann H (1990): OP-1 cDNA encodes an osteogenic protein in the TGF-β family. EMBO J 9:2085–2093.

Padgett RW, St. Johnston RD, Gelbart WM (1987): A transcript from a *Drosophila* pattern gene predicts a protein homologous to the transforming growth factor-β family. Nature 325:81–84.

Pelton RW, Dickinson ME, Moses HL, Hogan BLM (1990): In situ hybridization analysis of TGFβ3 RNA expression during mouse development: comparative studies with TGFβ1 and β2. Development 110:609–620.

Rosen V, Bauduy M, McQuaid D, Donaldson D, Thies S, Wozney J (1991): Expression of osteoblast-like phenotype in mouse embryo limb bud cell lines cultured in BMP-2 and retinoid acid. Proc Endocrine Soc 73:57.

Rosen V, Wozney JM, Wang EA, Cordes P, Celeste A, McQuaid D, Kurtzberg L (1989): Purification and molecular cloning of a novel group of BMPs and localization of BMP mRNA in developing bone. Connect Tissue Res 20:313–319.

Sampath TK, Reddi AH (1981): Dissociative extraction and reconstitution of extracellular matrix components involved in local bone differentiation. Proc Natl Acad Sci USA 78:7599–7603.

Takuwa Y, Ohse C, Wang EA, Wozney JM, Yamashita K (1991): Bone morphogenetic protein-2 stimulates alkaline phosphatase activity and collagen synthesis in cultured osteoblastic cells, MC3T3-E1. Biochem Biophys Res Commun 174:96–101.

Thies RS, Bauduy M, Ashton BA, Kurtzberg L, Wozney JM, Rosen V (1992): Recombinant human bone morphogenetic protein-2 induces osteoblastic differentiation in W-20-17 stromal cells. Endocrinology 130:1318–1324.

Urist MR (1965): Bone: Formation by autoinduction. Science 150:893–899.

Vukicevic S, Luyten FP, Reddi AH (1989): Stimulation of the expression of osteogenic and chondrogenic phenotypes in vitro by osteogenin. Proc Natl Acad Sci USA 86:8793–8797.

Wang EA, Rosen V, Cordes P, Hewick RM, Kriz MJ, Luxenberg DP, Sibley BS, Wozney JM (1988): Purification and characterization of other distinct bone-inducing factors. Proc Natl Acad Sci USA 85:9484–9488.

Wang EA, Rosen V, D'Alessandro JS, Bauduy M, Cordes P, Harada T, Israel D, Hewick RM, Kerns K, LaPan P, Luxenberg DP, McQuaid D, Moutsatsos I, Nove J, Wozney JM (1990): Recombinant human bone morphogenetic protein induces bone formation. Proc Natl Acad Sci USA 87:2220–2224.

Wang EA, Gerhart TN, Toriumi DM (1993): Bone morphogenetic proteins and development. In "Proceedings of the 4th International Symposium on Chemistry and Biology of Mineralized Tissues" (in press).

Wharton KA, Thomsen GH, Gelbart WM (1991): *Drosophila* 60A gene, another transforming growth factor β family member, is closely related to human bone morphogenetic proteins. Proc Natl Acad Sci USA 88:9214–9218.

Wlodarski K (1969): The inductive properties of epithelial established cell lines. Exp Cell Res 57:446–448.

Wozney JM (1989): Bone morphogenetic proteins. Prog Growth Factor Res 1:267–280.

Wozney JM, Rosen V, Celeste AJ, Mitsock LM, Whitters MJ, Kriz RW, Hewick RM, Wang EA (1988): Novel regulators of bone formation: Molecular clones and activities. Science 242:1528–1534.

Yamaguchi A, Katagiri T, Ikeda T, Wozney JM, Rosen V, Wang EA, Kahn AJ, Suda T, Yoshiki S (1991): Recombinant human bone morphogenetic protein-2 stimulates osteoblastic maturation and inhibits myogenic differentiation in vitro. J Cell Biol 113:681–687.

Yamaguchi A, Ikeda T, Katagiri T, Suda T, Yoshiki S (1992): BMP-2 induces differentiation of a non-osteogenic fibroblastic cell line (C3H10T1/2) into both osteoblasts and chondroblasts in vitro. Bone Miner 17S:191.

Molecular Basis of Morphogenesis, pages 231–240
© 1993 Wiley-Liss, Inc.

16. Homeobox Genes and Epithelial Patterning in *Hydra*

M. A. Shenk and R. E. Steele

Department of Biological Chemistry and the Developmental Biology Center, University of California, Irvine, California 92717-1700

INTRODUCTION

Studies of arthropods, nematodes, and vertebrates have provided strong evidence that a highly conserved group of genes, the HOM/HOX homeobox genes, is responsible for regulating the elaboration of body patterns during development. In *Drosophila,* genes of the homeotic (HOM) complex are responsible for defining the identities of the segments set up by the gap, pair-rule, and segment polarity genes during embryonic development (Akam, 1987). The vertebrate homologues of these genes, the HOX genes, have been shown to be arranged in an identical fashion to the HOM complex in flies and to be expressed in the embryo in a fashion consistent with establishment of the identities of axial regions (Graham et al., 1989). The findings that deletions of the HOX genes give developmental defects in the mouse embryo support these conclusions (Chisaka et al., 1992; Le Mouellic et al., 1992). Recent work indicates that a similar situation to that in flies and vertebrates exists in the nematode *Caenorhabditis elegans* (Burglin et al., 1991; Costa et al., 1988; Kenyon and Wang, 1991). Given the remarkable structural and functional conservation of the genes in the HOM/HOX class, we have begun to investigate whether such genes play a role in axial patterning in the simple metazoan *Hydra*. The results of molecular and developmental studies show that a gene of the HOM/HOX class is expressed in the epithelial cells of the *Hydra* polyp and that its pattern of expression in these cells is consistent with a role in pattern formation.

PATTERN FORMATION IN *HYDRA*

Hydra is a member of the phylum Cnidaria, which consists of radially symmetric diploblastic animals having a distinct but simple axial pattern. The adult *Hydra* polyp consists of a hollow tube formed by two concentric epithelia and closed at one end by a tentacle-encircled head and at the other end by an adhesive foot (Campbell and Bode, 1983). Pattern formation in *Hydra* thus refers to the establishment of the head at one end of the polyp and the foot at the other (reviewed by Bode and Bode, 1984). Patterning in the adult *Hydra* polyp is manifested in two ways. First, the polyp continually produces new cells; yet the size and form of the polyp do not change. Some of the new cells are used to make buds, the asexual progeny of the adult polyp. Other new cells are displaced toward either the head or the foot, where they are eventually sloughed off. Given the constant form of the polyp in the presence of continual cell proliferation and displacement, it is clear that maintenance of the head and foot must be a continual process. A second way in which pattern formation occurs in the adult polyp is in the process of regeneration of the head or the foot. Removal of the head, the foot, or both, from the polyp is followed by the regeneration of the appropriate structures at the ends where they previously existed. In the process of regeneration, the ends of the body column must be repatterend as head and foot.

It has been demonstrated convincingly that the axial pattern of *Hydra* is established and maintained through the action of gradients emanating from the head and the foot (Bode and Bode, 1984). An 11-amino-acid polypeptide, the head activator, has been shown to be present in a graded manner, with the highest concentration at the head (Schaller and Gierer, 1973). A second molecule, apparently also a small polypeptide, is present in a graded distribution, with the highest concentration at the foot (Grimmelikhuijzen and Schaller, 1977). These two molecules are thought to be components of the pathways by which the head and foot activation gradients are established and maintained.

The *Hydra* polyp is composed of about a dozen different cell types (Campbell and Bode, 1983), any of which could possibly play a role in patterning. Several studies have, however, led to the conclusion that axial patterning in the polyp requires only the epithelial cells (Rubin and Bode, 1982; Takano and Sugiyama, 1983; Takano and Sugiyma, 1984). For this reason we sought HOM/HOX genes which were expressed in the epithelial cells of the *Hydra* polyp.

HOM/HOX-RELATED GENES IN CNIDARIANS

Given the roles of the HOM/HOX class in patterning in other animals, we, and others, sought to identify such genes in cnidarians and then to test whether they play roles in the patterning process in these simple animals. Initial unsuccessful attempts to detect homeobox-containing genes in cnidarians and other simple metazoans, such as the nematode *C. elegans,* involved hybridization with probes containing the homeodomains of the *Antennapedia* and *Ultrabithorax* genes of *Drosophila* (Holland and Hogan, 1986; McGinnis, 1985). Subsequent application of genetic methods led to the identification of *mab-5,* a HOM/HOX class homeobox gene involved in patterning of the posterior of the nematode (Costa et al., 1988). This finding indicated that HOM/HOX genes must have been present in the common ancestor of nematodes and the other metazoa. Only a few phyla predate the divergence of the nematodes and the rest of the metazoa; one such phylum is Cnidaria. Thus, it seemed plausible that HOM/HOX class genes would be present in cnidarians. The polymerase chain reaction (PCR) has made possible the rapid screening of genomes for various classes of genes and was the method of choice for searching for HOM/HOX class genes in *Hydra*. The polymerase chain reaction has the advantage of potentially yielding many members of the class in a single screen. Initial use of PCR to identify homeobox genes of the HOM/HOX class in the marine hydrozoans *Eleutheria dichotoma, Sarsia sp.*, and *Hydractinia symbiolongicarpus* resulted in the isolation of several genes (Murtha et al., 1991; Schierwater et al., 1991). While the predicted amino acid sequences of the PCR products did not allow the assignment of these genes as the homologues of known homeobox genes, they were, as expected, most closely related to genes in the HOM/HOX class. Subsequent work from Schaller's lab using oligonucleotide screening (Schummer et al., 1992), and from our group using PCR (Shenk et al., 1993), has led to the isolation and characterization of homeobox genes from *Hydra vulgaris* and *Hydra viridissima*.[1] Among the genes isolated from cnidarians, one has been identified in all five species examined (Fig. 1A). This gene has been termed *Cnox-2*.[2] Clones containing all or nearly all of the coding region for *Cnox-2* have been isolated from *Hydra vulgaris* and *Hydra viridissima*. A comparison of the predicted amino acid sequences of the *Cnox-2* proteins from the two *Hydra* species is shown in Fig. 2B. The homeobox sequences of the two proteins are identical while the flanking sequences show significant divergence.

Comparison of the sequences of the *Hydra Cnox-2* proteins with the proteins of other members of the HOM/HOX class has revealed that *Cnox-2* is most closely related to the *Deformed* gene of *Drosophila* and its vertebrate homologues (Shenk et al., 1993). This finding suggests that some of the other cnidarian homeobox genes which have been isolated may be related to particular genes in higher organisms. Such comparisons will be possible, however, only after full-length sequences are available for the other genes. Of particular interest with regard to evolution of the HOM/HOX genes will be examination of the organization of HOM/HOX class genes in the *Hydra* genome. The clustering, and order of arrangement in the cluster, of members of the HOM/HOX class of genes has been strictly conserved in flies and vertebrates

---

[1]Schummer et al. (1992) refer to this species as *Chlorohydra viridissima*. Campbell (1983) has argued that this species should be referred to as *Hydra viridissima*.
[2]*Cnox* (*Cn*idarian homeob*ox*) has been agreed upon as the acronym for HOM/HOX-related homeobox genes in cnidarians.

A

```
Hydra vulgaris              NNRYLSRLRRIQIAAILDLTEKQVK
Hydra viridissima           -------------------------
Eleutheria dichotoma        ---------------M---------
Hydractinia symbiolongicarpus ---------------M---------
Sarsia sp.                  ---------------M---------
```

B

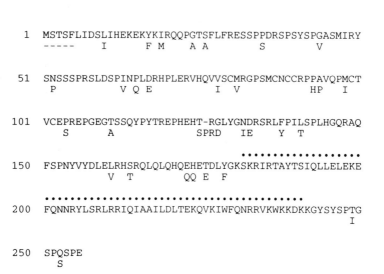

```
  1  MSTSFLIDSLIHEKEKYKIRQQPGTSFLFRESSPPDRSPSYSPGASMIRY
     -----     I   FM   AA          S         V

 51  SNSSSPRSLDSPINPLDRHPLERVHQVVSCMRGPSMCNCCRPPAVQPMCT
     P         VQE            I V             HP    I

101  VCEPREPGEGTSSQYPYTREPHEHT-RGLYGNDRSRLFPILSPLHGQRAQ
        S     A                SPRD    IE   Y  T
                                        ••••••••••••••••••
150  FSPNYVYDLELRHSRQLQLQHQEHETDLYGKSKRIRTAYTSIQLLELEKE
              V T       QQ E  F
     ••••••••••••••••••••••••••••••••••••
200  FQNNRYLSRLRRIQIAAILDLTEKQVKIWFQNRRVKWKKDKKGYSYSPTG
                                                  I

250  SPQSPE
        S
```

Fig. 1. Comparison of sequences of the protein encoded by the *Cnox-2* gene. **A:** Comparison of amino acid sequences of *Cnox-2* proteins from various cnidarians. Dashes indicate identities with the *Hydra vulgaris* sequence. The *Eleutheria*, *Hydractinia*, and *Sarsia* sequences were obtained from PCR products amplified using primers within the homeodomain (Murtha et al., 1991; Schierwater et al., 1991). **B:** Comparison of the longer sequences obtained from cDNA clones from the *Hydra vulgaris* and *Hydra viridissima* (Schummer et al., 1992) genes. The whole sequence of the *Hydra vulgaris* protein is shown. Only the amino acids for the *H. viridissima* sequence which differ from those of the *H. vulgaris* sequence are shown. Dashes at the amino-terminal end of the *H. viridissima* sequence indicate that the sequence of this portion of the protein is not known. Black dots overlie the amino acids of the homeodomain.

(Graham et al., 1989). The available data, though not yet complete, suggest a similar arrangement in nematodes (Burglin et al., 1991; Kenyon and Wang, 1991). As cDNA clones for the other HOM/HOX class genes from *Hydra* become available it should be possible to examine the arrangement of these genes in the genome by probing of Southern blots of large DNA fragments separated in pulsed field gels (Lai et al., 1989).

By probing of the blots of total DNA cut with various restriction enzymes, *Cnox-2* has been found to be a single-copy gene in both species of *Hydra*

Hydra Homeobox Genes   235

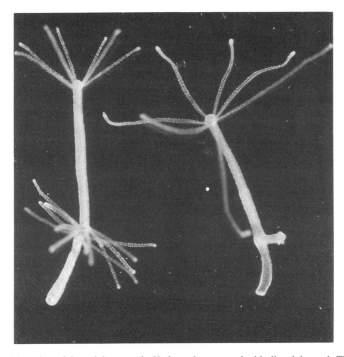

Fig. 2. Alteration of the axial pattern in *Hydra* polyps treated with diacylglycerol. The figure shows two *Hydra* polyps. The polyp on the right is an untreated control. The polyp on the left was given daily treatments with diacylglycerol. As a result of the diacylglycerol treatment, the polyp has generated a second ring of tentacles on the body column.

(Schummer et al., 1992; Shenk et al., 1993), a result which indicates that *Hydra* lacks the paralogous HOM/HOX gene sets that have been found in vertebrates (Duboule et al., 1990).

THE *CNOX-2* GENE ENCODES A NUCLEAR PROTEIN WHICH IS PRESENT IN A GRADED FASHION IN *HYDRA* EPITHELIA

To examine the expression of *Cnox-2* in the adult *Hydra* polyp, we generated antisera against a portion of the *Cnox-2* protein expressed in *Escherichia coli*. Staining of cells from adult polyps revealed that the protein product of the *Cnox-2* gene was, as expected, located in the nucleus (Shenk et al., 1993). While the nuclei of a variety of cell types were stained by the antibody, we have been most interested in the staining patterns of the epithelial cells, since it is the epithelial cells that are responsible for patterning (Rubin and Bode, 1982; Takano and Sugiyama, 1983; Takano and Sugiyama, 1984). Both endodermal and ectodermal epithelial cells were stained by the *Cnox-2* antibody, but the intensity of staining was not uniform. Some epithelial cells were not stained at all, others were stained lightly, and still others were stained intensely.

The *Hydra* polyp can be dissociated into individual fixed cells whose identities can be determined because they retain their normal distinctive morphologies (David, 1973). By dissecting the polyp into axial segments, then staining the dispersed cells of each segment with *Cnox-2* antiserum, one can obtain quantitative data regarding how many cells are expressing the *Cnox-2* gene at various positions along the polyp axis. Using this approach with *Hydra magnipapillata*, we have found that most of the epithelial cell nuclei in the body are intensely stained, while intensely stained nuclei are absent from the head (Shenk et al., 1993). Although this graded expression pattern is consistent with a role for *Cnox-2* in axial patterning, it does not provide proof for such a role.

EXPERIMENTAL ALTERATION OF THE AXIAL PATTERN IN *HYDRA* CHANGES THE DISTRIBUTION OF *CNOX-2* PROTEIN

If the *Cnox-2* protein is a component of the patterning machinery in the *Hydra* polyp, one would predict that altering the pattern of the polyp should cause concomitant changes in the pattern of *Cnox-2* expression. To test this hypothesis we have taken advantage of the finding that treatment of *Hydra* polyps with a protein kinase C-activating diacylglycerol results in an increase in the head activation level in the upper part of the body column (Müller, 1989; Müller, 1990; Müller, 1991). This increase in head activation results in the production of an ectopic ring of tentacles on the body column of the polyp (Fig. 2) and the regeneration of heads at both ends of an excised piece of body column (Fig. 3). Additional support for the idea that diacylglycerol treatment raises the head activation level in the body column comes from examination of the staining pattern obtained with an antibody which recognizes an FMRFamide-like peptide in the nerve cells of the *Hydra* head. Repeated treatment of a polyp with diacylglycerol results in the expansion of nerve staining by the antibody, down the body column, a result which is consistent with an increase of the head activation gradient (Fig. 4).

From the *Cnox-2* staining pattern in normal polyps (i.e., the absence of intensely stained nuclei in the head), we predicted that diacylglycerol treatment would result in the loss of intensely stained nuclei from epithelial cells in the upper body column as the head activation level rose in those cells. Staining of cells from axial segments of diacylglycerol-treated polyps gave the predicted result (Shenk et al., 1993). Schematic representations of the distributions of intensely stained nuclei in the epithelial cells of normal and diacylglycerol-treated polyps are shown in Fig. 5.

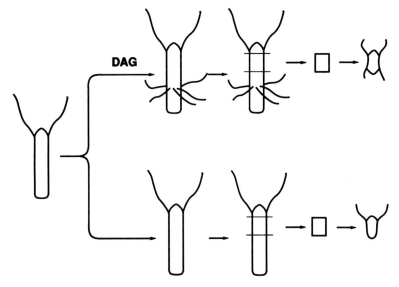

Fig. 3. Regeneration in diacylglycerol-treated *Hydra* polyps. A segment of body column excised from an untreated polyp will regenerate a head and a foot with the same polarity as the donor animal, as shown in the lower portion of the figure. In the upper portion of the figure, a comparable body column segment excised from a diacylglycerol-treated polyp regenerates a head at both ends.

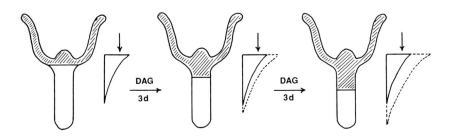

Fig. 4. Effect of diacylglycerol treatment of *Hydra* polyps on the head activation gradient, as measured by expression of a head-specific marker. Cross-hatching indicates the extent of staining of polyps with an antibody against an FMRFamide-like neuropeptide. In untreated polyps, staining is restricted to nerve cells in the hypostome (the dome containing the mouth) and the tentacles (Grimmelikhuijzen et al., 1982). Daily treatment of a polyp with diacylglycerol results in the spreading of staining down the body column (Bode, 1992a; Müller, 1991), as shown in the figure. Beside the diagrams of the polyps are shown plots of the head activation gradient in control (solid line) and diacylglycerol-treated polyps (dotted line) after 3 and 6 days of treatment. The arrow indicates a threshold in the gradient, above which activation of expression of the neuropeptide occurs. (Reprinted from Bode, 1992b, with permission of the publisher.)

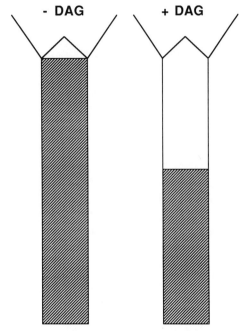

Fig. 5. Schematic of *Cnox-2* expression in the epithelial cells of normal (−DAG) and diacylglycerol-treated polyps (+DAG). Cross-hatching indicates regions in which the nuclei of virtually all of the epithelial cells contain high levels of *Cnox-2* protein, as assayed by immunocytochemical staining. In the regions without cross-hatching, none or few of the epithelial cells contain high levels of *Cnox-2* protein.

A MODEL FOR REGULATION OF *CNOX-2* EXPRESSION BY GRADIENTS EMANATING FROM THE HEAD AND FOOT OF THE *HYDRA* POLYP

The data we have obtained to date suggest that expression of *Cnox-2* is at least partly regulated by a signal emanating from the head, and that this signal acts through protein kinase C. Since treatment with diacylglycerol expands the zone in which cells lack high levels of *Cnox-2* protein, this signal must activate a pathway which inhibits *Cnox-2* expression. If the signal is graded from the head and has a threshold concentration below which it is not active, the pattern that is seen in the normal polyp would be obtained. In the normal polyp, the threshold would be at the head-body column border. The other component of the model for regulation of *Cnox-2* is a gradient which emanates from the foot and which has the capacity to activate high level expression of *Cnox-2* from the foot all of the way into the head. By this combination of an activating gradient arising from the foot and a counteracting inhibiting gradient coming from the head, the pattern of expression of *Cnox-2* which we see in the presence or absence of diacylglycerol can be explained. Since we know that patterning in *Hydra* is driven by gradients coming from the head and foot, this model is consistent with the *Cnox-2* gene product's playing a role in providing axial positional information in epithelia cells.

CONCLUSIONS

Using molecular approaches we have identified *Cnox-2*, a gene in the simple metazoan *Hydra vulgaris*, which is related to genes of the HOM/HOX class of homeobox genes. *Cnox-2* encodes a protein which is most closely related to the proteins encoded by the anterior members of the HOM/HOX class. The expression pattern of *Cnox-2* in the epithelial cells of the adult *Hydra* polyp is consistent with a role in defining the axial pattern of the polyp. Manipulation of the pattern of the polyp by treatment with diacylglycerol results in a change in the expression pattern of *Cnox-2* in keeping with its predicted role in patterning. Our findings with *Cnox-2* support the idea that the role of HOM/HOX genes in pattern formation was established very early in metazoan evolution. These findings also predict that addition HOM/HOX genes which have been identified in *Hydra* will be found to be components of the axial patterning mechanism.

ACKNOWLEDGMENTS

The work described here was supported by grants from the NSF and the March of Dimes Birth Defects Foundation. M.A.S. is supported by an NIH training grant to the UCI Developmental Biology Center. We thank Hans Bode for many helpful discussions and for supplying Figures 2, 3, and 4.

REFERENCES

Akam M (1987): The molecular basis for metameric pattern in the *Drosophila* embryo. Development 101:1–22.
Bode HR (1992a): Continuous conversion of neuron phenotype in hydra. Trends Genet 8:279–284.
Bode HR (1992b): Neuron determination in the ever-changing nervous system of *Hydra*. In Shankland SM, Macagno ER (eds): "Determinants of Neural Identity." San Diego: Academic Press, pp 323–357.
Bode PM, Bode HR (1984): Patterning in hydra. In Malacinski, GM, Bryant, SV (eds): "Pattern Formation, A Primer in Developmental Biology." New York: Macmillan, pp 213–241.
Burglin TR, Ruvkun G, Coulson A, Hawkins NC, McGhee JD, Schaller D, Wittmann C, Muller F, Waterston RH (1991): Nematode homeobox cluster. Nature 351:703.
Campbell RD (1983): Identifying hydra species. In Lenhoff, HM (ed): "Hydra: Research Methods." New York: Plenum Press, pp19–28.
Campbell RD, Bode HR (1983): Terminology for morphology and cell types. In Lenhoff, HM (ed): "Hydra: Research Methods." New York: Plenum Press, pp 5–14.

Chisaka O, Musci TS, Capecchi MR (1992): Developmental defects of the ear, cranial nerves and hindbrain from targeted disruption of the mouse homeobox gene *Hox-1.6*. Nature 355:516–520.

Costa M, Weir M, Coulson A, Sulston J, Kenyon C (1988): Posterior pattern formation in C. elegans involves position-specific expression of a gene containing a homeobox. Cell 55:747–756.

David CN (1973): A quantitative method for maceration of *Hydra* tissue. Wilhelm Roux Arch Entwicklungsmech Org 171:259–268.

Duboule D, Boncinelli E, DeRobertis E, Featherstone M, Lonal P, Oliver G, Ruddle FH (1990): An update of mouse and human HOX gene nomenclature. Genomics 7:458–459.

Graham A, Papalopulu N, Krumlauf R (1989): The murine and Drosophila homeobox gene complexes have common features of organization and expression. Cell 57:367–378.

Grimmelikhuijzen CJP, Schaller HC (1977): Isolation of a substance activating foot formation in hydra. Cell Differentiation 6:297–305.

Grimmelikhuijzen CJP, Dockray GJ, Schot LPC (1982): FMRFamide-like immunoreactivity in the nervous system of hydra. Histochemistry 73:499–508.

Holland PWH, Hogan BLM (1986): Phylogenetic distribution of *Antennapedia*-like homoeoboxes. Nature 321:251–253.

Kenyon C, Wang B (1991): A cluster of *Antennapedia*-class homeobox genes in a nonsegmented animal. Science 253:516–517.

Lai E, Birren BW, Clark SM, Simon MI, Hood L (1989): Pulsed field gel electrophoresis. Biotechniques 7:34–42.

Le Mouellic H, Lallemand Y, Brûlet P (1992): Homeosis in the mouse induced by a null mutation in the *Hox-3.1* gene. Cell 69:251–264.

McGinnis W (1985): Homoeobox sequences of the Antennapedia class are conserved only in higher animal genomes. Cold Spring Harbor Symp Quant Biol 50:263–270.

Müller WA (1989): Diacylglycerol-induced multihead formation in *Hydra*. Development 105:309–316.

Müller WA (1990): Ectopic head and foot formation in *Hydra*: Diacylglycerol-induced increase in positional value and assistance of the head in foot formation. Differentiation 42:131–143.

Müller WA (1991): Stimulation of head-specific nerve cell formation in *Hydra* by pulses of diacylglycerol. Dev Biol 147:460–463.

Murtha MT, Leckman JF, Ruddle FH (1991): Detection of homeobox genes in development and evolution. Proc Natl Acad Sci USA 88:10711–10715.

Rubin DI, Bode HR (1982): Both the epithelial cells and the nerve cells are involved in the head inhibition properties in *Hydra attenuata*. Dev Biol 89:332–338.

Schaller H, Gierer A (1973): Distribution of the head-activating substance in hydra and its localization in membranous particles in nerve cells. J Embryol Exp Morph 29:39–52.

Schierwater B, Murtha M, Dick M, Ruddle FH, Buss LW (1991): Homeoboxes in Cnidarians. J Exp Zool 260:413–416.

Schummer M, Scheurlen I, Schaller C, Galliot B (1992): HOM/HOX homeobox genes are present in hydra (*Chlorohydra viridissima*) and are differentially expressed during regeneration. EMBO J 11:1815–1823.

Shenk MA, Bode HR, Steele RE (1993): Expression of *Cnox-2*, a HOM/HOX homeobox gene in *Hydra*, is correlated with axial pattern formation. Development 117:657–667.

Takano J, Sugiyama T (1983): Genetic analysis of developmental mechanisms in hydra, VIII. Head-activation and head-inhibition potentials of a slow-budding strain (L4). J Embryol Exp Morph 78:141–168.

Takano J, Sugiyama T (1984): Genetic analysis of developmental mechanisms in hydra, XII. Analysis of a chimaeric hydra produced from a normal and a slow-budding strain (L4). J Embryol Exp Morph 80:155–173.

Molecular Basis of Morphogenesis, pages 241–254
© 1993 Wiley-Liss, Inc.

17. Human Epidermal Keratinocytes in Culture: Role of Integrins in Regulating Adhesion and Terminal Differentiation

Fiona M. Watt

Keratinocyte Laboratory, Imperial Cancer Research Fund,
London, England WC2A 3PX

INTRODUCTION

The epidermis is the stratified squamous epithelium that forms the outer covering of the skin. The most abundant cell type in the epidermis is the keratinocyte. The other permanent residents of the epidermis are melanocytes (pigment cells), Langerhans cells (macrophage-like) and Merkel cells (sensory cells associated with the nerve endings). In addition, cells of the immune system, such as T lymphocytes, migrate into and out of the epidermis.

The spatial organization of keratinocytes within the epidermis is relatively simple: the cells are arranged as multiple layers (Fig. 1) (Matoltsy, 1986). The basal (deepest) layer is attached to a basement membrane that separates the epidermis from the underlying connective tissue, the dermis. Melanocytes and Merkel cells are confined to the basal layer, whereas Langerhans cells can migrate through all the viable cell layers. It is within the basal layer that most keratinocyte proliferation takes place. Cells that leave the basal layer can no longer divide and they undergo a series of phenotypic changes, constituting their programme of terminal differentiation, as they move upwards towards the epidermal surface.

Three different types of suprabasal layer can be distinguished by the histological appearance of the constituent cells. Immediately above the basal layer lie about 5 to 10 spinous layers: spinous cells cannot divide, but are metabolically active and larger than basal cells; they are connected to one another by prominent desmosomal junctions. Above the spinous layers are one or two layers of granular cells, so named because their cytoplasm contains amorphous, electron-dense granules called keratohyalin granules. The outermost epidermal layers are the cornified layers, comprising cells that have lost their nucleus and intracellular organelles; their cytoplasm is filled with aggregated keratin filaments, and an insoluble protein envelope (cornified envelope) is closely apposed to the cytoplasmic face of the plasma membrane. The number

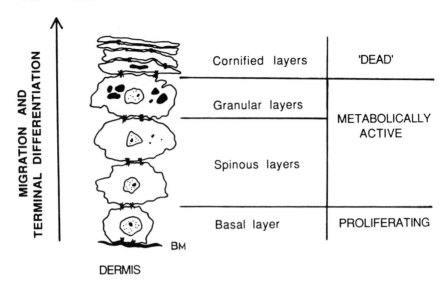

Fig. 1. The four types of cell layer that can be identified by morphology in human epidermis. The basal layer contains most of the proliferating cells and the suprabasal layers contain cells at different stages in the terminal differentiation process. Cells in the spinous and granular layers are metabolically active, but cornified cells are dead cells. Redrawn from Watt (1987).

of cornified cells varies, depending on body site, but is usually between 10 and 30 (Pinkus, 1951; Baker and Kligman, 1967). Although the cornified cells are dead, they have an important function in protecting the underlying living epidermal layers—and indeed the body as a whole—from desiccation, infection, and mechanical damage.

The epidermis is one of the tissues of the body in which proliferation continues throughout adult life. Cornified cells are continually shed from the cell surface and new cells are continually produced by mitosis in the basal layer. Under normal conditions, the rate at which new cells are produced balances the rate of shedding of cornified cells at the end of the terminal differentiation pathway, and so the total number of keratinocytes remains constant. The average time taken for a keratinocyte to move from the basal layer to the tissue surface is about 52 to 75 days, including an average transit time through the cornified layers of 14 days; however, there is regional variation between body sites and the transit time is decreased in psoriatic skin (Baker and Kligman, 1967; Halprin, 1972).

A number of features of the epidermis make it an attractive tissue to study. The simple spatial organization of the cells into layers can be mimicked in culture for studies of tissue assembly. The existence of a stem cell population that feeds a single terminal differentiation pathway offers the opportunity to analyze the factors that regulate stem cell behavior and the initiation of terminal differentiation. In addition, there is considerable clinical interest in wound healing; in benign hyperproliferative disorders, such as psoriasis; and in the high frequency of malignant conversion of normal keratinocytes to basal and squamous cell carcinomas. Finally, there are a number of practical applications for keratinocytes, including the use of the epidermis as a route for drug delivery and for gene therapy.

In this review I shall describe some of the methods that are available for culturing human keratinocytes and then discuss recent evidence that extracellular matrix receptors of the integrin family play a role both in determining the spatial organization of the cells and also in regulating the initiation of terminal differentiation.

TECHNIQUES FOR CULTURING HUMAN EPIDERMAL KERATINOCYTES

Human skin was one of the first tissues to be cultured in vitro. In 1897, C.A. Ljunggren reported that excised pieces of human skin maintained in ascites fluid at room temperature remained viable for several weeks and could be grafted onto wounds. Most of the early studies (to mid-twentieth century) were carried out on explants of skin and the epidermal sheets that migrated out from them (reviewed by Matoltsy, 1960). Later workers were able to establish cultures from disaggregated keratinocytes seeded at high density, but subculture was rarely successful (reviewed by Green, 1980).

A major advance came in 1975, when Rheinwald and Green showed that human epidermal keratinocytes isolated as a single-cell suspension could grow at clonal density if maintained in the presence of a feeder layer of irradiated 3T3 mouse embryo cells (Fig. 2). The 3T3 cells secrete factors that stimulate keratinocyte proliferation and also deposit extracellular matrix molecules that promote keratinocyte adhesion (Green, 1980; Rheinwald, 1980; Alitalo et al., 1982). The Rheinwald-Green technique is now widely used, the main refinements over the years being in the composition of the culture medium (Rheinwald, 1989). The most widely used medium formulation at present is a mixture of Dulbecco's modified Eagle's medium and Ham's F12, supplemented with adenine, fetal calf serum, hydrocortisone, insulin, cholera toxin, and epidermal growth factor. Hydrocortisone stimulates proliferation and gives the growing colonies a more orderly appearance (Rheinwald and Green, 1975). Cholera toxin also stimulates keratinocyte proliferation (Green, 1978). Epidermal growth factor increases the number of cell generations prior to senescence (Rheinwald and Green, 1977) and stimulates proliferation by stimulating the outward migration of the rapidly proliferating cells at the edges of growing colonies (Barrandon and Green, 1987).

In Rheinwald-Green cultures, keratinocytes from newborn foreskin epidermis have a replicative life span of 80 to 100 cell generations before senes-

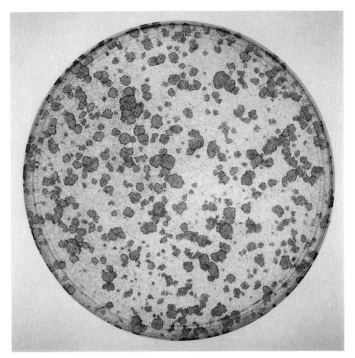

Fig. 2. Individual keratinocyte colonies grown on a feeder layer of 3T3 cells, using the Rheinwald-Green technique. Note the variation in colony size and morphology. Magnification × 0.98.

cence, whereas epidermal cells from children and adults have a life span of 40–70 cell generations (Rheinwald, 1989). Thus by freezing down keratinocytes at low passage numbers it is possible to work with the same strain of cells over several passage numbers and years.

In 1980 Hennings et al. described a technique for culturing newborn mouse keratinocytes in medium containing a lower concentration of calcium ions (0.05–0.1 mM) than standard medium formulations (1.2–1.8 mM). The low concentration of calcium ions inhibits desmosome formation and thus stratification (Hennings and Holbrook, 1983; Watt et al., 1984). Human keratinocytes can be grown in low-calcium medium under Rheinwald-Green conditions (Watt and Green, 1982), although a higher seeding density is required and the feeder layer must be replaced more frequently. Stratification is prevented, but cells can still initiate terminal differentiation (Watt and Green, 1982; Watt, 1983; Morrison et al., 1988). Thus the low-calcium cultures are useful for studying terminal differentiation in the absence of stratification (Fig. 3). Raising the calcium ion concentration to its normal level induces stratification within 24 hours and there is selective migration of terminally differentiating cells off

Role of Integrins in Human Keratinocytes 245

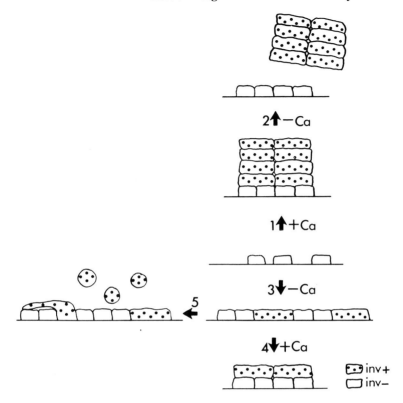

Fig. 3. Diagram illustrating the effects of changing the calcium ion concentration of the culture medium: − Ca, 0.1 mM calcium ions; + Ca, 1.8 mM calcium ions. Inv −, involucrin-negative cell; inv +, involucrin-positive cell. Data are from: 1, Banks-Schlegel and Green (1981); 2, Read and Watt (1988); 3,4,5, Watt and Green (1982). Drawing based on Watt (1987).

the culture dish (Watt and Green, 1982; Watt, 1984; Magee et al., 1987). The suprabasal layers can be detached from the basal layer in stratified cultures by a switch from standard to low-calcium medium (Jensen and Bolund, 1988; Read and Watt, 1988; Siegenthaler et al., 1988).

For some experiments, such as assays of potential mitogens, it is desirable to culture keratinocytes in the absence of a feeder layer, in serum-free medium. This has become possible through the development of a fully defined medium, MCDB 153 (Boyce and Ham, 1985). An extract of whole bovine pituitaries must be added for initiation of primary cultures and for serial culture, but can be omitted for clonal growth assays. The calcium ion concentration is lower than in standard media and the colony-forming efficiency is reported to be less than that obtained in Rheinwald-Green cultures (Rheinwald, 1989). More recently a defined medium containing 1.2 mM calcium ions has been developed that does not require addition of bovine pituitary extract (Morris et al., 1987, 1991).

The Rheinwald and Green cultures provide large numbers of keratinocytes from low seeding density; the cells stratify and express a range of terminal differentiation markers characteristic of spinous, granular, and cornified layers, including specific keratins and precursors of the cornified envelope (reviewed by Watt, 1989; Fuchs, 1990). Nevertheless, the histological appearance of the cultures is poor, since all cell layers are flattened, and granular and cornified layers do not accumulate. The poor morphology is a response of the cells to deficiencies in the culture environment, because within 1 week of grafting onto a suitable human recipient or onto nude mice the cells assume a normal epidermal morphology (Barrandon et al., 1988; Compton et al., 1989).

Techniques for improving the histological appearance of keratinocytes in culture are based on approximating a normal epidermal environment in vitro. The earliest attempts involved seeding keratinocytes on a type I collagen gel (Karasek and Charlton, 1971). This was later modified by seeding the gel with fibroblasts (Bell et al., 1981; Asselineau and Pruniéras, 1984) or 3T3 cells (Kopan et al., 1987). Another modification is to form copolymers of collagen and glycosaminoglycans (Yannas et al., 1989). The influence of intact dermis has been assayed by placing pieces of living or dead dermis underneath a collagen gel supporting the keratinocytes (Mackenzie and Fusenig, 1983), or seeding keratinocytes directly onto dermis from which the epidermis has been removed, but which retains the basement membrane (Régnier et al., 1981; Pruniéras et al., 1983). Cultures are supported on grids in chambers, on permeable membranes, or are floated, so that the keratinocytes are fed from below and exposed at the air-medium interface (see, for example, Mackenzie and Fusenig, 1983; Pruniéras et al., 1983).

In summary, there are a wide range of culture techniques available, and the method of choice depends on the relative importance of generating large numbers of cells or maximizing histological differentiation.

THE KERATINOCYTE INTEGRINS

Integrins are heterodimers of noncovalently associated transmembrane glycoproteins; each integrin consists of one α and one β subunit (Hynes, 1992). Keratinocytes express three integrins that share a common β subunit: $\alpha_5\beta_1$, which mediates adhesion to fibronectin; $\alpha_3\beta_1$, which mediates binding to laminin and epiligrin; and $\alpha_2\beta_1$ which is a collagen receptor that also participates in keratinocyte adhesion to laminin (Carter et al., 1990a,b, 1991; Adams and Watt, 1990, 1991). Keratinocytes also express a vitronectin receptor, $\alpha_v\beta_5$ (Adams and Watt, 1991; Marchisio et al., 1991), and $\alpha_6\beta_4$, which is a component of hemidesmosomes (Carter et al., 1990a; Stepp et al., 1990). The ligand for keratinocyte $\alpha_6\beta_4$ remains to be established unequivocally (Carter et al., 1990a; Staquet et al., 1990; Adams and Watt, 1991), but it may be laminin (De Luca et al., 1990; Lee et al., 1992).

All of the integrins expressed by keratinocytes in culture can also be detected in the epidermis, and expression of each integrin is largely confined to basal keratinocytes, both in vivo and in vitro (Nicholson and Watt, 1991; Adams and Watt, 1991; Hertle et al., 1991, 1992, and references cited therein). During epidermal development there are changes in the abundance and distribution of individual integrin subunits, particularly at the onset of stratification, and this provides indirect evidence that the receptors may play a role in establishing the spatial organization of keratinocytes in the epidermis (Hertle et al., 1991).

There are a number of situations in which aberrant integrin expression patterns are observed. During wound healing, and in psoriatic epidermis, integrins are expressed on the surface of keratinocytes that have left the basal layer and are undergoing terminal differentiation (Ralfkiaer et al., 1991; Hertle et al., 1992). In wound healing, the suprabasal cells are known to coexpress integrins and terminal differentiation markers, and suprabasal expression is not associated with the presence of suprabasal laminin, type IV collagen, or fibronectin, which are potential ligands for the receptors. In basal and squamous cell carcinomas of the epidermis, variable expression patterns are seen, including both increased and decreased expression of individual integrins (Peltonen et al., 1989; Stamp and Pignatelli, 1991; Sollberg et al., 1992).

ROLE OF INTEGRINS IN THE SELECTIVE MIGRATION OF KERATINOCYTES OUT OF THE BASAL LAYER

As keratinocytes undergo terminal differentiation, their ability to adhere to a range of extracellular matrix proteins is markedly decreased (Stanley et al., 1980; Toda and Grinnell, 1987; Adams and Watt, 1990). The adhesive changes precede the loss of β_1 integrins from the cell surface and, in the case of $\alpha_5\beta_1$, there is direct evidence that the receptors are downregulated in two stages (Adams and Watt, 1990). The first stage involves a reduction in the ability of $\alpha_5\beta_1$ to bind fibronectin without any reduction in the level of $\alpha_5\beta_1$ on the cell surface; this coincides with commitment to terminal differentiation (Adams and Watt, 1989, 1990). The second stage, coinciding with overt terminal differentiation, involves loss of $\alpha_5\beta_1$ from the cell surface (Adams and Watt, 1990) and a reduction in the levels of both subunit mRNAs (Nicholson and Watt, 1991) as a result of inhibition of transcription of the subunit genes (Hotchin and Watt, 1992).

There are a number of mechanisms by which the ability of $\alpha_5\beta_1$ to bind ligand could be decreased when basal keratinocytes become committed to terminal differentiation. Potential mechanisms fall into two categories: those that involve modulation of preexisting receptors in the plasma membrane, for example, through changes in receptor conformation, and those that involve synthesis of integrins that have been modified, for example, through glycosylation or alternative splicing (reviewed by Hynes, 1992). We have found that when keratinocytes are induced to differentiate, N-linked glycosylation of immature β_1 integrin subunits is inhibited and the immature subunits are not transported to the cell surface; thus functional downregulation must involve modification of preexisting receptors (Hotchin and Watt, 1992).

The functional downregulation of integrins on commitment to terminal differentiation must play a role in the selective migration of committed cells out of the basal layer and thus in determining the normal spatial organization of keratinocytes (Adams and Watt, 1990). However, terminal differentiation is associated not only with a reduction in keratinocyte–extracellular matrix adhesion, but also with changes in cell–cell adhesion (Watt, 1984, 1987). The integrins have a pericellular distribution in the basal epidermal layer (see, for example, Hertle et al., 1991, 1992) and there is some evidence from antibody blocking experiments that $\alpha_2\beta_1$ and $\alpha_3\beta_1$ may play a role in intercellular adhesion of keratinocytes in addition to their role in extracellular matrix adhesion (Carter et al., 1990b; Larjarva et al., 1990). The best-characterized of the keratinocyte intercellular adhesive mechanisms, however, involve desmosomes (Schwarz et al., 1990) and the nonjunctional cadherins, P- and E-cadherin (Takeichi, 1991). All keratinocytes express E-cadherin, but P-cadherin is only expressed by cells in the basal layer (Nose and Takeichi, 1986; Shimoyama et al., 1989; Nicholson et al., 1991; Wheelock and Jensen, 1992). The desmosomal glycoproteins also belong to the cadherin superfamily (Wheeler et al., 1991), and there are changes in both the number and composition of desmosomes during keratinocyte terminal differentiation (Skerrow, 1978; Parrish et al., 1986; King et al., 1991).

REGULATION OF TERMINAL DIFFERENTIATION BY INTEGRINS

When keratinocytes are disaggregated and placed as a single-cell suspension in medium made viscous by the addition of methyl cellulose, the cells are prevented from attaching to one another and to an adhesive substrate. Within 12 hours DNA synthesis is irreversibly inhibited; by 24 hours the majority of cells express involucrin, which is a precursor of the cornified envelope and a useful marker of the metabolically active phase of terminal differentiation; after 2 to 5 days the cells reach the final phase of terminal differentiation and

---

Fig. 4. Keratinocyte suspensions stained with antibodies to involucrin. **a:** Starting population. **b:** After suspension in methyl cellulose medium for 24 hours. **c:** Same as (b) except that the medium contained rabbit antiserum raised against the $\alpha_5\beta_1$ fibronectin receptor. See Adams and Watt (1989). Bar = 100 µm.

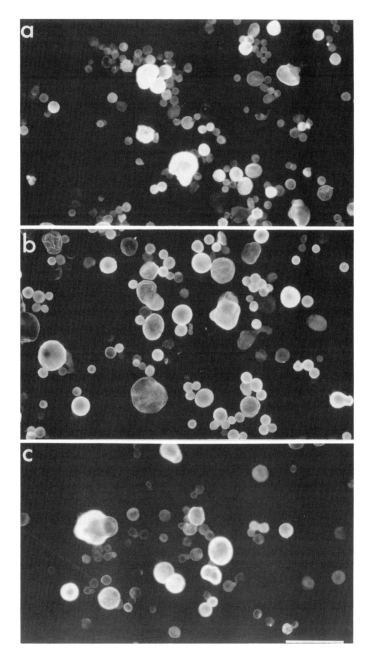

Figure 4.

assemble cornified envelopes (Green, 1977; Watt et al., 1988; Adams and Watt, 1989; Fig. 4). Under these conditions there are a number of potential stimuli for terminal differentiation, including the inhibition of DNA synthesis, loss of cell–cell contact, and loss of cell–substrate contact.

In order to investigate whether restricted substrate contact could act as a trigger for terminal differentiation, individual involucrin-negative keratinocytes were seeded on circular adhesive islands of varying diameter, under conditions in which cell–cell adhesion and proliferation were prevented (Watt et al., 1988). By varying the area of the islands, cell shape could be made to vary from fully spread to almost spherical. When substrate contact was restricted, DNA synthesis was inhibited and involucrin expression was induced. Thus, restricted spreading can act as a stimulus for terminal differentiation.

These results could either indicate that cell shape regulates the onset of terminal differentiation or that it is the proportion of the extracellular matrix receptors occupied by ligand that is important. In support of the latter idea, we found that it is possible to inhibit terminal differentiation in suspension by including fibronectin in the methyl cellulose, even though the cells remain rounded (Adams and Watt, 1989). Fibronectin inhibits involucrin expression, but does not prevent the inhibition of DNA synthesis, an observation which lends weight to the conclusion that although cell cycle withdrawal is a marker of commitment to terminal differentiation, proliferation and terminal differentiation can be regulated independently (Watt 1988; Watt et al., 1988; Adams and Watt, 1989).

The fibronectin effect is mediated by binding to the $\alpha_5\beta_1$ integrin (Adams and Watt, 1989). Fragments of fibronectin containing the RGD peptide that binds $\alpha_5\beta_1$ inhibit terminal differentiation whereas other regions of the molecule do not. Fab fragments prepared from a rabbit antiserum that binds to the β_1 integrins including $\alpha_5\beta_1$, are also capable of inhibiting terminal differentiation (Fig. 4). We have some evidence that laminin and type IV collagen can also participate in regulating terminal differentiation and that differentiation can either be regulated by occupancy of one specific integrin ($\alpha_5\beta_1$) or by the proportion of all β_1 integrins with bound ligand (J.C. Adams, N.A. Hotchin, and F.M. Watt, unpublished observations).

The inhibition of terminal differentiation by extracellular matrix proteins or anti-integrin antibodies is never complete, and this probably reflects the presence, in the starting population, of cells that are already committed to terminal differentiation and are thus unable to bind the proteins. In support of this, keratinocytes become insensitive to added fibronectin after 4 hours in suspension (Adams and Watt, 1989), a finding which fits with the kinetics of the functional downregulation of $\alpha_5\beta_1$ (Adams and Watt, 1990).

CONCLUSIONS

The ability to grow human keratinocytes in culture has made it possible to study the factors that control keratinocyte proliferation, terminal differentiation, and spatial organization. There is now good evidence that functional downregulation of β_1 integrins on commitment to terminal differentiation plays a role in ensuring the selective migration of committed cells out of the epidermal basal layer. Further, the proportion of β_1 integrins occupied by extracellular matrix proteins can regulate the initiation of terminal differentiation. Thus in keratinocytes, as in a number of other cell types, integrin-mediated signals can pass in two directions: signals from outside the cell (reduced contact with the extracellular matrix) can trigger differentiation, while on initiation of differentiation a signal from within the cell results in a decrease in ligand-binding ability by the receptors (see Hynes, 1992). One of our immediate challenges is to discover the nature of these signaling processes.

REFERENCES

Adams JC, Watt FM (1989): Fibronectin inhibits the terminal differentiation of human keratinocytes. Nature 340:307–309.

Adams JC, Watt FM (1990): Changes in keratinocyte adhesion during terminal differentiation: Reduction in fibronectin binding precedes $\alpha_5\beta_1$ integrin loss from the cell surface. Cell 63:425–435.

Adams JC, Watt FM (1991): Expression of β_1, β_3, β_4, and β_5 integrins by human epidermal keratinocytes and non-differentiating keratinocytes. J Cell Biol 115:829–841.

Alitalo K, Kuismanen E, Myllylä R, Kiistala U, Asko-Seljavaara S, Vaheri A (1982): Extracellular proteins of human epidermal keratinocytes and feeder 3T3 cells. J Cell Biol 94:497–505.

Asselineau D, Pruniéras M (1984): Reconstruction of 'simplified' skin: Control of fabrication. Br J Dermatol 111 (Suppl 27):219–222.

Baker H, Kligman AM (1967): Technique for estimating turnover time of human stratum corneum. Arch Dermatol 95:408–411.

Banks-Schlegel S, Green H (1981): Involucrin synthesis and tissue assembly by keratinocytes in natural and cultured human epithelia. J Cell Biol 90:732–737.

Barrandon Y, Green H (1987): Cell migration is essential for sustained growth of keratinocyte colonies: The roles of transforming growth factor-α and epidermal growth factor. Cell 50:1131–1137.

Barrandon Y, Li V, Green H (1988): New techniques for the grafting of cultured human epidermal cells onto athymic animals. J Invest Dermatol 91:315–318.

Bell E, Ehrlich HP, Buttle DJ, Nakatsuji T (1981): Living tissue formed in vitro and accepted as skin-equivalent tissue of full thickness. Science 211:1052–1054.

Boyce ST, Ham RG (1985): Cultivation, frozen storage, and clonal growth of normal human epidermal keratinocytes in serum-free media. J Tissue Cult Methods 9:83–93.

Carter WG, Kaur P, Gil SG, Gahr PJ, Wayner EA (1990a): Distinct functions for integrins $\alpha_3\beta_1$ in focal adhesions and $\alpha_6\beta_4$/bullous pemphigoid antigen in a new stable anchoring contact (SAC) of keratinocytes: Relation to hemidesmosomes. J Cell Biol 111:3141–3154.

Carter WG, Wayner EA, Bouchard TS, Kaur P (1990b): The role of integrins $\alpha_2\beta_1$ and $\alpha_3\beta_1$ in cell-cell and cell-substrate adhesion of human epidermal cells. J Cell Biol 110:1387–1404.

Carter WG, Ryan MC, Gahr PJ (1991): Epiligrin, a new cell adhesion ligand for integrin $\alpha_3\beta_1$ in epithelial basement membranes. Cell 65:599–610.

Compton CC, Gill JM, Bradford DA, Regauer S, Gallico GG, O'Connor NE (1989): Skin regenerated from cultured epithelial autografts on full-thickness burn wounds from 6 days to 5 years after grafting. A light, electron microscopic and immunohistochemical study. Lab Invest 60:600–612.

De Luca M, Tamura RN, Kajiji S, Bondanza S, Rossino P, Cancedda R, Marchisio PC, Quaranta V (1990): Polarized integrin mediates human keratinocyte adhesion to basal lamina. Proc Natl Acad Sci USA 87:6888–6892.

Fuchs E (1990): Epidermal differentiation. Curr Opin Cell Biol 2:1028–1035.

Green H (1977): Terminal differentiation of cultured human epidermal cells. Cell 11:405–416.

Green H (1978): Cyclic AMP in relation to proliferation of the epidermal cell: A new view. Cell 15:801–811.

Green H (1980): The keratinocyte as differentiated cell type. In "The Harvey Lectures, Series 74." New York: Academic Press, pp 101–139.

Halprin KM (1972): Epidermal "turnover time"—a re-examination. Br J Dermatol 86:14–19.

Hennings H, Holbrook KA (1983): Calcium regulation of cell-cell contact and differentiation of epidermal cells in culture. An ultrastructural study. Exp Cell Res 143:127–142.

Hennings H, Michael D, Cheng C, Steinert P, Holbrook K, Yuspa SH (1980): Calcium regulation of growth and differentiation of mouse epidermal cells in culture. Cell 19:245–254.

Hertle MD, Adams JC, Watt FM (1991): Integrin expression during human epidermal development in vivo and in vitro. Development 112:193–206.

Hertle MD, Kubler M-D, Leigh IM, Watt FM (1992): Aberrant integrin expression during epidermal wound healing and in psoriatic epidermis. J Clin Invest 89:1892–1901.

Hotchin NA, Watt FM (1992): Transcriptional and post-translational regulation of β_1 integrin expression during keratinocyte terminal differentiation. J Biol Chem 267:14852–14858.

Hynes RO (1992): Integrins: versatility, modulation, and signaling in cell adhesion. Cell 69:11–25.

Jensen PKA, Bolund L. (1988): Low Ca^{2+} stripping of differentiating cell layers in human epidermal cultures: An in vitro model of epidermal regeneration. Exp Cell Res 175:63–73.

Karasek MA, Charlton ME (1971): Growth of postembryonic skin epithelial cells on collagen gels. J Invest Dermatol 56:205–210.

King IA, Magee AI, Rees DA, Buxton RS (1991): Keratinization is associated with the expression of a new protein related to the desmosomal cadherins DGII/III. FEBS Letts 286:9–12.

Kopan R, Traska G, Fuchs E (1987): Retinoids as important regulators of terminal differentiation: Examining keratin expression in individual epidermal cells at various stages of keratinization. J Cell Biol 105:427–440.

Larjava H, Peltonen J, Akiyama SK, Yamada SS, Gralnick HR, Uitto J, Yamada KM (1990): Novel function for β_1 integrins in keratinocyte cell-cell interactions. J Cell Biol 110:803–815.

Lee EC, Lotz MM, Steele GD Jr, Mercurio AM (1992): The integrin $\alpha_6\beta_4$ is a laminin receptor. J Cell Biol 117:671–678.

Ljunggren CA (1897): Von der Fähigkeit des Hautepithels, ausserhalb des Organismus sein Leben zu behalten, mit Berücksichtigung der Transplantation. Dtsch Zeit Chir 47:608–615.

Mackenzie IC, Fusenig NE (1983): Regeneration of organized epithelial structure. J Invest Dermatol 81:189S–194S.

Magee AI, Lytton NA, Watt FM (1987): Calcium-induced changes in cytoskeleton and motility of cultured human keratinocytes. Exp Cell Res 172:43–53.

Marchisio PC, Bondanza S, Cremona O, Cancedda R, De Luca M (1991): Polarized expression of integrin receptors ($\alpha_6\beta_4$, $\alpha_2\beta_1$, $\alpha_3\beta_1$, and $\alpha_v\beta_5$) and their relationship with the cytoskeleton and basement membrane matrix in cultured human keratinocytes. J Cell Biol 112:761–773.

Matoltsy AG (1960): Epidermal cells in culture. Int Rev Cytol 10:315–351.

Matoltsy AG (1986): Structure and function in the mammalian epidermis. In Bereiter-Hahn J, Matoltsy AG, Richards KS (eds): "Biology of the Integument: Vol 2, Vertebrates" Berlin: Springer-Verlag, pp 255–271.

Morris RJ, Tacker KC, Baldwin JK, Fischer SM, Slaga TJ (1987): A new medium for primary cultures of adult murine epidermal cells: Application to experimental carcinogenesis. Cancer Lett 34:297–304.

Morris RJ, Haynes AC, Fischer SM, Slaga TJ (1991): Concomitant proliferation and formation of a stratified epithelial sheet by explant outgrowth of epidermal keratinocytes from adult mice. In Vitro Cell Dev Biol 27A:886–895.

Morrison AI, Keeble S, Watt FM (1988): The peanut lectin-binding glycoproteins of human epidermal keratinocytes. Exp Cell Res 177:247–256.

Nicholson LJ, Watt FM (1991): Decreased expression of fibronectin and the $\alpha_5\beta_1$ integrin during terminal differentiation of human keratinocytes. J Cell Sci 98:225–232.

Nicholson LJ, Pei XF, Watt FM (1991): Expression of E-cadherin, P-cadherin and involucrin by normal and neoplastic keratinocytes in culture. Carcinogenesis 12:1345–1349.

Nose A, Takeichi M (1986): A novel cadherin cell adhesion molecule: Its expression patterns associated with implantation and organogenesis of mouse embryos. J Cell Biol 103:2649–2658.

Parrish EP, Garrod DR, Mattey DL, Hand L, Steart PV, Weller RO (1986): Mouse antisera specific for desmosomal adhesion molecules of suprabasal skin cells, meninges, and meningioma. Proc Natl Acad Sci USA 83:2657–2661.

Peltonen J, Larjarva H, Jaakkola S, Gralnick H, Akiyama SK, Yamada SS, Yamada KM, Uitto J (1989): Localization of integrin receptors for fibronectin, collagen, and laminin in human skin. Variable expression in basal and squamous cell carcinomas. J Clin Invest 84:1916–1923.

Pinkus H (1951): Examination of the epidermis by the strip method of removing horny layers, I. Observations on thickness of the horny layer, and on mitotic activity after stripping. J Invest Dermatol 16:383–386.

Pruniéras M, Régnier M, Woodley D (1983): Methods for cultivation of keratinocytes with an air-liquid interface. J Invest Dermatol 81:28S–33S.

Ralfkiaer E, Thomsen K, Vejlsgaard GL (1991): Expression of a cell adhesion protein (VLA-β) in normal and diseased skin. Br J Dermatol 124:527–532.

Read J, Watt FM (1988): A model for in vitro studies of epidermal homeostasis: Proliferation and involucrin synthesis by cultured human keratinocytes during recovery after stripping off the suprabasal layers. J Invest Dermatol 90:739–743.

Régnier M, Pruniéras M, Woodley D (1981): Growth and differentiation of adult human epidermal cells on dermal substrates. Front Matrix Biol 9:4–35.

Rheinwald JG (1980): Serial cultivation of normal human epidermal keratinocytes. Methods Cell Biol 21A:229–254.

Rheinwald JG (1989): Methods for clonal growth and serial cultivation of normal human epidermal keratinocytes and mesothelial cells. In Baserga R (ed): "Cell Growth and Division. A Practical Approach" Oxford: IRL Press, pp 81–94.

Rheinwald JG, Green H (1975): Serial cultivation of strains of human epidermal keratinocytes: The formation of keratinizing colonies from single cells. Cell 6:331–344.

Rheinwald JG, Green H (1977): Epidermal growth factor and the multiplication of cultured human epidermal keratinocytes. Nature 265:421–424.

Schwarz MA, Owaribe K, Kartenbeck J, Franke WW (1990): Desmosomes and hemidesmosomes: Constitutive molecular components. Annu Rev Cell Biol 6:461–491.

Shimoyama Y, Hirohashi S, Hirano S, Noguchi M, Shimosato Y, Takeichi M, Abe O (1989): Cadherin cell-adhesion molecules in human epithelial tissues and carcinomas. Cancer Res 49:2128–2133.

Siegenthaler G, Saurat J-H, Ponec M (1988): Terminal differentiation in cultured human keratinocytes is associated with increased levels of cellular retinoic acid-binding protein. Exp Cell Res 178:114–126.

Skerrow CJ (1978): Intercellular adhesion and its role in epidermal differentiation. Invest Cell Pathol 1:23–37.

Sollberg S, Peltonen J, Uitto J (1992): Differential expression of laminin isoforms and β_4 integrin epitopes in the basement membrane zone of normal human skin and basal cell carcinomas. J Invest Dermatol 98:864–870.

Stamp GWH, Pignatelli M (1991): Distribution of β_1, α_1, α_2 and α_3 integrin chains in basal cell carcinomas. J Pathol 163:307–313.

Stanley JR, Foidart J-M, Murray JC, Martin GR, Katz SI (1980): The epidermal cell which selectively adheres to a collagen substrate is the basal cell. J Invest Dermatol 74:54–58.

Staquet MJ, Levarlet B, Dezutter-Dambuyant C, Schmitt D, Thivolet J (1990): Identification of specific human epithelial cell integrin receptors as VLA proteins. Exp Cell Res 187:277–283.

Stepp MA, Spurr-Michaud S, Tisdale A, Elwell J, Gipson IK (1990): $\alpha_6\beta_4$ integrin heterodimer is a component of hemidesmosomes. Proc Natl Acad Sci USA 87:8970–8974.

Takeichi M (1991): Cadherin cell adhesion receptors as a morphogenetic regulator. Science 251:1451–1455.

Toda K, Grinnell F (1987): Activation of human keratinocyte fibronectin receptor function in relation to other ligand-receptor interactions. J Invest Dermatol 88:412–417.

Watt FM (1983): Involucrin and other markers of keratinocyte terminal differentiation. J Invest Dermatol 81:100S–103S.

Watt FM (1984): Selective migration of terminally differentiating cells from the basal layer of cultured human epidermis. J Cell Biol 98:16–21.

Watt FM (1987): Influence of cell shape and adhesiveness on stratification and terminal differentiation of human keratinocytes in culture. J Cell Sci Suppl 8:313–326.

Watt FM (1988): Keratinocyte cultures: An experimental model for studying how proliferation and terminal differentiation are co-ordinated in the epidermis. J Cell Sci 90:525–529.

Watt FM (1989): Terminal differentiation of epidermal keratinocytes. Curr Opin Cell Biol 1:1107–1115.

Watt FM, Green H (1982): Stratification and terminal differentiation of cultured epidermal cells. Nature 295:434–436.

Watt FM, Mattey DL, Garrod DR (1984): Calcium-induced reorganization of desmosomal components in cultured human keratinocytes. J Cell Biol 99:2211–2215.

Watt FM, Jordan PW, O'Neill CH (1988): Cell shape controls terminal differentiation of human epidermal keratinocytes. Proc Natl Acad Sci USA 85:5576–5580.

Wheeler GN, Buxton RS, Parker AE, Arnemann J, Rees DA, King IA, Magee AI (1991): Desmosomal glycoproteins I, II and III: Novel members of the cadherin superfamily. Biochem Soc Trans 19:1060–1064.

Wheelock MJ, Jensen PJ (1992): Regulation of keratinocyte intercellular junction organization and epidermal morphogenesis by E-cadherin. J Cell Biol 117:415–425.

Yannas IV, Lee E, Orgill DP, Skrabut EM, Murphy GF (1989): Synthesis and characterization of a model extracellular matrix that induces partial regeneration of adult mammalian skin. Proc Natl Acad Sci USA 86:933–937.

Molecular Basis of Morphogenesis, pages 255–266
© 1993 Wiley-Liss, Inc.

18. Differentiation of the Endodermal Epithelium During Gastrulation in the Sea Urchin Embryo

Gary M. Wessel

Division of Biology and Medicine, Brown University, Providence, Rhode Island 02912

INTRODUCTION

Gastrulation is a complex series of cell and tissue rearrangements used by most metazoans to generate the basic body plan of the embryo: ectoderm on the outside, a tube of endoderm inside, and mesoderm cells scattered in between. This basic body plan is generated by diverse strategies throughout phylogeny, although many of the underlying cellular changes are well conserved. The sea urchin embryo has long been used to study the morphogenetic mechanisms of gastrulation, including ingression of the primary mesenchyme cells and invagination of the endodermal tissue. All cells are readily observable in this transparent embryo, and since the culturing conditions are simple, many investigators, both classical and contemporary, have capitalized on surgical manipulations to examine its embryogenesis. More recently these studies have been complemented by the identification of cell-type-specific molecular probes (Cameron and Davidson, 1991). In this chapter, I will discuss the development of the endoderm lineage of the sea urchin and focus on the activation of new gene activity in endoderm cells during gastrulation.

MORPHOGENESIS OF THE ENDODERM AT GASTRULATION

Gastrulation in the sea urchin embryo begins in the vegetal plate of the embryo, the region of slightly thickened cells (Fig. 1A). Cells in the vegetal plate are direct descendents of the vegetal pole of the egg. This lineage and others of the early embryo are known precisely because the embryo exhibits invariant cleavage and there are no cell migrations or positional changes prior to gastrulation, (Cameron and Davidson, 1991).

Cells of the future endodermal epithelium of the sea urchin are distinguishable first during primary invagination (Fig. 1B), an inward buckling of the vegetal plate epithelium that displaces the blastocoel. The motive force to initiate this buckling and the directionality for primary invagination is still unclear but the

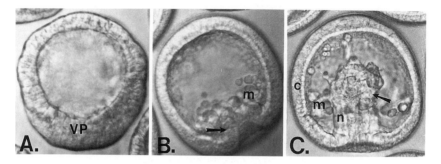

Fig. 1. Gastrulation in the sea urchin *Lytechinus variegatus*. **A:** The vegetal plate (VP) thickens at the blastula stage shown here and contains the future endoderm and mesoderm cell populations. **B:** After primary mesenchyme cell (m) ingression, the vegetal plate epithelium begins to buckle inward (arrow) signifying primary invagination. **C:** During secondary invagination, the endodermal epithelium (arrow) extends across the blastocoel. Primary mesenchyme cells undergo a characteristic migration behavior, and the aboral ectoderm begins to form a squamous epithelium. Embryos are approximately 100 microns in diameter.

underlying mechanisms are probably shared by many developing animals. Among the hypotheses to explain the motive force of primary invagination in the sea urchin are: changes in cell shape that result from differential cell adhesion (Gustafson and Wolpert, 1967), "tractoring" of cells adjacent to the vegetal plate towards the vegetal pole to force a buckling of the central tissue mass (Burke et al., 1991), and a polarized, unidirectional secretion of molecules to modify the basal and/or apical cellular environment to facilitate the tissue bending (Solursh, 1986; C. Lane, Young Investigator Abstract, this volume). The various mechanisms proposed to contribute to the motive force of primary invagination are certainly not mutually exclusive and each may contribute collectively to the observed morphogenesis.

Secondary invagination (Fig. 1C) refers to the series of events which result in the extension of the endodermal gut tube, the archenteron, across the blastocoel. This dramatic elongation results largely from a progressive repacking of neighboring endodermal epithelial cells, followed by filopodial contraction from secondary mesenchyme cells (SMC) bridging the tip of the archenteron to the overlying ectodermal roof. The proposed mechanism(s) and regulation of the target site for (SMC) contraction has been reviewed in an earlier volume of this series (McClay et al., 1991). Following gastrulation, the larval digestive system is morphologically separated into the foregut (prospective esophagus and pharynx), midgut (prospective stomach), and hindgut (prospective intestine) domains so that the free-swimming larvae are able to feed and prepare for metamorphosis.

Microsurgical experiments showed clearly that the force(s) necessary for endoderm invagination are inherent within cells of the vegetal plate and their immediate neighbors. Global forces generated by the embryo, like pressure within the epithelium from mitotic cell activity (Ettensohn, 1984; Stephens et al., 1986), or cell spreading or epiboly (Ettensohn and Ingersoll, 1992), although they may contribute to certain features of morphogenesis during gastrulation, are not necessary for invagination. Examining the differentiation of cells in the vegetal plate should, therefore, reveal some of the features necessary for invagination.

ENDODERM-SPECIFIC GENE EXPRESSION DURING GASTRULATION

The vegetal plate of the embryo contains a heterogeneous population of cells that includes presumptive endoderm cells, primary mesenchyme cells (PMC), and secondary mesenchyme cells. Recently several gene products were identified that distinguish various descendents of the vegetal plate (Coffman and Davidson, 1992), including cells of the endoderm. The endoderm lineage was shown to express a unique set of genes at gastrulation, which coincide with the changes in cell shape and behavior that accompany invagination. Although the function of these gene products is not known, they have been used to examine various features of differentiation in this lineage and may be useful to examine how in the early, cleaving embryo, an endodermal fate is assigned. Three genes specific to the endoderm lineage have been identified: Endo 16, Endo 1, and LvN1.2 (Fig. 2).

Endo 16 was identified by cDNA cloning and its RNA is expressed in cells of the endoderm during primary and secondary invagination (Fig. 2). After gastrulation, Endo 16 mRNA is degraded, first in the foregut and then in the hindgut, so that high levels of expression remain only in the midgut, the future stomach of the larva. The Endo 16 protein, whose molecular weight is over 300-kd, is concentrated both in the basal region of endodermal cells and in the extracellular matrix (Nocente-McGrath et al., 1989). A partial sequence of the Endo 16 cDNA suggests that the protein has a highly repetitive, acidic composition and contains the R-G-D amino acid sequence, known to be important in the cell binding domains of many extracellular matrix (ECM) proteins (Ruoslahti and Pierschbacher, 1987). Deposition of this protein into the extracellular matrix (ECM) have been shown to be important for endoderm differentiation at gastrulation (see below). Perhaps preendoderm cells synthesize and secrete their own ECM upon which subsequent morphogenesis depends. This hypothesis is readily testable in vivo in this embryo by injecting into the blastocoel reagents that react with specific constituents of the ECM (Ettensohn and Ingersoll, 1992). Such an autocrine effect by the ECM is known to occur in certain tissue culture cells after exposure to, for example, growth factor β (Massague, 1990).

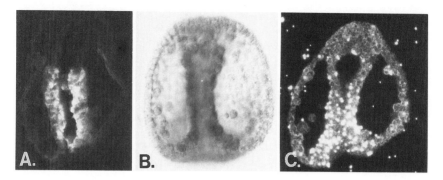

Fig. 2. Genes specific to endoderm lineage. **A:** Endo 1 seen by immunofluorescence. Endo 1 accumulates in the posterior two-thirds of the developing gut. **B:** Endo 16 mRNA detected by whole-mount RNA in situ hybridization. Endo 16 mRNA accumulates in all cells of the endoderm during gastrulation and then is restricted to the cells of the future stomach (not shown). **C:** LvN1.2 mRNA localization shown by autoradiographic in situ RNA hybridization. Like Endo 1, LvN1.2 accumulates only in the posterior two-thirds of the gut. (Photograph B is courtesy of A. Ransick and E. Davidson.)

Two other endoderm-specific products, Endo 1 and LvN1.2, have been identified (Fig. 2A,C). These molecules accumulate during primary invagination in only a subset of endoderm cells: those that form the posterior two-thirds of the digestive tract (the future stomach and intestine), but not in cells of the foregut, the future esophagus. Although the functions of Endo 1 and LvN1.2 are unknown, their expression is associated with the acquisition of a specific cell phenotype. Midgut and hindgut cells (Endo 1-positive) have extensive apical convolutions, are highly vesiculated, and are cuboidal, whereas foregut cells (Endo 1-negative) have smooth apical surfaces, and are ciliated. Indeed, foregut cells appear more similar to cells of the oral ectoderm, on the basis of several criteria, even though the two lineages are derived from distinct founder cells (Cameron and Davidson, 1991). Characteristics of the foregut that make it more similar to the oral ectoderm than the mid- or hindgut are its phenotype, a prolonged mitotic activity, and a retention of several "housekeeping" gene products (Kingsley et al., 1993), presumably for continued rapid growth.

Endo 1 was first identified by a monoclonal antibody screen to an embryo extract of membranes and extracellular matrix (Wessel and McClay, 1985). Endo 1 is a cell surface 320-kd glycoprotein found on both the apical and basolateral surfaces of endoderm cells. The gene encoding Endo 1 has not been identified but it is believed to be activated zygotically, just prior to gastrulation.

LvN1.2 was identified by a differential cDNA screen comparing RNA isolated from the ectoderm and from the endoderm at gastrulation (Wessel et al., 1989). The encoded 25-kd protein is concentrated at the apical region of endoderm cells of the hindgut and midgut, in the area that is highly vesiculated. The cytological association of LvN1.2 with vesicles and the similarity of its amino acid sequence to those of GTP-binding proteins of the small G-protein family (Balch, 1990; Bourne et al., 1991), suggest that LvN1.2 may be involved with vesicular transport during secretion and endocytosis for which endoderm cells are specialized.

The boundaries of expression between the foregut and midgut, and between the hindgut and ectoderm, are very sharp for the endoderm genes described above (Fig. 2A,C). Within one to two cell diameters, mRNA and protein accumulation goes from maximum levels to background levels. This sharp boundary is likely the result of the cells having arisen from different founder precursor cells. However, territories of cells can be experimentally induced to change their fates, as for example with the "vegetalizing" effect of LiCl or micromeres, so that a refinement of sharp boundaries may also result from cell interactions (Horstadius, 1972; Livingston and Wilt, 1990; Nocente-McGrath et al., 1991).

The expression of Endo 1 and LvN1.2 clearly distinguish midgut from foregut during gastrulation, but it is not clear when these cells are committed to their fate. Perhaps cell fate commitment occurs in the vegetal plate prior to gastrulation and then sort to their ultimate destination based on their previously assigned fate. Alternatively, cell fate may be determined during gastrulation as a result of a positional difference in the archenteron of "pluripotent" endoderm cells. Cells that find themselves within the leading edge of the archenteron may see a different environment and become foregut cells with a distinct gene program. It should be noted that in embryos induced to exogastrulate, where the tissue orientation is reversed with respect to the blastocoel, endoderm cells still retain their normal apical/basal polarity and their histospecific marker expression (i.e. Endo 1 and Endo 16), and the foregut cells remain histologically distinct from the hindgut and midgut regions.

In addition to the lineage-specific genes, the endoderm lineage also shows enriched activity of structural genes during gastrulation. For example, one member of the actin gene family, CyIIa (Coffman and Davidson, 1991) and the actin-binding protein α-spectrin (Wessel and Chen, 1993) show increased expression in the endoderm during gastrulation. Also the β-4 tubulin family member was shown to be enriched in endoderm cells during gastrulation (Harlow and Lane, 1987). Perhaps these proteins are selectively recruited for the extensive cell shape changes that occur during gastrulation.

Following gastrulation, cells of the gut tube may refine their phenotype and acquire distinct subpopulations. Myoepithelial cells for instance, develop characteristics both of muscle cells (muscle myosin) and epithelial cells (Endo 1 and LvN1.2) and constitute the sphincters separating the different gut domains of esophagus, stomach and intestine (Burke, 1981).

Endoderm-specific gene expression appears to be tightly coupled to endoderm morphogenesis at gastrulation. Whenever the endoderm invaginates, Endo 1 and LvN1.2 are expressed, whereas experimental inhibition of invagination by disruption of the extracellular matrix leads to an inhibition of Endo 1 and LvN1.2 expression. The two phenomena are inseparable. That is not to say that Endo1 is required for invagination but, experimentally, the signals to begin the cell shape changes for invagination might be the same as or similar to those involved in activating new gene expression. Alternatively, endoderm gene activation might result from changes in cell shape, similar to the activation seen in several mammalian cell lines (Werb et al., 1989). Increasing evidence shows that preendoderm cells respond to specific constituents in the extracellular matrix by activating and perpetuating endoderm differentiation so as to result in invagination and archenteron extension. Before discussing cell-ECM interactions that lead to endoderm differentiation at gastrulation, I will outline some of our recent findings on the composition and ontogeny of the ECM compartment.

THE EXTRACELLULAR MATRIX IN EARLY EMBRYOGENESIS

The early sea urchin embryo is well endowed with diverse ECM molecules in the blastocoel and basal lamina. The conserved members of the ECM family that have been identified include laminin (McCarthy et al., 1987), fibronectin, proteoglycans, (Wessel et al., 1984; DeSimmone et al., 1985) and collagens (Venkatesan et al., 1986; Angerer et al., 1988; Wessel et al., 1991; and Esposito et al., 1992). Several collagen genes have been identified in the sea urchin embryo, and it is likely that this embryo will contain much of the functional diversity seen for the collagen gene superfamily in vertebrates.

It is apparent that the embryonic blastocoel is not a homogeneous mixture of molecules. Instead the blastocoel has microenvironments, presumably as a result of differential synthesis, deposition, and degradation of its constituents. The result of this heterogeneous compartment is that cells traversing the blastocoel, that is, primary mesenchyme cells, secondary mesenchyme cells, and perhaps the invaginating endoderm cells during archenteron elongation, would experience different ECM environments. Such microenvironments may influence the direction of cell migration or the activation of specific gene expression.

Many proteins of the early ECM are contributed by maternal stores in the egg (Wessel et al., 1984; Figs. 3 and 4). Apparently, oocytes synthesize in these ECM proteins during oogenesis and store the proteins in distinct membrane-bound vesicles. After fertilization, and during early cleavages, these vesicles polarize to either the basal and/or apical cell surfaces of the blastomeres, where they are secreted. Extracellular matrix deposition into the

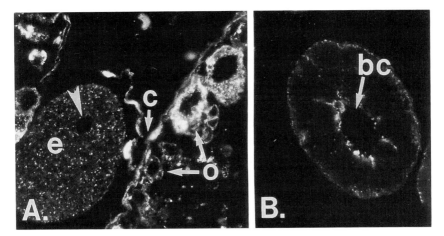

Fig. 3. Maternal contribution to the embryonic extracellular matrix (ECM) represented here by immunofluorescence localization of ECM 3, a recently identified molecule of the extracellular matrix with unknown function. **A:** An ovary section is shown with two opposed lobules separated by an egg. The egg contains numerous ECM granules dispersed throughout the cytoplasm. Note the unlabeled pronucleus (arrowhead). Oocytes at various stages of development are adjacent to the ovarian capsule, which consists of connective tissue sandwiched between two basal laminae. Extracellular matrix granules are present in these germ-line cells from early in oogenesis through maturation, but not in the surrounding somatic cells. ECM 3 is also present in the basal lamina of adult somatic tissue, that is, in the ovarian capsule that has peeled away from the germinal epithelium. For scale, the egg is 100 microns in diameter. **B:** Fertilization activates the redistribution of ECM granules during early cleavages. Shown is an early embryo (32 cells) during blastulation. The majority of ECM granules have polarized to the basal surface of the blastomeres and have begun to release their contents into the presumptive blastocoel. e, egg; o, early oocytes; c, ovarian capsules; bc, presumptive blastocoel.

nascent blastocoel begins during cleavage stages, and it is possible that this initial wave of maternal ECM participates in the process of blastulation.

The biosynthetic routing of the ECM vesicles in oogenesis is unknown, but the vesicles show additional regulatory steps beyond that seen in somatic cells. During oogenesis the newly synthesized ECM proteins enter only one type of a variety of secretory vesicles. Thus the "default" secretory pathway of somatic cells has "choices" in the egg. The ECM vesicles are then stalled during their routing and accumulate in the vicinity of the Golgi apparatus. However, another major secretory vesicle, the cortical granule, is instead routed to the surface of the egg and is secreted at fertilization as part of the fertilization reaction. The routing of ECM vesicles is reactivated at fertilization with most vesicles being polarized to the basal surface. Thus the developing egg is not only a warehouse of organelles, nutrients, and RNA, but must also act as a "control tower" for three-dimensional transportation.

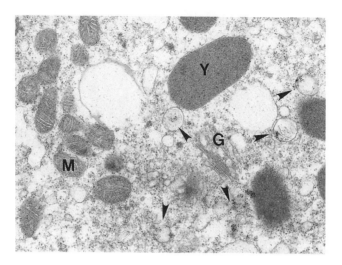

Fig. 4. Electronmicroscopic immunolabeling of extracellular matrix proteins stored in oocytes. Thin sections (80 nm) were immunolabeled with anti-fibronectin conjugated to colloidal gold (15 nm) and then visualized under an electron beam at 80 keV. Signals were concentrated to membrane-bound vesicles (arrows) in the vicinity of Golgi organelles. No significant label was seen in the cytoplasm, or in other organelles. Y, yolk platelet; G, Golgi; M, motichondria.

Once secreted into the extracellular environment, the dynamics of ECM proteins in the blastocoel are not understood. For instance, once secreted into the blastocoel, is each constituent confined to limited quadrants of the blastocoel defined by the location of the secreting cell? As the blastocoel expands, are new molecules added in layers as in onion skin, or is expansion (blastulation) a constant wedgelike insertion? With the recent cDNA cloning of ECM molecules (Angerer et al., 1988; G.M. Wessel and L. Berg, in preparation) and generation of antibody probes, we should be able to follow the construction and dynamics of this ubiquitous extracellular compartment.

ENDODERM INTERACTIONS WITH THE EXTRACELLULAR MATRIX

Several lines of evidence indicate that interactions of the preendodermal cells with the basal lamina-blastocoel matrix are important for primary invagination. The approaches used to examine this interaction are diverse, and include biochemical inhibitions, pharmacological treatments, and immunological reagents (reviewed in Ettensohn and Ingersoll, 1992). We have focused on the synthesis and processing of collagens, where there are several targeted steps for inhibition (Prockop, 1990). It is likely that in the extracellular matrix, collagen forms a structural backbone for other constituents, such that elimination of the collagen element leads to abnormal deposition, or turnover, of other ECM molecules.

Endoderm differentiation appears to be particularly sensitive to the extracellular matrix. In vivo inhibition of either glycosaminoglycans or collagen synthesis eliminates invagination and endoderm gene expression, whereas removal of the agents results in a rapid deposition of the affected molecules and restoration of complete endoderm development (Fig. 5). The normal recovery in these inhibited embryos is intriguing since, while the endoderm is apparently stalled in its developmental program, other cell types differentiate at least partially. Yet when gastrulation is restored, the developmentally "older" ectoderm has remained competent to interact with the invaginating endoderm to form a normal functioning mouth, a feeding larva, and a normal skeletal system.

Recently we have begun an in vitro approach to examine the dependence of endoderm differentiation on the extracellular environment. We have started by dissociating embryos prior to gastrulation and plating the single cells in culture. Preendoderm cells in these cultures were then challenged to "differentiate" in the presence of a variety of different substrates or additions to the media, and their success was measured by specific antibody and cDNA probes that could distinguish endoderm differentiation independent of morphogenesis. The question addressed here is whether the endoderm cells could differentiate in vitro, independent of their normal tissue contacts, and if so, could we identify what is important for the activation of the endodermal program. From these results, perhaps, we could also then learn how, and when, endoderm cells are determined in development.

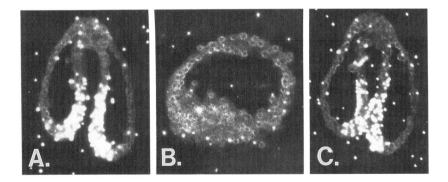

Fig. 5. Extracellular matrix is required for endoderm gene expression. In situ RNA hybridizations of LvN1.2 **A:** Embryos cultured 18 hours in seawater show normal accumulation patterns of LvN1.2 **B:** Sibling embryos to (A), but cultured in β-aminopropionitrile (β-APN) to disrupt the deposition of collagen and perturb normal extracellular matrix development. These embryos develop normally until gastrulation, but the endoderm does not invaginate and does not express LvN1.2. **C:** Culture mates of embryos in (B); these were allowed to recover by removal of β-APN. These embryos begin to deposit collagens and then invaginate, and express LvN1.2.

When embryonic cells were plated in the presence of an extracellular matrix made from EHS (Engelbreth-Holm-Swarm) cells (Kleinman et al., 1986), the number of cells expressing was about half that when Endo 1 or LvN1.2 cultured in vivo. Culturing cells in horse sera or fetal calf sera resulted in nearly complete endoderm differentiation, whereas cells plated on collagen, fibronectin, or laminin alone, or in the presence of various growth factors alone (EGF, basic-FGF), were insufficient to stimulate endoderm differentia-

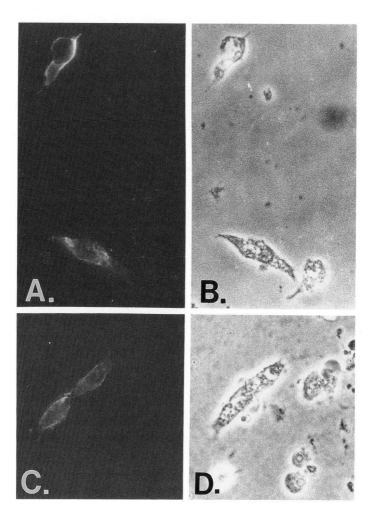

Fig. 6. Cells isolated from embryos prior to gastrulation and cultured for 18 hours in vitro in the presence of 5% horse sera. After fixation in methanol, cultures were stained with antibody to Endo 1 (see also Fig. 2) to identify "differentiated" endoderm cells. Endo 1–positive cells are elongated and distinct from other, nonendodermal cells of the culture. **A, C:** Immunofluorescence images. **B, D:** Phase photomicrographs of A and C, respectively.

tion (i.e., Endo 1 expression). Cells acquired at the mesenchyme blastula stage showed the best ability to differentiate, whereas younger cells from the blastula stage were almost completely incompetent. Thus, our initial conclusion is that endoderm cells are determined within the vegetal plate just prior to gastrulation and that these cells are competent to respond to in vitro environmental cues for activation of their endodermal program.

Clearly, gene expression has been separated from invagination under these conditions, yet the cultured endoderm cells do acquire distinct cell shapes. Many of these cells were spread with multiple cellular extensions (Fig. 6). Their shape is distinct from that of other cells in the culture, particularly from that of primary mesenchyme cells and ciliated ectoderm cells. Thus, these cells have acquired new cell phenotypes, along with new cellular proteins. Although cell shape in vitro was distinct relative to in vivo morphology, the protrusions from the endodermal cell bodies seen in vitro may correlate with the basal protrusions seen from endoderm cells during archenteron elongation in vivo (Ettensohn and Ingersoll, 1992). With the reagents and the tools to examine endoderm differentiation, we can begin to dissect the regulative capacity and the morphogenetic mechanisms used by this lineage during its impressive morphogenesis at gastrulation.

ACKNOWLEDGMENTS

I am grateful to Dr. Andrew Ransick and Dr. Eric Davidson for providing Figure 2B and to Dr. William H. Klein for providing helpful comments on the manuscript. This research was supported by grants from the NIH, the NSF, and the March of Dimes.

REFERENCES

Angerer LM, Chambers SA, Yang Q, Venkatesan M, Angerer RC, Simpson RT (1988): Expression of a collagen gene in mesenchyme lineages of the *Strongylocentrotus* embryo. Genes Dev 2:239.

Balch WE (1990): Small GTP-binding proteins in vesicular transport. Trends Biochem Sci 15:473–477.

Bourne HR, Sanders DA, McCormick F (1991): The GTPase superfamily: Conserved structure and molecular mechanisms. Nature 349:117–127.

Burke RD (1981): Structure of the digestive tract of the pluteus larva of *Dendraster excentricus*. Zoomorphology 98:209–225.

Burke RD, Myers RL, Sexton TL, Jackson C (1991): Cell movements during the initial phase of gastrulation in the sea urchin embryo. Dev Biol 146:542–557.

Cameron A, Davidson EH (1991): Cell-type specification during sea urchin development. Trends Genet 7:212–218.

Coffman J, Davidson EH (1992): Expression of spatially regulated genes in the sea urchin embryo. Curr Opin Genet Dev 2:260–268.

Coffman J, McClay DR (1990): A hyaline layer protein that becomes localized to the oral ectoderm and foregut of sea urchin embryos. Dev Biol 140:93–104.

DeSimmone DN, Spiegel E, Spiegel M (1985): The biochemical identification of fibronectin in the sea urchin embryo. Biochem Biophys Res Commun 133:183.

Ettensohn CA (1984): Primary invagination of the vegetal plate during sea urchin gastrulation. Am Zool 24:571–588.

Ettensohn CA, Ingersoll E (1992): Morphogenesis of the sea urchin embryo. In Rossomando E, Alexander S (eds): "Morphogenesis: An Analysis of the Development of Biological Form." New York: Marcel Dekker, pp 189–262.

Esposito JY, Dalessio M, Solursh M, Ramirez F (1992): Sea urchin collagen is evolutionarily homologous to vertebrate pro-alpha2(I) collagen. J Biol Chem 267:15559–15562.

Gustafson T, Wolpert L (1967): Cellular movement and contact in sea urchin morphogenesis. Biol Rev 42:441–498.

Harlow P, Nemer M (1987): Developmental and tissue-specific regulation of beta-tubulin gene expression in the embryo of the sea urchin *Strongylocentrotus purpuratus*. Genes Dev 1:147–160.

Horstadius S (1972): "Experimental Embryology of Echinoderms." Oxford: Clarendon Press.

Kingsley PD, Angerer LM, Angerer RC (1993): Major temporal and spatial patterns of gene expression during differentiation of the sea urchin embryo. Dev Biol 155:216–234.

Kleinman HK, McGarvey ML, Hassell JR, Star VL, Cannon FB, Laurie GW, Martin GR (1986): Biochemistry 25:312.

Livingston BT, Wilt FH (1990): Determination of cell fate in sea urchin embryos. Bioessays 12:115–119.

Massague J (1990): The transforming growth factor-beta family. Annu Rev Cell Biol 6:597–642.

McCarthy RA, Beck K, Burger MM (1987): Laminin is structurally conserved in the sea urchin basal lamina. EMBO J 6:1587.

McClay DR, Morrill J, Hardin J (1991): Archenteron morphogenesis in the sea urchin. In Gerhardt J, (ed): New York: "Cell-Cell Interactions in Early Development" Wiley-Liss, pp 15–29.

Nocente-McGrath C, Brenner CA, Ernst SG (1989): Endo 16, a lineage-specific protein of the sea urchin embryo, is first expressed just prior to gastrulation. Dev Biol 136:264–272.

Nocente-McGrath C, McIsaac R, Ernest SG (1991): Altered cell fate in LiCl-treated sea urchin embryos. Dev Biol 147:445–450.

Prockop DJ (1990): Mutations that alter the primary structure of type I collagen. J Biol Chem 265:15349–15352.

Ruoslahti E, Pierschbacher MD (1987): New perspectives in cell adhesion: RGD and integrins. Science 238: 491–497.

Solursh M (1986) Migration of sea urchin primary mesenchyme cells. In Browder L (ed): "Developmental Biology: A Comprehensive Synthesis," Vol. 2. New York: Plenum Press, pp 394–431.

Stephens L, Hardin J, Keller R, Wilt F (1986): The effects of aphidicolin on morphogenesis and differentiation in the sea urchin embryo. Dev Biol 118:64–69.

Van Katesan J, De Pablo F, Vogeli G, Simpson RT (1986): Structure and developmentally regulated expression of a *Strongylocentrotus purpuratas* collagen gene. Proc Natl Acad Sci USA 83:3351–3355.

Werb Z, Tremble PM, Behrendtsen O, Crowley E, Damsky CH (1989): Signal transduction through the fibronectin receptor induces collagenase and stromalysin gene expression. J Cell Biol 109:877–889.

Wessel GM, Chen SW (1993): Transient, localized accumulation of α-spectrin during sea urchin morphogenesis. Dev Biol 155:161–171.

Wessel GM, McClay DR (1985): Sequential expression of germ-layer specific molecules in the sea urchin. Dev Biol 111:451–463.

Wessel GM, Goldberg L, Lennarz WJ, Klein WH (1989): Gastrulation in the sea urchin is accompanied by the accumulation of an endoderm-specific mRNA. Dev Biol 136:526–536.

Wessel GM, Etkin M, Benson S (1991): Primary mesenchyme cells of the sea urchin embryo require an autonomously produced, nonfibrillar collagen for spiculogenesis. Dev Biol 148:261–272.

Wessel GM, Marchase RB, McClay DR (1984): Ontogeny of the basal lamina in the sea urchin embryo. Dev Biol 103:235–245.

Index

A 60A gene, as BMP homologue, 223
A-CAM, in adherens junctions, 50-51
ACK2 monoclonal antibody, *c-kit* and, 5, 10
Actin gene, activation of, 159
Activin, *Wnts* and, 38, 40-42, 44-45
Adherens junctions, in morphogenesis, 49-51
Adrenals, *mZP3* in, 29
Ag gene, in flower development, 94, 100-102, 105
Alkaline phosphatase, germ cell expression in, 2
Allantois
  c-kit in, 3-4, 7
  W mutations in, 1
Amacrine cells, dopaminergic, induction and, 123
Ambystoma mexicanum, *cardiac lethal* mutation in, 87
Amino acids
  in cadherins, 53, 55, 57
  in *mZP3*, 22-24, 30
β-Aminopropionitrile, differentiation during gastrulation and, 263
Anaphase, in developing *Drosophila* retina, 127
Anemia, *W* allele in, 1-2
Ankyrin, cadherins and, 56
Antennapedia gene, homeobox domain in, 232
Anteroposterior axis
  positional specification in muscle development and, 178-179, 183-184
  retinoic acid and, 211-213, 215
  in zebrafish, 79-90
Antibodies
  boss protein, 115
  cardiac morphogenesis in zebrafish and, 83
  c-kit, 5, 10, 13

Cnox-2, 235-236
FMRFamide-like neuropeptide, 237
retinoid, 211
sevenless protein, 115
vimentin, 149
Antirrhinum spp., floral control genes in, 103-104
Aorta, dorsal, *c-kit* in, 4
Ap1 gene, in flower development, 94, 96, 99-102, 105
Ap2 gene, in flower development, 94, 96, 99-102, 105
Ap3 gene, in flower development, 94, 96, 98-103, 105
Apical ectodermal ridge
  bone morphogenetic proteins in, 224-225
  in limb development, 208, 214-216
Arabidopsis spp., flower development in, 93-105
Archenteron, *Wnts* in, 36-37
argos gene, in developing *Drosophila* retina, 122
armadillo gene, plakoglobin family and, 58, 60-61
Asparagine-oligosaccharides, in mZP3 glycoprotein, 19
Atresia, in primordial oocytes, 14
Autocrine developmental signaling molecules, *Wnts* and, 39
Axial pattern, *Cnox-2* protein and, 236-238

Basal layer, keratinocyte migration from, 247-248
Basic fibroblast growth factor (bFGF)
  myogenic regulation and, 161
  Wnts and, 41
bFGF. *See* Basic fibroblast growth factor
Bicoid protein, in determining cell fate, 35
Birthdays, cell, 135, 137-139
Blastocyst, *mZP3* in, 21
Blastomere, *Wnts* in, 37, 39, 42-43

267

Blastopore, *Wnts* in, 37
Blastula, *Wnts* in, 35-38, 40
BMP-2. *See* Bone morphogenetic protein 2
BMP-4. *See* Bone morphogenetic protein 4
BMPs. *See* Bone morphogenetic proteins
Bone morphogenetic protein 2 (BMP-2), ridge signal and, 215
Bone morphogenetic protein 4 (BMP-4), as ventral mesodermalizing factor, 44
Bone morphogenetic proteins (BMPs), in cartilage and bone development, 221-228
boss gene. *See bride of sevenless* gene
brachyury gene, *Xenopus* homolog of, 44
Brain
   int-2 in, 74
   mZP3 in, 29
   Wnts in, 37, 43-44
BrDU
   in cerebral cortex, 145
   in *Drosophila* retina, 124-126
bride of sevenless (boss) gene, in developing *Drosophila* retina, 113-121, 129

22C10 neuronal epitope, in developing *Drosophila* retina, 128
C3H10T1/2 cell line
   bone morphogenetic proteins in, 223
   myogenic regulation in, 155
Cadastral genes, in flower development, 94, 100, 104-105
Cadherins
   in keratinocytes, 248
   in morphogenesis, 49, 50-51, 52-58, 61
Caenorhabditis elegans
   cell lineage in, 138
   homeobox genes in, 231-232
Calcium-binding pockets, in cadherins, 53
Calcium ion concentration, of culture medium, in culturing keratinocytes, 245-246
cal gene, in flower development, 94, 99, 105
Calvarial cells, bone morphogenetic proteins in, 223
cAMP. *See* Cyclic adenosine monophosphate
cardiac lethal mutation, in axolotl, 87
Cartilage
   bone morphogenetic proteins in development of, 221-228
   collagen fibrils in, 189-191

Catenins, in junction formation, 50-51, 57-59
CDC25 protein, in p21 activation, 120
cDNA. *See* Copy DNA
Cell cycle, induction in, 123-128
Cellular adhesion, integrins in, 241-251
Cellular differentiation
   cadherins in, 56-57
   cell division in, 128
   during gastrulation, 255-265
Cellular recognition, cadherins in, 54-55
Cellular retinal binding protein (CRBP), in limb development, 211
Cellular retinoic acid binding protein (CRABP), in limb development, 211
Cerebellum, *mZP3* in, 29
Cerebral cortex
   development of, 135-151
   organization of, 136
   sagittal view of, 137
Chemotactic factors, c-*kit* and, 4
Chicks
   collagen genes in, 192-199
   limb morphogenesis in, 208-215, 217
   myogenic regulation in, 158, 166-174, 179
Chondrocytes, collagenous proteins and, 196-197
Chordate structure, somite as fundamental unit of, 165
Chromatids, *mZP3* and, 21
Chromosome 5, PDGF receptor on, 2
Chromosomes, *mZP3* and, 21
cis-acting elements, in *mZP3* expression, 20, 25, 27-29, 31
Citric acid extraction, of desmosomes, 51
c-*kit* protooncogene, in primordial germ cell development, 1-8, 14-15
Cleavage-stage embryos
   mZP3 in, 21-22, 25
   Wnts in, 37-40, 42
clv genes, in flower development, 94
Cnidarians, HOM/HOX-related genes in, 232-235
cnox-2 gene, in *Hydra*, 233-239
Collagenous proteins, role of, 189-202
Colony stimulating factor 1 (CSF-1), c-*kit* and, 2
Condensation, in cartilage differentiation, 217
Cone cells, in developing *Drosophila* retina, 110-114, 119, 124

constans gene, in flower development, 94
Copy DNA (cDNA)
   cadherin, 56
   α-catenin, 57
   myogenic factor, 155-156
Cortical granule enzymes, in zona pellucida, 19
Cortical rotation
   in embryonic patterning, 35-36
   Wnts in, 45
COUP steroid receptor, *sevenup* gene and, 123
CRABP. *See* Cellular retinoic acid binding protein
CRBP. *See* Cellular retinal binding protein
c-ros gene, in R7 specification, 113
CSF-1. *See* Colony stimulating factor 1
c-src gene, in adherens junctions, 59
CV1 cell line, myogenic regulation in, 156
Cyclic adenosine monophosphate (cAMP), in spatial segregation, 110
Cyclin B, in developing *Drosophila* retina, 124-125
c-yes gene, in adherens junctions, 59
Cysteine
   mZP3 and, 23-24
   Wnts and, 39
Cytoplasmic domain
   cadherin-mediated adhesion in, 57-58
   c-kit in, 10

decapentaplegic (dpp) gene
   as BMP homologue, 223-224
   in developing *Drosophila* retina, 122, 125
def gene
   in flower development, 103
   homeobox gene homology to, 233
DER1 gene, in developing *Drosophila* retina, 120
Desmocalmin, in desmosomes, 51-52
Desmocollins, in desmosomes, 51-52, 55
Desmogleins, in desmosomes, 51-53, 55
Desmoplakins, in desmosomes, 51-52
Desmosomal band 6, in desmosomes, 51-52
Desmosomes, in morphogenesis, 49-52
Desmoyokin, in desmosomes, 51
Diacylglycerol, *Hydra* treated with, 235-239
Dictyate stage. *See* Diplotene stage
Dictyostelium spp., differentiation in, 128
Diffusible factors, in local induction, 122-123
Diplotene stage
   c-kit in, 8, 11, 14
   mZP3 in, 20-21
Disulfides, intermolecular, *mZP3* and, 23
Dlx gene, in zebrafish embryo, 69-73
Dorsal mesentery
   c-kit in, 3, 8
   W mutations in, 1
dpp gene. *See decapentaplegic* gene
Drosophila spp.
   armadillo gene in, 58, 60-61
   BMP homologues in, 223
   cadherins in, 55-56
   fat gene in, 54, 56
   homeobox genes in, 231-233
   induction in retina of, 109-129
   intercellular junctions in, 54-56, 58, 60-61
   myogenic factors in, 158
   wingless gene in, 60-61
Dynamin, homologue to, 118

E12 protein, in myogenic regulation, 156
Ear, zebrafish, homeobox genes in, 71-75
E box, in myogenic regulation, 156
echinoid mutation, in developing *Drosophila* retina, 128
echinus mutation, in developing *Drosophila* retina, 128
Ectoderm, *Wnts* in, 36-38, 40, 44-45
EGF. *See* Epidermal growth factor
Eleutheria dichotoma, homeobox genes in, 233-234
ellipse gene
   in developing *Drosophila* retina, 128
   in R7 specification, 126-127
Embryoids, *Wnts* in, 40
Embryonic patterning
   mZP3 in, 24-27
   Wnts in, 35-46
Endo 1 protein, differentiation during gastrulation and, 258-260, 264-265
Endo 16 protein, differentiation during gastrulation and, 257-259
Endoderm
   c-kit in, 3-4, 6-7
   differentiation of during gastrulation, 255-265
   Wnts in, 36-37
Endoplasmic reticulum, in R8, 115

Engelbreth-Holm-Swarm cells, extracellular matrix made from, 264
Epaxial domain, developing somite and, 171-174
Epiblast, germ cell generation from, 2
Epidermal growth factor (EGF)
  cadherins and, 54
  c-*kit* and, 11
  in developing *Drosophila* retina, 118, 126
Epidermis
  desmosomes in, 51
  integrins in, 241-251
  Wnts in, 38
Epididymis, *mZP3* in, 29
Epithelial patterning, homeobox genes in, 231-239
Erythroid progenitors, *W* mutations in, 1
Escherichia coli, *Cnox-2* protein expressed in, 235
Extracellular matrix
  differentiation during gastrulation and, 257-258, 260-264
  terminal differentiation by, 250-251
Eye, zebrafish, homeobox genes in, 69-71, 74-75

fat gene, cadherin repeats in, 54, 56
fca gene, in flower development, 94, 105
fd gene, in flower development, 94, 105
fe gene, in flower development, 94
Feline sarcoma virus, c-*kit* and, 2
Ferrets, cerebral cortex in, 139-140, 146, 148
Fertility, reduced, *W* allele in, 2
Fertilization, zona pellucida in regulation of, 19-21
FGF. *See* Fibroblast growth factor
FGF-4. *See* Fibroblast growth factor 4
fha gene, in flower development, 94
fibrils unbundled mutation, in heart development, 90
Fibroblast growth factor (FGF)
  int-2 and, 74
  Wnts and, 38, 45
Fibroblast growth factor 4 (FGF-4), ridge signal and, 215
Fibroblasts
  cadherins in, 54, 56-57
  collagenous proteins in, 196-197
Fibronectin
  spatial segregation and, 110
  terminal differentiation and, 250
Filaments, of zona pellucida, 19
First polar body, *mZP3* in, 21
fl genes, in flower development, 94
flo gene, in flower development, 103-104
Floorplate, c-*kit* in, 8
Floral control genes, evolutionary conservation of, 102-104
Floral transition, genes affecting, 95-96
Flower development
  genetic control of, 93-105
  wild-type, 95-96
FMRFamide-like neuropeptide, antibody to, 237
Fodrin, cadherins and, 56
Fogo selvagum, desmoglein I in, 52
Follicles
  c-*kit* in, 8-11
  mZP3 in, 21, 22, 26-27
Forebrain
  mZP3 in, 29
  Wnts in, 43-44
fpa gene, in flower development, 94
Frogs
  development of, 35-46
  limb morphogenesis in, 216
ft gene, in flower development, 94
fve gene, in flower development, 94
fwa gene, in flower development, 94
fy gene, in flower development, 94

G protein receptor family, boss protein in, 115
Gametes, c-*kit* encoded at *W* locus of mice and, 1-15
Gap junctions, formation of, 54
GAP1 gene, in developing *Drosophila* retina, 120-122
Gastrula, *Wnts* in, 36-37, 39, 42-43, 45
Gastrulation, cell differentiation during, 255-265
GATA family proteins, *mZP3* and, 30-31
Genetic hierarchy, controlling flower development, 93-105
gigantea gene, in flower development, 94
Ginkgo spp., floral control genes in, 104
Glycoproteins
  desmosomal, 51, 55
  mZP3 and, 19-31
  Wnts and, 39

Glycosylation sites, *mZP3* and, 23-24
Golgi apparatus, *mZP3* in, 21-22
Gonadal ridge
  c-*kit* in, 3-4, 7-8
  W mutations in, 1
Gonads, fetal, c-*kit* in, 3
Gonial cell cycle, c-*kit* in, 14
goosecoid gene, *Wnts* and, 42, 45
Graafian follicles, *mZP3* in, 21
Growth cones, cortical, 146
Gut
  c-*kit* in, 8
  mZP3 in, 29
  Wnts in, 37

hairy gene, in developing *Drosophila* retina, 128
Half-somite transplant experiments
  characterization of, 170-171
  epaxial and hypaxial domains, 171-174
Hamsters, *hZP3* and, 28
HAV tripeptide, in cadherins, 55-56
HDGC desmoglein, in desmosomes, 51, 55
Heart
  c-*kit* in, 4
  mZP3 in, 29
  tube, in zebrafish, 79-90
Helix-loop-helix domain
  myogenic regulation and, 156, 161
  mZP3 and, 30
Hematopoietic cells, *W* mutations in, 1
Hensen's node, in limb development, 210
Herculin, in myogenic regulation, 155, 169
Hindbrain
  int-2 in, 74
  Krox-20 in, 74
  Wnts in, 43
Hindgut
  c-*kit* in, 3, 5, 7
  W mutations and, 1
Homeobox (*Hox*) genes
  in *Hydra*, 231-239
  in limb development, 212-216, 218
  in zebrafish embryos, 69-75
Homeotic genes, in flower development, 94, 101-102, 104-105
Hox genes. *See* Homeobox genes
Hydra spp., homeobox genes in, 231-239
Hydractinia symbiolongicarpus, homeobox genes in, 233-234

Hypaxial domain, developing somite and, 171-174
hZP3, in transgenic mice, 28

Id protein, in myogenic regulation, 161
Induction, in developing *Drosophila* retina, 109-129
int-2 gene, in zebrafish embryo, 74
Integrins, in human keratinocytes, 241-251
Intercellular adhesive junctions, in morphogenesis, 49-61
Interleukins, spatial segregation and, 110
Intermediate filament proteins, desmoplakins and, 52

Keratinocytes, integrins in, 241-251
Keratohyalin granules, in epidermis, 241
Kidney, *mZP3* in, 29
Krox-20 gene, in hindbrain, 74

Laminar identity, determination of, 139-146
Lamin B-like protein, in desmosomes, 51-52
late flowering genes, in flower development, 94-96, 98-99, 105
L-CAM, in adherens junctions, 50-51
lfy gene, in flower development, 94, 96-104
LH. *See* Luteinizing hormone
Limb buds
  developmental expression of myogenic factors in, 159-161
  myogenic factor genes in, 155-162
Limb morphogenesis, pattern formation and, 207-218
Liver
  c-*kit* in, 4
  mZP3 and, 29
  W mutations in, 1
Lizards, myogenic regulation in, 173
Luciferase gene, firefly, *mZP3* and, 25, 27-29
Lumbar vertebrae, collagenous proteins in, 201
Lung, *mZP3* in, 29
Luteinizing hormone (LH)
  c-*kit* and, 11
  mZP3 and, 21
LvN1.2, differentiation during gastrulation and, 258-260, 263-264
Lytechinus variegatus, differentiation during gastrulation in, 255-265

mab-5 gene, in pattern formation, 232
Madin-Darbin canine kidney (MDCK) cells, cadherins in, 56
MADS box, in flower development, 94, 103
Mandible, c-*kit* in, 4
Mast cells
  c-*kit* in, 9
  W mutations in, 1
Master regulatory genes, myogenesis and, 169
MCDB medium, in culturing keratinocytes, 245
MDCK cells. *See* Madin-Darbin canine kidney cells
MEF-2 factor, in myogenic regulation, 159
Meiosis
  c-*kit* in, 9, 11
  mZP3 in, 20-21, 31
Meiotic prophase, c-s*kit* in, 3, 8, 11, 14
Melanoblasts, c-*kit* in, 4-5
Melanocytes
  c-*kit* in, 8
  W mutations in, 1
Meninges, c-*kit* in, 7
Meristem identity genes, in flower development, 94, 97-100, 104-105
Mesenchyme
  bone morphogenetic proteins in, 224, 226, 228
  c-*kit* in, 4, 7
  differentiation during gastrulation and, 256-257
  in limb development, 208, 211
Mesoderm
  c-*kit* in, 3
  germ cells in, 3
  Wnts in, 35-46
Mesodermal patterning, synergistic model for, 45-46
Mesonephric region, c-*kit* in, 7
Messenger RNA (mRNA)
  ap3, 98, 101-102
  bone morphogenetic protein, 224-228
  c-*kit*, 6-8, 11-12, 14
  collagen, 195
  Endo 16, 258
  goosecoid, 42
  lfy, 97, 100
  muscle creatinine kinase, 159
  myogenic factor, 157-160, 167
  mZP3, 22, 25-27, 31

noggin, 41
pi, 101
string, 125, 127
Wnts, 39-40
Metaphase
  in developing *Drosophila* retina, 127
  mZP3 in, 21
[^{35}S]-Methionine, *mZP3* and, 22
MF20 monoclonal antibody, cardiac morphogenesis in zebrafish and, 83, 87
Mice
  bone morphogenetic proteins in, 224-228
  c-*kit* encoded at *W* locus of, 1-15
  collagen genes in, 190-191, 199-202
  limb morphogenesis in, 210-211, 213-214
  myogenic factor genes in, 155-162
  mZP3 in, 19-31
Midblastula transition, *Wnts* in, 36-37, 41-45
Migration, neuronal, development of neocortical areas and, 146-150
Missense mutations, *W* mutations as, 2
Mitochondria, *mZP3* in, 21
Mitosis
  in cerebral cortex, 141-142, 144-145
  in developing *Drosophila* retina, 125, 127-128
MLC-CAT transgene, in positional specification of muscle precursors, 180-185
Molecular classes, in cell induction, 110
Morphogenesis, intercellular adhesive junctions in, 49-61
Morphogenetic furrow, in developing *Drosophila* retina, 111-112, 114, 116-117, 122-128
Morphogen gradients, in determining cell fate, 35
Motility, stimulation of, c-*kit* in, 2-3
MRF4 gene, in myogenic regulation, 155, 157-161, 169
mRNA. *See* Messenger RNA
Msh gene, in zebrafish embryo, 69, 71-73
Multivesicular bodies, in developing *Drosophila* retina, 115, 117
Muscle creatinine kinase gene, activation of, 159
Muscle development, positional specification in, 177-185
myd gene, in myogenic regulation, 159

myf-5 gene, in myogenic regulation, 155, 157-161, 166-167, 169
myf-6 gene, in myogenic regulation, 155
MyoD, in myogenic regulation, 155-161, 166-167, 169
Myogenic factor genes, in mouse embryos, 155-162
Myogenic lineages, positional differences in, 178-180
Myogenic linkages, within developing somites, 165-175
Myogenin, in myogenic regulation, 155, 157-161, 166-167, 169
Myosin gene, activation of, 159
Myosin heavy chain, myogenic regulation and, 158
mZP3, regulation of expression of
  characteristics of mZP3 gene and polypeptide, 22-24
  cis-acting elements and, 25-30
  introduction to, 19-20
  oogenesis in mice and production of zona pellucida, 20-21, 23
  pattern of expression during mouse development, 24-27
  synthesis and secretion of mZP3 glycoprotein, 21-22
  trans-acting elements and, 30

Na,K-ATPase, cadherins in, 56
Nasal epithelium, *c-kit* in, 4
N-cadherin, in adherens junctions, 50-51
NC4 globular domain, encoding of, 191
NDF, in developing *Drosophila* retina, 118
ned-1 mutant embryos, cell degeneration in CNS of, 88
Neocortex, development of, 146-150
Neural crest, *W* mutations in, 1
Neural tissue, *Wnts* in, 37, 40
Neurons, in cerebral cortex, 135-151

Neurula, *Wnts* in, 36
Nicotiana spp., floral control genes in, 104
Nieuwkoop center, *Wnts* and, 37-42, 45
Noggin gene, *Wnts* and, 41, 45
Notch gene, in developing *Drosophila* retina, 112, 118-119
Notochord, *Wnts* in, 43-44
Null mutation, *W* mutation as, 1

Oligosaccharides, in mZP3 glycoprotein, 19, 22, 30
Ommatidia, in developing *Drosophila* retina, 110-112, 114, 117, 119, 122-123, 129
Oocytes
  c-kit in, 6, 8-11, 14
  extracellular matrix and, 260-262
  mZP3 in, 20-23, 25-28, 30-31
OSP-1 protein, *mZP3* and, 30-31
Osteocalcin, BMPs in induction of, 223
Ovaries
  c-kit in, 5, 7, 11
  mZP3 in, 21-23, 25-27, 29-31
Ovulation, *mZP3* in, 21, 31

p21, in developing *Drosophila* retina, 120-122
p120, phosphorylation of, 60
Paracrine developmental signaling molecules, *Wnts* and, 39
Parietal endoderm, *c-kit* in, 3, 6
Pattern formation
  homeobox genes and, 231-239
  limb morphogenesis and, 207-218
Pax gene, in zebrafish embryo, 69-74
PC9 lung carcinoma cells, α-catenin in, 57
PC12 pheochromocytoma cells
  cadherins in, 56
  Wnt-1 in, 61
PDGF. *See* Platelet-derived growth factor
pebbled mutation, in developing *Drosophila* retina, 128
Pemphigus foliaceus, desmoglein I in, 52
Pemphigus vulgaris antigen
  in desmosomes, 51
  in morphogenesis, 55
Peptide growth factors, *Wnts* and, 38
perianthia genes, in flower development, 94
Periosteum, bone morphogenetic proteins in, 224-228
pGEM-72f vector, *mZP3* and, 28
Phosphorylation
  regulation by, 59-60
  of sevenless receptor, 115
Photoreceptor cells, in developing *Drosophila* retina, 109-129
Pi gene, in flower development, 94, 101-102, 105
Pituitary gland, *mZP3* in, 29
Plakoglobin, in intercellular adhesive junctions, 49-52, 58
Plaques, in intercellular adhesive junctions, 50-51

Plasma membrane, *mZP3* in, 23
Platelet-derived growth factor (PDGF), c-*kit* and, 2, 11
Plectin, in desmosomes, 51-52
Polarizing region, in limb development, 208-215, 217
Poly(A) RNA, *mZP3* and, 25
Polydactyly, prepattern mechanism and, 215-216
Polyspermia, zona pellucida in prevention of, 19
Positional information, recording, 212-214
Positional specification, in muscle development, 177-185
Preimplantation embryos, *mZP3* in, 27, 31
Prepattern mechanism, in limb development, 215-216
Primitive streak, germ cells in, 3
Primordial germ cells
  c-*kit* in, 2-8, 14-15
  mZP3 in, 20
Programmed cell death
  cerebral cortex and, 140
  c-*kit* and, 14
  eye disk development and, 112
Progress zone, in limb development, 208-211, 214-215
Proline, *mZP3* and, 22
Prosencephalon, *Pax* gene in, 69
Protein kinase C, *Cnox-2* protein and, 236, 238
Protein kinases, in adherens junctions, 50, 59-60
Proximal-distal axis, in limb development, 212

Quails, myogenic regulation in, 158, 167, 170-172, 174, 179

R7 photoreceptor cell pathway, in cell induction, 109-129
Radial glial cells, in cerebral cortex, 136, 149
Radixin, in adherens junctions, 50-51
ras genes, in R7 development, 120-122, 129
Ratchet mechanism, in limb development, 214
Receptor-mediated endocytosis, in developing *Drosophila* retina, 118
Reporter genes, *mZP3* and, 25
Retina, *Drosophila*, induction in, 109-129

Retinoic acid
  heart tube polarity and, 81-90
  homeobox genes and, 73
  signal from polarizing region and, 210-212
Rostrocaudal axis, positional specification in muscle development and, 177-180, 185
rough gene, in developing *Drosophila* retina, 111, 118-119, 128

S2 cell lines, clonal, boss and sevenless protein expression in, 115
S46 monoclonal antibody, cardiac morphogenesis in zebrafish and, 83, 87
Saccharomyces cerevisiae
  CDC25 protein in, 120, 125
  differentiation in, 128
Sarsia sp., homeobox genes in, 233-234
scabrous gene, in developing *Drosophila* retina, 122, 127-128
Schizosaccharomyces pombe, differentiation in, 128
Sea urchin, differentiation during gastrulation in, 255-265
Second messengers, cadherins and, 49, 56
Second polar body, *mZP3* in, 21
Secretory vesicles, *mZP3* in, 22-23
Segment polarity genes, in intercellular junctions, 60
Self differentiation, *Wnts* in, 38
Seminiferous epithelium, c-*kit* in, 11
Serine, in mZP3 glycoprotein, 22-23
Sertoli cells, c-*kit* in, 11-12
sevenless gene, in R7 specification, 113-122, 129
sevenup gene, in developing *Drosophila* retina, 111, 123
sextra gene, in developing *Drosophila* retina, 121-122
SH2 domains, *GAP1* and, 122
Sharks, myogenic regulation in, 173
shibire gene, in developing *Drosophila* retina, 118
Siamese twin embryos, *Wnts* in, 39
Signal transduction
  cell induction and, 109
  c-*kit* in, 2
  in developing *Drosophila* retina, 120, 122, 129

Wnts in, 41, 46
silent heart mutation, in heart development, 90
Simian virus 40 (SV40), *mZP3* and, 28
sina gene, in R7 development, 121
Sinapus spp., floral induction in, 99
Skeletal muscles, embryonic origins of, 156-157, 178
Solanum spp., floral control genes in, 104
Somite
   c-kit and, 5, 8
   development of, 165-169
   as fundamental unit of chordate structure, 165
   half-somite transplant experiments and, 170-171
   myogenesis and, 155-162, 169-170
son of sevenless (sos) gene, in R7 specification, 120
sos gene. *See son of sevenless* gene
sparkling poliert mutation, in developing *Drosophila* retina, 128
Spatial segregation, in cell induction, 110
α-Spectrin, differentiation during gastrulation and, 259
Spemann organizer, *Wnts* and, 37-39, 42-43, 45-46
Spermatogonia, *c-kit* in, 11-15
Sperm receptor gene, regulation of expression of, 19-31
Spinal cord
   c-kit in, 7
   Wnts in, 37
Spleen, *mZP3* in, 29
split mutation, in developing *Drosophila* retina, 128
Spontaneous cell death, *c-kit* and, 14
src family, protein kinases of, in adherens junctions, 50, 59-60
Star mutation, in developing *Drosophila* retina, 128
steel (Sl) mutation
   c-kit and, 1-3, 5, 7, 10-11, 13
   in developing *Drosophila* retina, 118
Stomach, *c-kit* in, 4
string gene, in developing *Drosophila* retina, 125, 127-128
Stromal cell lines, bone morphogenetic proteins in, 223
Subcortical proteins, in morphogenesis, 49
sup gene, in flower development, 94, 96, 100, 105
SV40. *See* Simian virus 40
Swine, collagen genes in, 193

Tadpoles, *Wnts* in, 36
Tailbud, *Wnts* in, 36
talpid mutant, limb morphogenesis in, 213
Temporal competence, in cell induction, 110
Terminal differentiation, integrins in, 241-251
Testis
   c-kit in, 11-14
   mZP3 in, 29
tfl gene, in flower development, 94, 96, 99-100, 105
TGF-α. *See* Transforming growth factor α
TGF-β. *See* Transforming growth factor β
Threonine-oligosaccharides, in mZP3 glycoprotein, 19
[^{3}H]-Thymidine, in developing *Drosophila* retina, 125
Thymus, *mZP3* in, 29
Tibia, collagenous proteins in, 202
Tight junctions, formation of, 54
TNF. *See* Tumor necrosis factor
Tongue, desmosomes in, 51
Tractoring, of cells, 256
trans-acting elements, in *mZP3* expression, 20, 30-31
Transcription factors, *Wnts* and, 42
Transforming growth factor α (TGF-α), in developing *Drosophila* retina, 118
Transforming growth factor β (TGF-β)
   bone morphogenetic proteins and, 44, 223, 227-228
   in developing *Drosophila* retina, 125-126
   spatial segregation and, 110
Trans-Golgi network, cadherins in, 53
Transmembrane protein, in developing *Drosophila* retina, 113, 115
Tubulin β-4, differentiation during gastrulation and, 259
Tumor necrosis factor (TNF), in developing *Drosophila* retina, 118
Tyrosine kinases
   c-kit and, 2, 11
   in developing *Drosophila* retina, 113, 115, 120, 122

Ultrabithorax gene, homeobox domain in, 232

Ultraviolet radiation
  cortical rotation blocked by, 37
  Wnts and, 37-38, 40
unusual floral organs genes, in flower development, 94
Uterus, c-*kit* in, 3
Uvomorulin, in adherens junctions, 50

Ventricular zone, cortical neurons in, 136, 138-139, 142, 146-149
Vimentin, antibodies to, 149
Vinculin
  in adherens junctions, 50-51
  α-catenin and, 58
v-*src* gene, in adherens junctions, 59

W-20 cells, bone morphogenetic proteins in, 223
white spotting locus (W) mutation, c-*kit* encoded at, 1-15
Wild Fire, desmoglein I in, 52
wingless gene, plakoglobin family and, 60-61
W locus. *See white spotting locus* mutation
Wnt-1 gene, in intercellular junctions, 60
wnts gene, in mesodermal patterning of *Xenopus* embryo, 35-46
α1X collagen gene
  mutations of in transgenic mice, 199-202
  structure and regulation of, 191-199

Xbra, as ventral mesodermalizing factor, 44
Xenopus laevis
  cadherins in, 57
  early development of, 36
  myogenic factors in, 158
  Wnts in, 35-46
XFKH-1 transcription factor, *Wnts* and, 42
Xlim 1 transcription factor, *Wnts* and, 42
Xwnt genes, in early mesodermal patterning, 38-41, 45-46

Yolk sac, *W* mutations in, 1

Zea spp., floral control genes in, 104
Zebrafish
  cardiac morphogenesis in, 79-90
  eyes and ears in, 69-75
  homeobox genes in, 69-75
Zona pellucida
  c-*kit* in, 10
  mZP3 in, 19-21, 23, 28, 31
Zonula adherens, in epithelial tissues, 59
Zygote, *Wnts* in, 35-36
Zyxin, in adherens junctions, 51

THE LIBRARY
UNIVERISTY OF CALIFORNIA, SAN FRANCISCO
(415) 476-2335

THIS BOOK IS DUE ON THE LAST DATE

Books not ret